Rheinwerk

Computing

Have you been to our website?

For code downloads, print and e-book bundles, extensive samples from all books, special deals, and our blog, please visit us at:

www.rheinwerk-computing.com

Rheinwerk Computing

The Rheinwerk Computing series offers new and established professionals comprehensive guidance to enrich their skillsets and enhance their career prospects. Our publications are written by the leading experts in their fields. Each book is detailed and hands-on to help readers develop essential, practical skills that they can apply to their daily work.

Explore more of the Rheinwerk Computing library!

Jürgen Wolf
HTML and CSS: The Comprehensive Guide
2023, 814 pages, paperback and e-book
www.rheinwerk-computing.com/5695

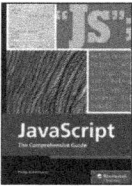

Philip Ackermann
JavaScript: The Comprehensive Guide
2022, 982 pages, paperback and e-book
www.rheinwerk-computing.com/5554

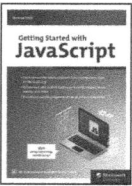

Thomas Theis
Getting Started with JavaScript
2025, 456 pp, paperback and e-book
www.rheinwerk-computing.com/5875

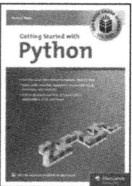

Thomas Theis
Getting Started with Python
2024, 437 pages, paperback and e-book
www.rheinwerk-computing.com/5876

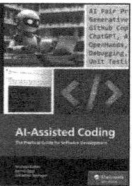

Michael Kofler, Bernd Öggl, Sebastian Springer
AI-Assisted Coding: The Practical Guide for Software Development
2025, 395 pages, paperback and e-book
www.rheinwerk-computing.com/6058

www.rheinwerk-computing.com

Kerem Koseoglu

Swift

The Practical Guide

Rheinwerk
Computing

Editors Rachel Gibson, Meagan White
Acquisitions Editor Hareem Shafi
Copyeditor Melinda Rankin
Cover Design Graham Geary
Photo Credits iStockphoto: 1341709750/© 123ducu; The Swift logo is a trademark of Apple Inc.
Layout Design Vera Brauner
Production Hannah Lane
Typesetting SatzPro, Germany
Printed and bound in the United States of America, on paper from sustainable sources

ISBN 978-1-4932-2718-1
1st edition 2026

© 2026 by:
Rheinwerk Publishing, Inc.
2 Heritage Drive, Suite 305
Quincy, MA 02171
USA
info@rheinwerk-publishing.com
+1.781.228.5070

Represented in the E.U. by:
Rheinwerk Verlag GmbH
Rheinwerkallee 4
53227 Bonn
Germany
service@rheinwerk-verlag.de
+49 (0) 228 42150-0

Library of Congress Cataloging-in-Publication Control Number: 2025046147

Contents at a Glance

Contents

3   Collections

4   Control Flow

5   Functions

6   Optionals

7   Enumerations

8　Structs

9　Classes

10 Protocols

16 Conclusion

Appendices

Dedication

To my beloved wife, Ozge—
my partner, my love, my constant in every journey,
standing by me through every chapter.

Chapter 1
Introduction

This chapter introduces the Apple development ecosystem and the role of the Swift programming language. It sets the stage for the rest of the book by explaining who the book is for, what you will learn, and how you might approach the topics covered in the book.

Hello, and welcome to the wonderful world of Swift! I feel honored to have been picked as your introductory guide to Swift development. Together, we will cruise through the features offered by the language; by the end of the book, you will be ready to jump into Apple development, with a strong technical infrastructure in place so that you can build any app you want.

This book is for everyone with basic computer literacy. If you don't have any former programming experience, don't fret! Swift has a very approachable syntax, making it an ideal candidate for a debut programming language. If you conceptually know the basics of programming, that should be enough to follow through the book. If you already have experience with another programming language, all the better: You'll cruise through the content like a breeze!

Now that we've defined the target audience, let's talk about what this book has to offer.

1.1 About this Book

As noted, the purpose of this book is to teach you how to program in the Swift language. And that's already a tall order—as is evident by the number of pages contained in this book.

Swift, on its own, is a general-purpose programming language. You can learn Swift and use it to develop command-line tools, mobile/desktop applications, backend services, libraries, and more. In practice, though, Swift is most commonly used in Apple-related development. The Swift ecosystem mostly revolves around development for Apple devices.

Any serious native Apple developer would be expected to be fluent in Swift, simply because the core language is the fundamental framework of the entire development stack.

However, Swift alone is not enough to develop the next big app. On higher levels of the development stack, there are Apple frameworks and libraries. Table 1.1 showcases some Apple frameworks you can use in your projects.

Framework	Purpose
Accessibility	Make apps accessible to people with disabilities
AVFoundation	Audio and video content
CloudKit	Cloud storage and sync
Core Data	Persistence using SQLite
Core Location	GPS and location services
Core ML	Machine learning
Foundation	Core functionality
GameKit	Multiplayer games and leaderboards
HealthKit	Health and fitness data
HomeKit	Smart home automation
MapKit	Apple Maps
Notifications	Local and remote push notifications
RealityKit	AR experiences
SceneKit	3D game and animations
SiriKit	Voice commands and automation
SpriteKit	2D game and animations
SwiftUI	User interfaces
TVUIKit	Apple TV apps
Vision	Image analysis, face recognition
WatchKit	watchOS apps

Table 1.1 Some Apple Frameworks

In a typical project, an Apple developer would use Swift language to code the app and use the appropriate Apple libraries as necessary to achieve specific goals. For instance, if you are developing a mobile chess game, you could do the following:

- Build a custom chess AI in standard Swift
- Show graphics using Swift plus the SpriteKit framework
- Handle data storage using Swift plus the CoreData framework

As you can see, the actual coding is done in Swift in each case. You use frameworks as helpers supporting your Swift code.

Those frameworks can be imagined as preprogrammed mature libraries of Swift code, which are kindly supplied by Apple. Instead of reinventing the wheel, you can include them in your projects and perform the necessary function calls to easily fulfill tasks such as displaying a UI, storing data, popping up notifications, and so on. Naturally, you can also pull in third-party frameworks or libraries if the need arises.

Swift Is the Central Language
In a typical Apple project, the only language you would use would be Swift. Additional frameworks are included in the project as needed for your convenience.

However, don't fall into the trap of underestimating those frameworks. Many frameworks contain vast amounts of functionality and will require you to dedicate time and practice to master them. Some frameworks are comprehensive enough to warrant a book of their own.

That said, each good book should have a clearly defined scope; no book can be everything to everyone. In this book, we will focus on the Swift programming language itself. You will discover different aspects of the language, and by the end of the book, you will have a strong foundation in the language—one that you can easily keep building further by learning about the frameworks you're interested in.

That's the natural bottom-up learning order. We don't recommend jumping to frameworks without a strong Swift foundation. That would be like trying to dance without learning how to walk first, or composing music without learning the notes and chords first. You get the idea: Build a strong foundation first, then you can tackle any requirement coming your way—like a tree with strong roots.

Now that we have defined the scope of the book, let's officially welcome you to the world of Swift.

1.2   Welcome to Swift!

Because you purchased this book and are reading these lines, you should already have a basic understanding of Swift and where it stands within the Apple ecosystem. Still, a formal introduction is in order.

Swift is a modern programming language developed by Apple. It was introduced in 2014 as a replacement for Objective-C, which had been the primary language for Apple development until then. Although legacy code chunks of Objective-C are still around here and there, Swift is the common choice for any fresh native development today.

Swift was designed with modern programming requirements in mind. It supports both functional and object-oriented programming paradigms as well as advanced features like generics, closures, and error handling.

Its readability is very high, and Swift is an expressive and fun language: It's a joy to code in Swift! Because it's a beginner-friendly language, you can invoke many strong features with simple statements. But don't let its simplicity fool you: Swift is an industrial-quality language with impressive performance.

Compiled Language Performance

Programming languages generally fall into two categories. In *interpreted languages*, such as Python or JavaScript, the source code is read and executed line by line at runtime. In *compiled languages*, such as C or Rust, the code is translated into machine code ahead of time, producing an executable that runs directly.

Although interpreted languages are arguably more comfortable to work with, that convenience usually comes at the cost of performance. Compiled languages, on the other hand, offer superior speed. Because Swift is a compiled language, your code is translated into machine code, leading to performance comparable to that of other compiled languages like C, C++, or Rust.

Swift is an open-source project, which makes it a transparent language. If you're interested, you can learn more at *https://www.swift.org*.

The Swift syntax leads programmers toward a cleaner architecture and helps prevent common programming errors thanks to elements such as the following:

- Variable initialization before usage
- Automatic overflow checks
- Nil value handling over optionals
- Automatic memory management
- Error handling

If you are unfamiliar with some of those concepts, don't worry! That's why you're here.

1.3   Why Learn Swift?

As an open-source general-purpose language, Swift can theoretically be used to tackle any programming task. Although the natural habitat for Swift is the Apple ecosystem, installers and development tools for Windows and Linux are also available. This means that you can develop and run Swift applications on all major operating systems.

That being said, most members of the Swift community are focused on Apple development. Most resources and libraries also target apps on Apple devices.

Practically speaking, the main goal of learning Swift is therefore using it for development on/for Apple devices. You may have a different sort of main goal for learning Swift, though!

You might be a complete beginner to programming and may want to start with Swift. Or you may be a student who is required to learn Swift as part of your courses. Even if you aren't sure if Swift development is going to be the pinnacle of your future career, it is a fun and useful language to learn due to its simplicity and its secure and powerful features. The architecture and good habits of the language will surely improve the quality of your approach to any programming area.

You may want to learn Swift alongside your primary programming language to broaden your horizons. That's another perfectly valid goal! If you are coming from a low-level language like C, C++, or Rust, you will probably feel like a first-class passenger due to the modern, powerful, yet easy-to-invoke features of the language, which nonetheless don't compromise performance too much. If you are coming from a high-level language like Python, Java, or C#, the English-like syntax of Swift will be familiar to you, and your know-how from your former language will be mostly applicable to the new environment.

High- Versus Low-Level Languages

In programming terminology, we often make the distinction between high- and low-level languages.

High-level languages are closer to human language and provide an approach closer to human logic and real-world concepts. They are typically easier to understand and code in, but they come at the cost of slower execution, higher memory usage, and less control over hardware. Such languages are used for a wide variety of tasks, such as web, mobile, and desktop applications, as well as data science, enterprise software, and automation.

Low-level languages are closer to machine code and offer direct hardware control and efficiency. They are typically harder to understand and code in, simply because you need to understand the architecture of the underlying hardware. That effort pays off with the highest degree of control and performance. Such languages are typically used for operating systems, embedded programming, device drivers, and game development.

Career advancement is another popular reason to learn Swift. You may already be an experienced programmer considering switching to Apple-focused development, or you may be an analyst, tester, manager, or the like who won't necessarily hand-code applications but needs to unbox some mysteries and understand the code structures a little better. Learning Swift would be the core place to start!

Some of you might have a good idea for an app and want to develop it yourself. This idea may even have a possible long-term agenda of turning the app into a business, maybe with a little help from investors. If you have experience in other programming languages already, why not learn Swift and develop your own codebase instead of rely-ing on (and paying) third parties?

Learn Swift First

As stated before, it makes sense to start with core Swift first before jumping to higher-level libraries and frameworks. No matter your background or goal, you won't get too far with Apple development if you don't have a sound grasp of the underlying program-ming language—which is Swift.

You might get some quick tasks done using code samples and online tutorials, such as a simple Mac calculator, iPhone shopping list, or Apple Watch chronometer. But as requirements become more complex, you will most likely regret not learning Swift bet-ter beforehand. Or you might have to reinvent the wheel in some cases, without even knowing that Swift offers a much better built-in mechanism for what you need.

In any case, beginning your journey with a book focusing on core Swift will surely let you start with a solid framework.

Those are some of the typical cases to learn Swift. If your motivation is completely dif-ferent, don't fret! Due to our inclusive approach, you should be able to follow through this book, no matter your background and goal!

Most programming languages are more similar than different. Although each language is invented with a specific mindset, they all share a vast common ground: variables, conditions, iterations, subroutines, and so on. Your background in other languages will surely aid you in your Swift journey—or in contrast, the concepts you'll learn in Swift will be transferrable to other languages in the future.

Now that we've discussed the reasons to learn Swift, let's help you get to know the lan-guage a little better by comparing it to some other languages. This should be useful especially for readers with programming experience in other environments.

1.4   Swift Versus Other Programming Languages

In this section, we will compare Swift with some other popular programming lan-guages. This will give you a better idea of where the language fits in the ecosystem as well as its advantages and disadvantages for your use case. This will also be useful for those who are trying to build a learning path of multiple languages, in which Swift is merely one of the target languages.

Motivation Alert

The language comparisons ahead are targeted to readers with programming back-grounds in other languages. If you are new to programming and don't understand some

---

of the terms, don't be disheartened. It is definitely not a prerequisite to know all of these concepts and understand the terminology in advance. Quite the contrary: You are here to learn!

1.4.1 Python

Python is a high-level, interpreted, cross-platform language known for its simplicity. It was created by Guido van Rossum in 1991 and is accepted as one of the most beginner-friendly languages. It is a dynamically typed language supporting multiple programming paradigms, like object-oriented, functional, and procedural programming. Some use cases for Python include web development, data science, machine learning, scripting, and fintech. It has a huge community and rich set of libraries ready to tackle most high-level tasks.

Cross-Platform

Saying that an application is *cross-platform* means that it can run on various operating systems or devices.

Table 1.2 showcases three key feature differences between Swift and Python.

Feature	Swift	Python
Type system	Statically typed	Dynamically typed
Execution	Compiled, fast	Interpreted, slower
Runtime target	Single platform	Cross-platform

Table 1.2 Swift Versus Python

Swift is a *statically typed language*, meaning that you have to determine the type of each variable in advance most of the time. Python instead is a *dynamically typed language*; that is, the types of variables are determined during runtime based on the values assigned to them. Although type hints are available, they don't enforce type checking during runtime.

Whether dynamic typing is a strength or weakness depends on the use case. It allows more flexibility and rapid development, but if you aren't careful, you'll get errors during runtime instead of compile time, which is bad for your production system. Using type check libraries or unit tests is essential to guard against that risk.

On the other hand, the static typing of Swift makes it easier to catch some simple (but fatal) errors during compilation. That helps with building more robust applications—but at the cost of slower development and reduced flexibility.

Performance is another significant difference. As a compiled language, Swift code is expected to run much faster than interpreted Python code. But don't let that statement fool you: You'll need to have CPU-heavy computations to feel the difference. In a simple, everyday application, Python may not feel slow at all: quite the contrary! Computers tend to be very fast nowadays, and even unoptimized systems seem to perform well—assuming you aren't pushing their limits.

Besides, although the interpreted nature of Python comes with a performance penalty, it pays off in some other ways:

- The source code of each application is open and transparent; there are no hidden surprises.
- Code on the production system can be debugged directly.
- In emergencies, quick fixes can be applied by simply changing the code on the production system.
- Interactive execution enables you to use tools like Jupyter Notebook, in which code can be executed line by line.

Both Swift and Python are easy to learn languages with a simple and readable syntax for any English speaker. However, dynamic typing, a reduced number of syntax artifacts, and its lack of compilation make Python code a little more approachable for complete beginners.

A big advantage of Python over Swift is its cross-platform nature. If you aren't relying on external compiled libraries and have followed best practices, then a pure Python project can simply be copied and pasted to any OS (with Python installed) and executed directly.

1.4.2   Java

Java is a high-level, object-oriented, and cross-platform language developed by Sun Microsystems in 1995. Java follows the "write once, run anywhere" philosophy, meaning that Java code can run on any platform with a Java Virtual Machine (JVM). Some use cases include enterprise applications, backend development, web development, and financial and banking applications.

Table 1.3 showcases three key differences between Swift and Java.

Feature	Swift	Java
Compilation	Native machine code	Bytecode for Java Virtual Machine (JVM)
Execution	High native performance	Reduced due to JVM overhead
Runtime target	Single platform	Cross-platform

Table 1.3 Swift Versus Java

The most significant difference between these languages is the way they compile. On an Apple device, Swift code is compiled directly to native machine code, which translates into a very fast application. In a Java environment, however, code is compiled to bytecode, which runs on the intermediary JVM platform preinstalled on the target machine. That may require more preparation and possible license considerations on the target platform.

The native nature of its compiled code leads to a performance advantage for Swift in some cases—but Java applications are not slow by any means. The difference may be tangible for heavy calculations, though.

In terms of syntax, Swift is a clean and brief language in which syntax clutter is mostly avoided. In Java, a higher degree of clutter is arguably present.

Like Python, Java has the undisputed advantage of being a cross-platform language. A pure Java application is compiled once and can be executed on any OS with Java support. However, JVM version differences can be a source of headaches. Managing multiple JVM versions and handling corresponding environment variables for different Java apps may not be your favorite part of the Java deployment process.

1.4.3   C#

C# arguably started as Microsoft's take on Java, but it has improved over time. It was published in 2000 as part of the .NET framework. Some typical use cases include desktop applications, web backends, and game development targeting the Unity engine.

The comparison between Swift and C# is very similar to the comparison between Swift and Java as C# is largely inspired by Java in the first place. Overlooking their syntax differences, C# is a similar technology that runs on the .NET runtime instead of JVM. Due to this similarity, we won't duplicate the Java comparison information here.

1.4.4   JavaScript

JavaScript—not to be confused with Java—is a high-level, interpreted programming language that is arguably the backbone of web development. It was created by Brendan Eich in 1995 and is the standard language for use with web browsers, enabling interactive and dynamic websites. That being said, it outgrew that initial purpose over the years, broadening its usage to areas like backend servers and cross-platform mobile and desktop apps.

Table 1.4 showcases three key differences between Swift and JavaScript.

Feature	Swift	JavaScript
Type system	Statistically typed	Dynamically typed
Execution	Compiled to machine code; fast	Interpreted; slower
Reusability	Low	High

Table 1.4 Swift Versus JavaScript

Like Python, JavaScript is a dynamically typed language—and this feature offers the same advantages and disadvantages to the language. In terms of execution, JavaScript is similar to Python once again; it is an interpreted language, which offers some advantages at the cost of performance. Refer to Section 1.4.1 for a refresher on those points.

One of the key differences between Swift and JavaScript is in their reusability. As a developer, it is natural to target the highest return on your time investment while learning a new technology. The scope of Swift is more or less defined within the Apple ecosystem, whereas JavaScript has a much broader scope—for example:

- Frontend development using vanilla JavaScript or certain frameworks
- Backend development using frameworks like Node.js
- Cross-platform desktop development using frameworks like Electron
- Cross-platform mobile development using frameworks like React Native

This doesn't mean that JavaScript is the best choice for any given development task; like any other choice, picking JavaScript for a certain task will surely have advantages and disadvantages. However, the wide horizon of its applicability makes it an attractive choice for developers considering shipping a cross-platform application with a single code base.

1.4.5   Dart

Dart is a modern, object-oriented language released by Google in 2011. It is best known for powering Flutter, which is Google's cross-platform framework for mobile, web, and desktop applications.

Table 1.5 showcases three key differences between Swift and Dart.

Feature	Swift	Dart
Vendor	Apple	Google
Main target	Apple devices	Cross-platform
UI framework	SwiftUI (native)	Flutter (custom)

Table 1.5 Swift Versus Dart

Swift and Dart offer similar features and a modern syntax. Swift is invented and supported by Apple, and its main purpose is app development for Apple devices, whereas Dart is invented and supported by Google. Its main purpose is to power Flutter, which is a cross-platform UI framework for building natively compiled applications for various targets, like iOS, Android, web, Windows, macOS, and Linux.

In that regard, Dart in combination with Flutter offers an alternative for cross-platform development similar to JavaScript in combination with React Native. You get to have a single codebase that can be published in distinct apps for each target platform. Advantages and disadvantages of this approach will be discussed in Section 1.4.9.

1.4.6   Kotlin

Kotlin is a modern, statically typed programming language developed by JetBrains in 2011, arguably as an effort to improve some shortcomings of Java for Android development. It is fully interoperable with Java and designed to be concise and expressive, reducing boilerplate code and enhancing productivity. Today, it is fully supported by Google for Android development.

Table 1.6 showcases three key differences between Swift and Kotlin.

Feature	Swift	Kotlin
Primary platform	Apple	Android
Interoperability	Objective-C	Java
Compilation	Native machine code	Java-compatible bytecode

Table 1.6 Swift Versus Kotlin

First and foremost, Swift and Kotlin were born with similar stories but targeted different purposes. Swift was born as a modernized replacement of Objective-C for Apple platforms, whereas Kotlin was born as a modernized replacement of Java for Android platforms.

This similarity and contrast between Swift and Kotlin is reflected in their features as well. For instance, Swift can interoperate with legacy Objective-C code, while Kotlin can interoperate with legacy Java code.

With the rise of Kotlin Native, though, Kotlin became a player in the cross-platform development game too. Beyond those core differences, both are modern languages with expressive syntax.

1.4.7   C++

C++ is a high-performance, object-oriented programming language developed by Bjarne Stroustrup in 1985 as an extension of C. It enables programmers to work close to the hardware via low-level memory control. It is widely used for system programming, game development, and high-performance computing.

Table 1.7 showcases three key differences between Swift and C++.

Feature	Swift	C++
Level of abstraction	High	Low; closer to hardware
Memory safety	High	Low
Syntax	Clean	Complex

Table 1.7 Swift Versus C++

First and foremost, Swift provides a high level of abstraction. It acts like an intermediary layer between the hardware and the programmer, enabling the programmer to focus on their algorithms instead of orchestration of the underlying hardware elements. C++, meanwhile, is a lower-level language, which enables a higher degree of hardware manipulation.

This makes C++ an attractive choice for low-level frameworks in which performance is the top priority, such as game engines, system programming, operating systems, and such.

However, such power comes at the cost of safety risks. The requirement of manual memory management in C++ has been the source of countless application crashes and security risks in the history of computing—simply because programmers are human and make mistakes.

Another significant difference is the syntax. Swift is a young language built from the ground up for the modern programmer of our age, which translates into a clean and approachable syntax. C++, on the other hand, is an old language that has evolved over the years, adding new elements as the need emerged. Its syntax is very complex and probably scary for many beginners.

1.4.8   Rust

Rust is a high-performance, system-level programming language developed by Mozilla and initially released in 2010. It was arguably invented to overcome some shortcomings of C/C++, providing an alternative with similar performance plus increased safety and readability.

Table 1.8 showcases three key differences between Swift and Rust.

Feature	Swift	Rust
Memory safety	Automatic Reference Counting (ARC)	Ownership model
Main focus	App-level	System-level
Learning	Easier	Harder

Table 1.8 Swift Versus Rust

Memory safety is one of the main reasons that Rust emerged: to solve the issues of C++ highlighted in Section 1.4.7. Rust uses a system of ownership and borrowing and lifetimes to ensure memory safety without a garbage collector, whereas Swift uses a system called Automatic Reference Counting (ARC) to manage memory automatically—meaning that developers don't need to manage object lifetimes manually.

Garbage Collector

A *garbage collector* is a feature in some programming languages that automatically manages the memory of the application, freeing the developer from that responsibility and reducing the risk of crashes due to human errors. It brings simplicity and security features to the table but inevitably adds runtime overhead and reduces control over memory.

Swift and Rust were born with different purposes in mind. The main focus of Swift is application development on consumer devices (with an imprint of a certain fruit). Rust, meanwhile, is a system-level language like C/C++, providing a closer relationship with the hardware. Even WebAssembly development is possible with Rust.

WebAssembly

Often shortened as WASM, WebAssembly is a low-level binary format designed to run code inside web browsers at near-native speeds. Unlike JavaScript, which is text-based and interpreted, WASM is a compiled format, making it much faster for performance-critical tasks like games and multimedia processing.

The raw power of Rust comes at the cost of a steeper learning curve. The syntax of the language is certainly much more approachable than C++, but still, programmers need to understand many low-level concepts before developing competent, robust applications. Swift instead offers a lower entry barrier and an easier jump start.

1.4.9 Picking the Ideal Language

If you are trying to decide between Swift and another language for your software project, there are naturally many factors to consider. This section offers some thoughts to point you in the right direction.

If you are targeting Apple devices, Swift is the natural choice for native development. It ensures the best performance and reliability in most cases, and you get access to the full power of the device plus the earliest access to new features. In addition, the footprint of your app will be minimal.

If you are targeting multiple platforms, there are two typical approaches:

- **Native**
  With the native approach, you can use a native language for each platform. For instance, you can use Swift for Apple development and Kotlin for Android development. That brings the advantages of native development to the table, but your development time and maintenance costs will probably increase because you'll have to build and manage different codebases for different target platforms.

- **Cross-platform**
  With the cross-platform approach, you can use an OS-independent platform to build your app—like React Native, Electron, Tauri, or Flutter. In that case, you most likely won't need to handwrite much Swift code; most of the development will be in a middle ground language like JavaScript. That brings the advantage of reduced development costs and complexity, but you'll have to endure performance overhead, a larger app size, limited access to native features, and an unavoidable vendor lock-in. It is also not uncommon to find yourself in a desperate situation if the app on the device spits out an error rooted in the cross-platform framework; those errors may be hard to solve without vendor support and can cause delays.

For projects without an Apple focus, Swift won't be the first choice of the typical programmer. If you are stepping outside of Apple's ecosystem, you'll probably find languages with more mature and vibrant communities, resources, and readily available libraries targeting your task at hand.

For the lowest level of development, such as embedded or kernel programming, the lowest-level programming languages would be preferred over others anyway—such as C, C++, Rust, or even Assembly.

There may be some other, more powerful factors affecting your language choice; for example, the experience of the development team at hand or the availability of ready-to-use libraries on a certain platform might be leading factors of your decision. For instance, Python is one of the exclusive choices for machine learning due to the high availability of libraries and resources on that topic.

1

Avoid Tech Fanaticism

We recommend being careful to avoid fanaticism. Some communities can be very vocal about their technological choices, creating an impression of universal superiority.

However, forks, spoons, and knives were invented for different purposes, and they reside peacefully on the same table. The same applies to development technologies such as programming languages. There are good and not-so-good use cases for each language; no programming language can do everything equally well.

A healthy approach is to understand the strengths and weakness of each option and to pick the best tool accordingly. It doesn't make sense to use a fork to eat soup, right?

Now that you understand Swift's place within the programming industry, you can take your first steps by preparing your development environment before jumping into our debut "Hello World" app!

1.5   Setting Up the Development Environment

In this section, you will prepare your development environment for Swift development, which is a rather easy task on a Mac. Yes, you read that right: It is highly recommended to have a Mac for Swift development—which makes sense as the primary goal of Swift is development for Apple devices. It's therefore logical to do the development itself on an Apple device and OS as well, where the infrastructure and emulators of all kinds are readily available. During the learning phase, any Mac will do so long as it can run Xcode. Don't feel obliged to make a huge investment for your debut.

If you don't have a Mac, don't fret: The situation is not hopeless. We'll offer alternative solutions.

1.5.1   Preparing Your Mac

In this section, we'll guide you through the simple steps to prepare your Mac for Swift development.

Installing Xcode

Unlike some other languages with complex installation and configuration steps, all you need to do is to install the latest IDE on your Mac, which is Xcode. You might already have Xcode installed on your Mac; if that's the case, fire it up and complete any necessary updates. Otherwise, simply find it in the App Store and install it like any other application. That's it!

Once you have it up and running, Xcode should greet you with its initial welcome window, shown in Figure 1.1.

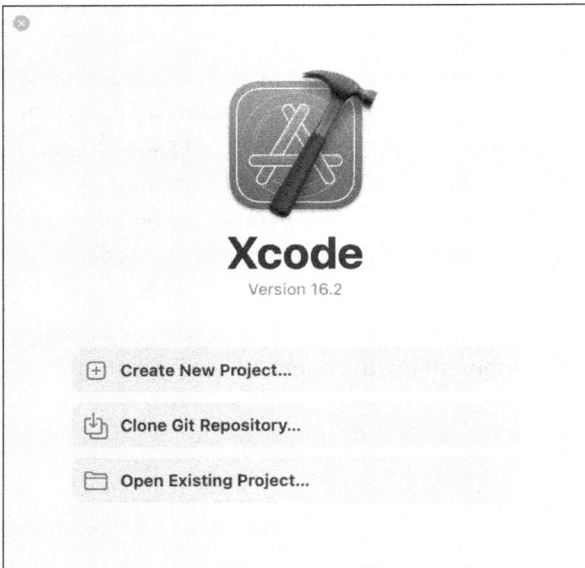

Figure 1.1 Welcome Window of Xcode

Xcode Playground Versus Projects

In your Swift learning journey, the first thing to know is the difference between a playground and a project. A *playground* can be considered a quick and dirty scratchpad file for the Swift language in which you can write and execute simple Swift code snippets to test things out. Although it has limited functionality for serious software projects, it is a nice platform to discover the language and try out features.

A *project* is a fully featured development package. If you are developing a new app or library, you ought to create a new project for it. However, this doesn't mean that you can't create educational projects during your learning phase: quite the contrary! Because creating a project costs nothing, you can open up as many projects as you like and play around with them.

Now let's see both options in action!

"Hello, World!" in a Playground

Let's begin with the simpler option and write a "Hello, World!" code snippet in a playground. On the Xcode menu, select **File · New · Playground**, as shown in Figure 1.2.

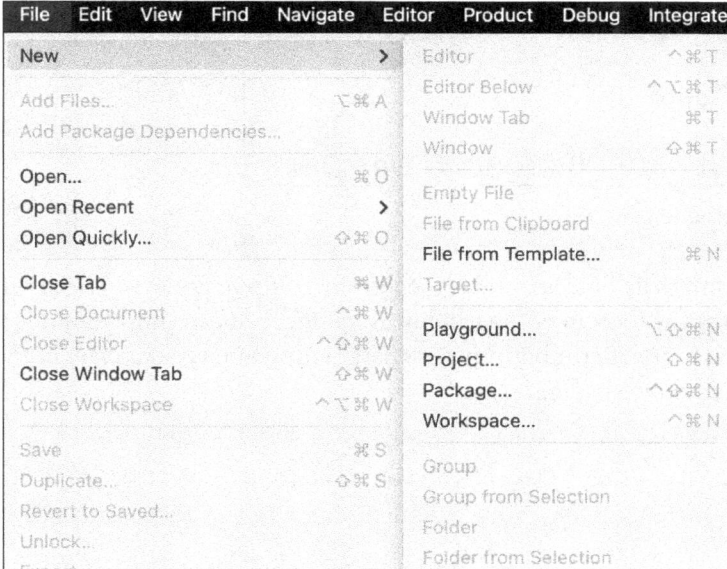

Figure 1.2 Starting New Playground in Xcode

In the popup window, browse to the **macOS** tab and select **Blank**, as shown in Figure 1.3.

Figure 1.3 Creating Blank Playground

Now Xcode will save your playground file to the disk. Pick an appropriate folder, then save your playground as "HelloWorld.playground", as shown in Figure 1.4.

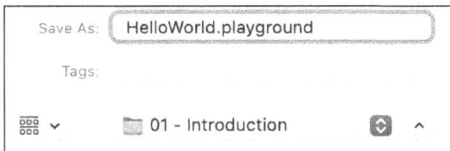

Figure 1.4 Saving Playground to Disk

That's it! Now Xcode will bring up the playground editor, in which you can write Swift code. There may even be some initial code in it already, as shown in Figure 1.5.

Figure 1.5 Initial Playground Code

The first thing you want to do here is to show the terminal output as you will need it. If the terminal section at the bottom of the window is hidden, you can simply unhide it by clicking the **Expand** button in the bottom-right part of the window. Your playground window should look as shown in Figure 1.6.

Figure 1.6 Visible Terminal Section of Playground

Now let's write some Swift code! Delete the initial playground code and replace it with the content of Listing 1.1.

```
print("Hello world!")
```

Listing 1.1 "Hello World" Code for Playground

Click the **Run** button at the top of the terminal, and voilà! Your Swift code is running! Your playground should look as shown in Figure 1.7.

You can close and reopen your playground file via the intuitive **File · Open** menu anytime you want. You can simply think of it like any other text editor.

Next, let's do the same thing, but in a project.

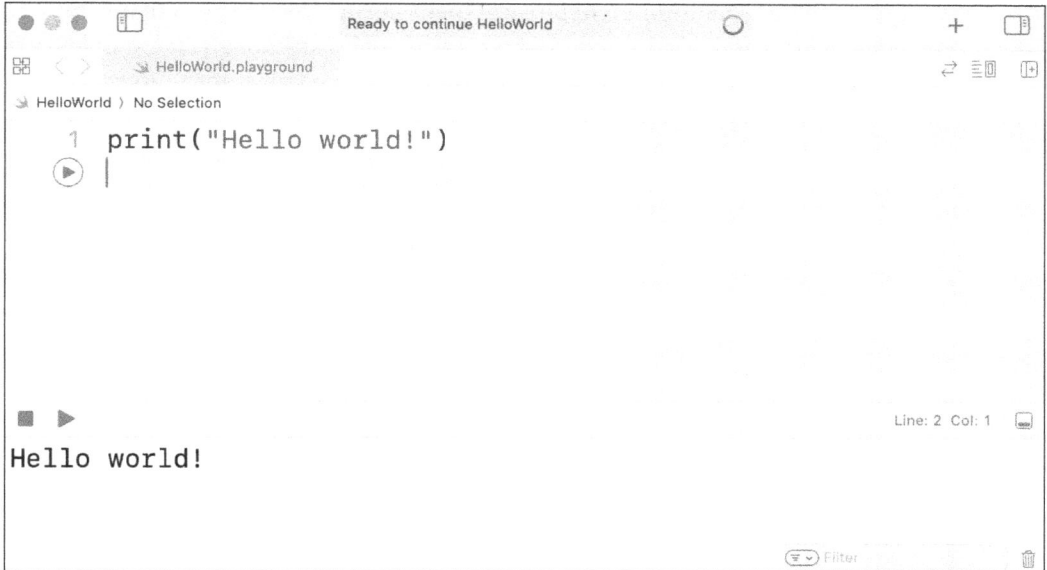

Figure 1.7 Playground Output for "Hello World"

"Hello, World!" in a Project

To create a new Xcode project, follow the **File · New · Project** menu path, as shown in Figure 1.8.

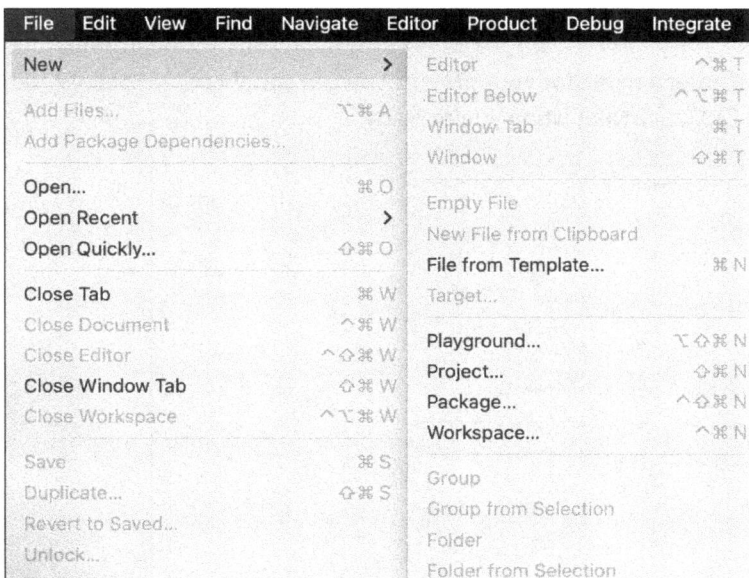

Figure 1.8 Starting New Project in Xcode

Although there are many different project types available, a simple command line project is enough to learn Swift basics—and certainly more than enough for a simple "Hello World" app! Select **macOS · Command Line Tool**, as shown in Figure 1.9, and click **Next**.

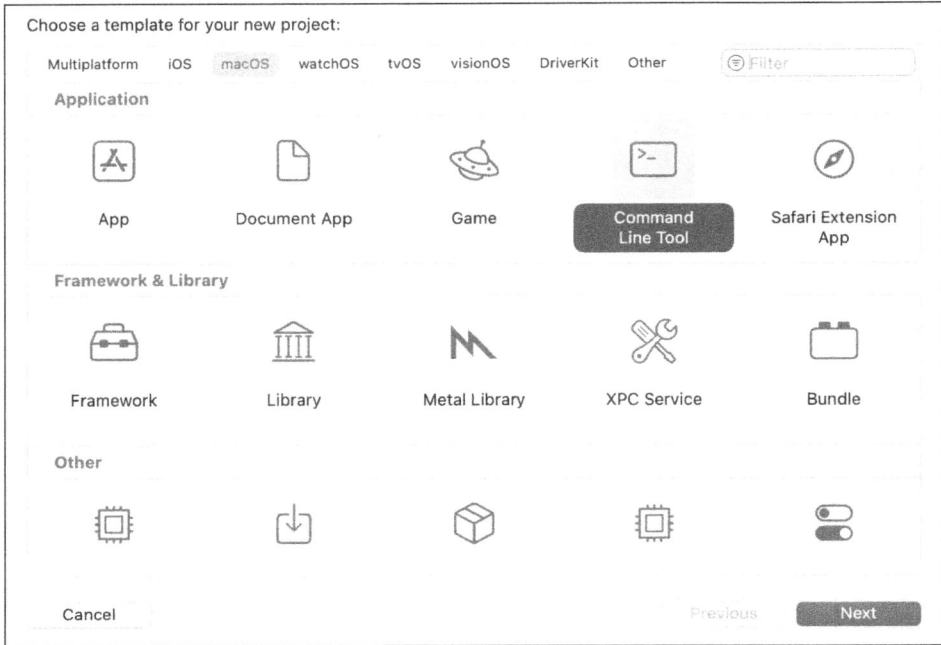

Figure 1.9 Picking Project Type

In the next window, enter a name for your project. You can simply call it "HelloWorld", as shown in Figure 1.10. Click **Next** when you're ready.

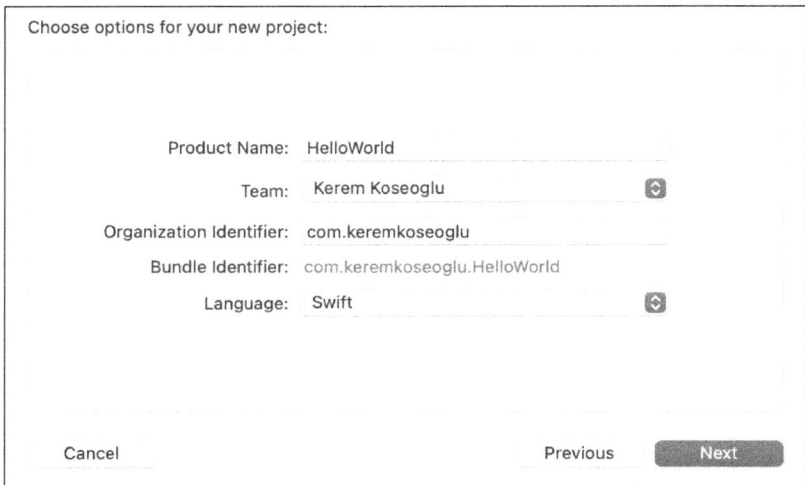

Figure 1.10 Naming Project

Select an appropriate folder for your project, as shown in Figure 1.11, and save it. You don't need to worry about further options (like Git) at this point.

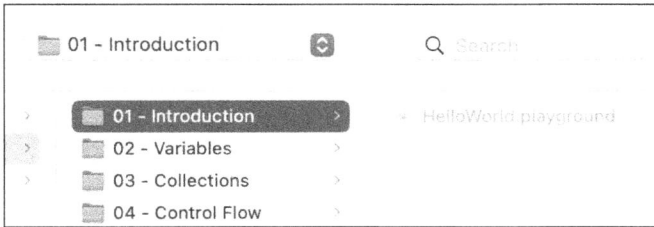

Figure 1.11 Saving Project to Disk

And voilà! The project is saved to your disk. You can close or open the project in Xcode via its intuitive **File • Open** menu any time you want, just like any other file.

On the left side of Xcode, you will see your project structure, as shown in Figure 1.12. Here, click **main.swift**, which is the main file in which to write Swift code. Xcode will show the editor, which contains some initial autogenerated code.

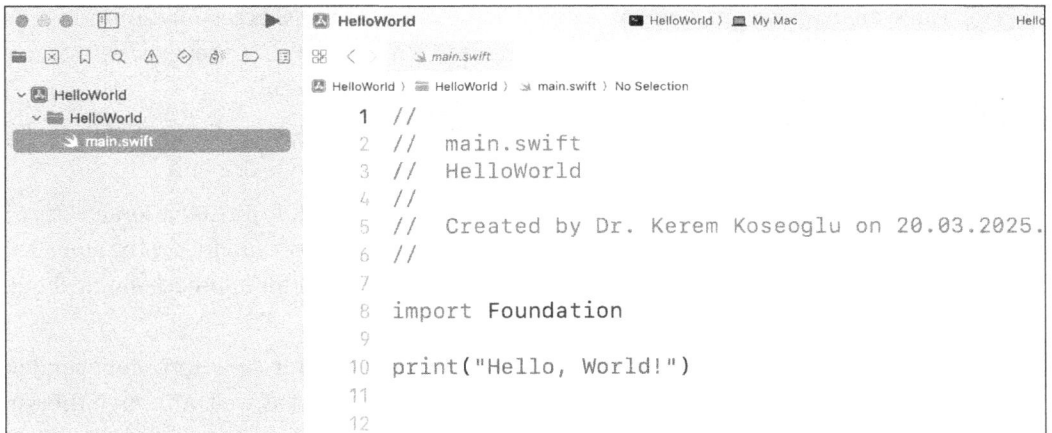

Figure 1.12 Main Swift Editor

You can replace the entire code with the content of Listing 1.2 to simplify it even further.

```
print("Hello, World!")
```

Listing 1.2 "Hello World" Code Snippet

Now click the **Run** button at the top of your editor. Alternatively, you can press ⌘+R (on your keyboard or select **Product • Run** from the Xcode menu. In any case, the simple code should run successfully and generate the output shown in Figure 1.13.

```
Hello, World!
Program ended with exit code: 0
```

Figure 1.13 Xcode Project "Hello World" Output

Voilà! Now you know how to study Swift over an Xcode project as well.

Note that all code samples in this book are available online! You can find all the Swift code within this book at *https://github.com/keremkoseoglu/swift/*, which can make it easier to follow through the book. Instead of manually writing code samples into Xcode, you can simply copy and paste from the repository.

1.5.2   Alternative Options for Swift Development

For those of you who don't possess a Mac, here are some alternative paths you can follow.

There are commercial cloud services offering remote macOS access. By subscribing to such a service, you can basically have access to a remote Mac. However, the costs may exceed the cost of a Mac in the long run, so do your math well!

In the Apple ecosystem, there is an iPad app called Swift Playgrounds. If you own an iPad (and preferably an external keyboard), you can study Swift via this app.

For Windows and Linux, Apple provides online guidelines for installation at *https://www.swift.org/install/*. You can pick your OS and distribution and follow the steps provided here. There might be missing or differing features in this approach, but it should be enough to start with Swift.

If you don't want to install anything at all or are looking for a platform-independent solution, there are browser-based online Swift playgrounds as well. Although they are not officially supported by Apple, they can be used for Swift studies. You can find them via your favorite search engine.

Hackintosh

You may have heard this term before, so here is the explanation. Apple's end user license agreement (EULA) contains license terms that state that macOS is only authorized to run on Apple-branded hardware. Despite that, some users misbehave and manage to install a copy of macOS on a PC or virtual machine. That way, they are able to run Mac apps—including Xcode.

We neither suggest nor support this kind of contract violation. We are merely explaining the term.

1.6 Summary

1

In this section, we took a bird's-eye look at the Swift language and its place in the programming ecosystem. Although Swift is a general-purpose language, its primary focus is development for Apple devices. It features a modern and approachable syntax while providing a strong and secure framework.

We explained the difference between playgrounds and projects while having you write your first "Hello World" apps via both options. Having a Mac and using Xcode is the best way to write Swift code, but there are alternative options if you don't.

Now that you've gotten your feet in the water, we can advance a little further and start your Swift coding journey! In the next chapter, you'll learn about variables.

Chapter 2
Variables

This chapter goes through the very basics of Swift, covering variables, basic data types, and operators.

In this chapter, you will make your debut into hands-on Swift adventures by learning about one of the most fundamental programming concepts: variables! For novices, we'll start by explaining the basics of variables, then we'll follow that up by covering different data types offered by Swift. We'll finally conclude with some tips and tricks. Buckle up, and let's go!

Xcode Screenshots

Because this chapter is your hands-on debut to Swift and Xcode, we'll provide screenshots of Xcode outputs for nearly every example. This will ensure that you get used to the IDE and understand how to see the results of Swift statements. In other chapters, the volume of screenshots will be reduced as appropriate.

2.1 Variables

So, what is a variable? We could approach this question from a hardware point of view, explaining complex details about RAM, stack, heap, and how ones and zeros dance on those floors. Or we could approach it from a math point of view, browsing our way through formulas and equations.

But because Swift is a high-level language with a focus on simplicity for developers, this Swift book can follow the same path. We can take a human-oriented approach to make concepts more understandable—which is going to be the main compass of this book whenever possible.

Imagine that a friend tells you to memorize the number 42 for a while. That's an easy task, right? What you do semantically is create a little box in your head and put the value 42 in there. Then, the same friend tells you to memorize the name Pam. What do you do? Create another little box in your head and put that name there.

The values *42* and *Pam* are stored in those boxes, and you can recall them when needed.

The operation of creating boxes to store values is present in programming languages as well, meaning that you can do that in Swift, as shown in Listing 2.1.

```
var theNumber = 42
var theName = "Pam"
```

Listing 2.1 Creating Boxes (Variables) in Memory to Store Values

As a result of the statements in Listing 2.1, the computer will happily create two boxes in its memory. One box will be called theNumber and will store the value 42. The other box will be called theName and will store the value Pam. This state is represented visually in Figure 2.1.

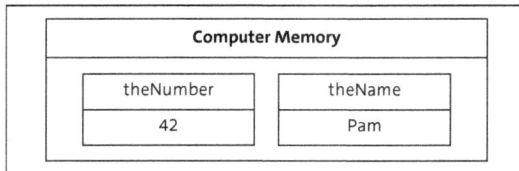

Figure 2.1 Variables in Computer Memory

In this simple case, theNumber and theName are variables. That's clear as day, right? They are little boxes used by the computer to hold/remember values to be used later.

Now that you've learned what a variable is, we need to differentiate between two core concepts—variables and constants—which we'll discuss in the following sections.

2.1.1 About Variables and Constants

Basically speaking, variables and constants are almost the same thing. They both are memory boxes that contain values. The distinction lies in their mutability. In a nutshell, you can change the value of a variable multiple times, while the value of a constant can only be assigned once; it can't ever be changed again.

You may be wondering why constants are ever needed if they're so inflexible. You're about to discover the answer in the next section.

2.1.2 Declaring Simple Variables and Constants

Let's start with variables. As stated, a *variable* is the flexible option. You can assign different values to a variable at different points in the program flow, as demonstrated in Listing 2.2.

```
var myAge = 33
print(myAge)

myAge = 34
print(myAge)
```

Listing 2.2 Assigning Different Values to Variable

In the first part of the code, you declare the `myAge` variable with the statement `var myAge`, which instructs the computer to create a box in its memory and call it `myAge`. Then, you immediately put the value 33 into that box. You follow that with `print(myAge)`, which prints the value of `myAge`.

That's followed by `myAge = 34`, which assigns a new value to the variable. So starting from this point, `myAge` no longer holds the old value, 33. Instead, it holds the new value, 34. When the last statement, `print(myAge)`, is executed, the program will now print the new value, 34.

You can type this code snippet into a playground and execute it! Figure 2.2 shows the output of this code snippet.

```
88   < >        002.playground
 002 ) No Selection
   1   var myAge = 33
   2   print(myAge)
   3
   4   myAge = 34
   5   print(myAge)

33
34
```

Figure 2.2 Output of Age Code Snippet

So far, so good! You've learned that you can change the value of a variable any time you want. But what makes a constant different? As its name implies, a *constant* is a strict kind of variable. Once you assign a value to a constant, you can never change it again. In programming terms, it is an *immutable* type of variable.

Listing 2.3 contrasts how variables and constants are declared.

```
let birthYear = 1993

var mood = "happy"
mood = "tired"
mood = "calm"
```

Listing 2.3 Constant Versus Variable Declaration

As shown in Listing 2.3, constants are defined with the `let` keyword. When you run the statement `let birthYear = 1993`, you're creating an immutable variable in the memory. You can no longer change the value of `birthYear`; even if you try, the compiler will generate an error.

This makes sense: The birth year of a person won't change after their birth. Such variables are preferred to be declared as constants for the following reasons:

- The intention of the programmer is clear.
- The purpose of the variable is made clearer.
- Accidental changes are prevented.
- Readability and maintainability are improved.

The mood of a person can change, however. If you declare a memory box to hold the mood of a user, you had better declare it as a variable instead of a constant. Which you did here, using the var mood statement. As shown in Listing 2.3, you could therefore change the value of mood whenever necessary.

That's clear, right? You could use variables all throughout your program, never declaring constants, and call it a day—but the distinction surely helps.

Naming Variables

Variable and constant names are case-sensitive in Swift. This means that myVariable, MyVariable, and MYVARIABLE are treated as different variables. In terms of standardization, the general approach is to use camelCase standard for variable names. This is not a technical requirement but just the general habit of the community.

camelCase is a naming convention in which each word in the name starts with a capital letter—except the very first word, which starts with a lowercase letter. Some examples include invoiceAmount, buttonIsEnabled, and expirationDate.

Constants follow the same convention for the most part. However, global constants are typically written in capitals and separated with underscores (*snake_case*), as in PI_VALUE or MAX_LIMIT.

It is also good to pick meaningful variable names. Instead of picking generic names like x, n1, or zAmount, self-explanatory names like studentIndex, firstName, and accountBalance will make your code easier to understand. You can use shorter names in very small local code blocks, though.

2.1.3   Type Annotations

In the introduction, we said that Swift is a statically typed language—which means that every variable in Swift must have a fixed type at compile time. Once you determine the data type of a variable, you can't change it thereafter.

Now, what is a data type? A *data type* indicates what kind of data can be stored in that variable. Let's walk you through Listing 2.4 for a better understanding.

```
var today = "Wednesday"
var hour = 19
```

```
hour = 20
hour = "abc" // ERROR!
```

Listing 2.4 Assigning Different Values to Variables

In the first line, you make the assignment var today = ."Wednesday". Swift is clever enough to understand that "Wednesday" is a string. Therefore, it assumes the variable today to be a *string* type of variable, which can be used to store texts.

You continue with the assignment var hour = 19. Once again, Swift has cleverly recognized the value 19 as a number (integer). Therefore, it assumed the variable hour to be an *integer* type of variable, which can be used to store numbers.

> **Data Types in Swift**
>
> In Swift language, there are various types of variables beyond strings and integers. You will discover these in the following sections of this chapter.

Now, pay attention to how the flow continues. The hour = 20 statement is perfectly fine; it changes the value of hour from 19 to 20. Because the new value, 20, is also a number, the assignment is valid.

But the last line is not perfectly fine. Here, hour = "abc" will inevitably fail because hour is an integer. You can't assign a string value to an integer variable—as the compiler will tell you (see Figure 2.3).

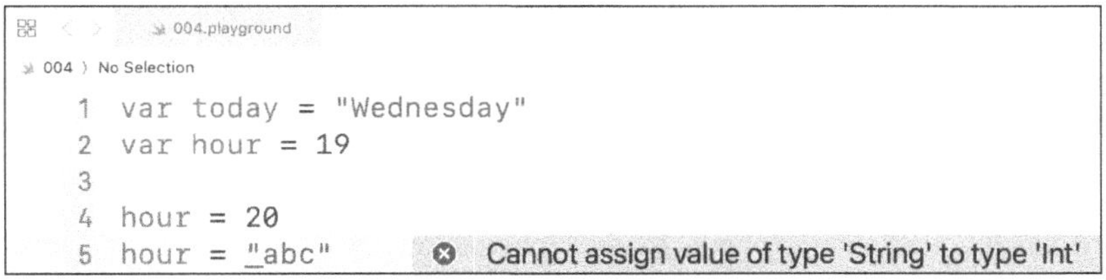

Figure 2.3 Compiler Error Due to Wrong Variable Assignment

This is basically what we mean by a *statically typed language*: The type of each variable is predetermined and won't be changed again. Swift is often clever enough to guess the type of the variable—especially for basic types like string and integers. This is called *type inference*.

But in many cases, you may want take full control and explicitly determine the type of the variable. Why? Some reasons are as follows:

- Some types are alike (e.g., integer, double, float), and Swift can make a wrong guess.
- You may want to declare the variable early without assigning a value yet.
- Swift can't infer the type of some empty variables, such as arrays or dictionaries.
- Code clarity to prevent programmer confusion is a good reason on its own.

With such motivations, let's show you how to explicitly determine the type of Swift variables. Listing 2.5 demonstrates the syntax of type annotations with direct value assignment.

```
var today: String = "Wednesday"
var hour: Int = 19
```

Listing 2.5 Type Annotations with Direct Value Assignment

In the first line, the today variable is explicitly typed as a String and its value is assigned immediately. In the second line, the hour variable is explicitly typed as Int and its value is assigned immediately as well. That's easy, right? There's no more room for confusion and type guessing!

Now what if you wanted to declare the variables first and assign their values later? It's easy! Check Listing 2.6, in which you simply split the declaration from the assignment.

```
var today: String
var hour: Int

today = "Wednesday"
hour = 19
```

Listing 2.6 Delaying Value Assignments

We advise that you become familiar with both direct and delayed assignments as both are going to be needed in different contexts.

Before moving on to the data types Swift has to offer, let's go over a small but useful coding feature: comments.

2.1.4   Comments

Programming is a hybrid task. In part, you're writing code that computers can understand. But never forget that humans need to understand your code as well. If the code is unclear and hard to understand, maintaining it through its lifecycle is going to be a difficult task. It can even be argued that if the code is clean enough, it should explain itself without needing any extra aid.

However, despite your best efforts, your code may still lack clarity in some cases, or some parts may be potentially ambiguous. In such cases, you have the option to include free text amid your Swift code, in the form of *comments*. Comments help humans understand the code a little better, but they are ignored by the compiler—so you can type in anything you want in plain English—or any human language you like!

In Swift, you have some different options for comments. Listing 2.7 demonstrates the single-line option, in which any line starting with // will be ignored by the compiler and can be used to contain free text.

```swift
// Define variables
var name: String
var surname: String
let birthYear: Int

// Assign dummy values to variables
name = "John"
surname = "Doe"
birthYear = 1990
```

Listing 2.7 Single-Line Comments

Listing 2.8 demonstrates the in-line option. At the end of any Swift statement, you can enter // and write free text after that in the same line—as in line 3 of the listing.

```swift
var name: String
var surname: String
let birthYear: Int // Because birth year never changes
```

Listing 2.8 In-Line Comments

Listing 2.9 demonstrates the multiline option. Here you're free to write any text you want between the /* and */ tokens. Those text inputs will be ignored by the compiler and will hopefully help humans follow through the code a little easier.

```swift
/*
 Define variables.
 We set birthYear as a constant, because
 the birth year of a person never changes.
 */
var name: String
var surname: String
let birthYear: Int

/*
 Now that we have the variables in place,
 let's assign some dummy values to them.
 */
name = "John"
surname = "Doe"
birthYear = 1990
```

Listing 2.9 Multiline Comments

You can even get a little adventurous and start an in-line comment using /*, continuing it on the following lines and ending with */ , as demonstrated in Listing 2.10.

```
var name = "John" /* This is a multiline comment
                    starting from the end of a
                    Swift command */
```

Listing 2.10 In-Line Multiline Comments

Are Comments Bad?

Although comments can be occasionally useful, relying on comments for code clarity is often considered a bad practice for a few reasons:

- If code needs comments to explain itself, it's likely not well structured.
- Comments can become outdated and misleading.
- Excessive commenting makes the code harder to read.

Some good reasons to write comments include the following:

- You can mark a temporary fix, indicating that it will be improved later.
- You can explain why code was written in case the business logic is not obvious.
- You can clarify complex algorithms or design patterns by naming them.

Although comments are not inherently bad, you should consider them as added spice, not the main dish. This will become clearer as you advance through your programming path.

Now that you know about comments, we can start applying them to the code samples in due course as needed.

2.2   Boolean

Although we took a sneak peek at strings (text) and integers (numbers) previously, we'll mark your official entry into Swift data types by introducing the most basic type of them all: Booleans. Following that, we'll cruise toward more advanced types.

2.2.1   About Booleans

Boolean data types exist in nearly all programming languages. It is a black-or-white kind of data type: It can be either true or false. This also represents one of the most fundamental concepts of human logic: Some statements are either correct or incorrect.

Listing 2.11 demonstrates how to declare Boolean variables and directly assign values to them.

```
var swiftIsEasy: Bool = true
var catsCanFly: Bool = false

var birdsCanFly: Bool
birdsCanFly = true
```

Listing 2.11 Direct Value Assignment to Boolean Variables

A Boolean data type in Swift is called `Bool`. Its possible values are `true` or `false`. That's easy, right?

On the hardware level, computers store information using a binary system of ones and zeros because digital circuits operate on electrical charges in memory cells. The presence of charge (high voltage) represents the number 1 and no charge (low voltage) represents the number 0. A Boolean is the most basic type of variable because only one cell is needed to represent/remember/store its value. `swiftIsEasy = true` would charge its memory cell with voltage, while `catsCanFly = false` would discharge its memory cell.

When the computer needs to remember the value of either variable, it simply checks the electrical charge of that variable's memory cell. Is there a charge present? Then the value is `true`. Otherwise, the value is `false`.

This was a simplified explanation of how memory works to emphasize the fact that every variable operation causes an electronic correspondence in your computer.

Let's get back to Booleans now. Although they look simple, they are extremely useful—and we'll use them extensively in Chapter 4 when dealing with `if` conditions and control flow.

For now, let's discover some Boolean-related operators. These operators are fun because they correspond directly to human logic!

2.2.2 AND Operator

AND is a logical operator, which states that all of two (or more) given conditions must be true. For example, a customer in an apparel store may make a statement like, "The shirt I'm looking for must be green and XL." That sentence includes an AND operator already: Color must match *AND* size must match. Both conditions must be fulfilled to set the customer's buying decision to `true`.

Now, let's express this customer statement in Swift! Check Listing 2.12 to see the code snippet.

```
// Variable declarations
var colorOK: Bool
var sizeOK: Bool
var willPurchase: Bool
```

```
// Case 1: Both OK
colorOK = true
sizeOK = true
willPurchase = colorOK && sizeOK

// Case 2: Invalid color
colorOK = false
sizeOK = true
willPurchase = colorOK && sizeOK
```

Listing 2.12 AND Operator in Action

Wow, that's a lot of code! But no worries; we'll cruise through the code together. In the first part, you simply declare three variables—nothing fancy. Table 2.1 explains the purpose of each variable.

Variable	Purpose
colorOK	Indicates if the color matches
sizeOK	Indicates if the size matches
willPurchase	Indicates if the customer will buy the shirt

Table 2.1 Variables in Code Snippet

In case 1, you set colorOK = true and sizeOK = true, which means that the perfect shirt has been found.

In the following line, you make use of the AND operator, which is represented as && in the Swift language. The statement willPurchase = colorOK && sizeOK means that if both colorOK is true and sizeOK is true, then willPurchase will also be true. In any other case, willPurchase will be false.

When you execute this code snippet in the playground, you should get the result shown in Figure 2.4. Because both colorOK and sizeOK are true, willPurchase is also calculated as true. Expressed in pseudocode, "colorOK is true AND sizeOK is true" is true!

```
6  // Case 1: Both OK
7  colorOK = true                        ▦ true
8  sizeOK = true                         ▦ true
9  willPurchase = colorOK && sizeOK      ▦ true
```

Figure 2.4 Positive Purchase Decision with AND

Following the same reasoning, let's move forward to case 2, in which the size is OK, but the color is not OK. Naturally, the purchase decision will be negative in that case—as demonstrated in Figure 2.5. Expressed in pseudo-code, "colorOK is false AND sizeOK is true" is false!

```
11  // Case 2: Invalid color
12  colorOK = false                           false
13  sizeOK = true                             true
14  willPurchase = colorOK && sizeOK          false
```

Figure 2.5 Negative Purchase Decision with AND

Congratulations on learning about the AND operator in Swift! Now, let's move on to an equally popular operator, called *OR*.

2.2.3 OR Operator

OR is a logical operator that states that any of two (or more) given conditions must be true. To continue the previous example, the customer in the apparel store may also be looking for a sweatshirt—but they are fine with either a hoodie or a zippered sweatshirt. Either option works for them. In that case, their statement would be, "I want to buy a sweatshirt with a hood or with a zipper." You see? The OR operator exists in your daily life already! All you need to do is to learn how it's represented in Swift.

You learned that the AND operator is represented as && in Swift. In contrast, the OR operator is represented as ||. To see the OR operator in action, examine Listing 2.13.

```
// Variable declarations
var hasHood: Bool
var hasZipper: Bool
var willPurchase: Bool

// Case 1: Shirt with hood
hasHood = true
hasZipper = false
willPurchase = hasHood || hasZipper

// Case 2: Unrelated shirt
hasHood = false
hasZipper = false
willPurchase = hasHood || hasZipper
```

Listing 2.13 OR Operator in Action

Understanding the OR operator is very easy after learning about the AND operator. In case 1, hasHood || hasZipper semantically means "hasHood is true OR hasZipper is true." So either option is fine for the customer. The output of running this in the playground is shown in Figure 2.6.

```
6  // Case 1: Shirt with hood
7  hasHood = true                              ▣ true
8  hasZipper = false                           ▣ false
9  willPurchase = hasHood || hasZipper         ▣ true
```

Figure 2.6 Positive Purchase Decision with OR

Having hasHood = true is enough to calculate willPurchase as true due to the OR operator.

In case 2, however, both hasHood and hasZipper are false. Therefore, willPurchase will also be calculated as false—indicating that the customer won't be buying a sweatshirt without a hood or zipper. The playground output of that case is shown in Figure 2.7.

```
13  // Case 2: Size not OK, with both hood and zipper
14  sizeOK = false                                   ▣ false
15  hasHood = true                                   ▣ true
16  hasZipper = true                                 ▣ true
17  willPurchase = sizeOK && (hasHood || hasZipper)  ▣ false
```

Figure 2.7 Negative Purchase Decision with OR

Got it? Good! Now, let's see how to combine multiple conditions.

2.2.4 Nested Conditions

In the examples so far, you've seen single AND or OR conditions. But it's also possible to combine these and build nested conditions. To demonstrate this in action, let's extend the request of the imaginary customer. Now, so long as the size is right, the customer will buy any sweatshirt with a hood or zipper. This new request can be expressed in pseudocode as "willPurchase = sizeOK AND (hasHood OR hasZipper)." That makes sense, right? It's just a semantic formulation of the customer request.

Careful readers have already noticed that this semantic formulation already features a nested form of the AND and OR conditions! Now all you need to do is to write this in Swift language: Check Listing 2.14.

```
// Variable declarations
var sizeOK: Bool
var hasHood: Bool
var hasZipper: Bool
var willPurchase: Bool

// Case 1: Size OK with hood
sizeOK = true
hasHood = true
hasZipper = false
```

```
willPurchase = sizeOK && (hasHood || hasZipper)

// Case 2: Size not OK, with both hood and zipper
sizeOK = false
hasHood = true
hasZipper = true
willPurchase = sizeOK && (hasHood || hasZipper)
```

Listing 2.14 Nested Condition Example

In case 1, the size of the sweatshirt is right and it has a hood (but no zipper). This is good enough for the customer and they will buy this shirt. When you run this in the playground, you see the same result—as shown in Figure 2.8.

```
 7  // Case 1: Size OK with hood
 8  sizeOK = true                                          true
 9  hasHood = true                                         true
10  hasZipper = false                                      false
11  willPurchase = sizeOK && (hasHood || hasZipper)        true
```

Figure 2.8 Positive Result of Nested Condition

In case 2, the size of the sweatshirt is wrong. Although it has both a hood and zipper, the customer won't buy it. The same conclusion is also reached in the playground—as shown in Figure 2.9.

```
13  // Case 2: Size not OK, with both hood and zipper
14  sizeOK = false                                         false
15  hasHood = true                                         true
16  hasZipper = true                                       true
17  willPurchase = sizeOK && (hasHood || hasZipper)        false
```

Figure 2.9 Negative Result of Nested Condition

Is that clear? Good! Needless to say, you can go further than a single set of parentheses and make your nested conditions as complex as needed, just like in a math formula. However, keep in mind that code readability is always a high priority. If your statements become too complex, you can always break them down.

Listing 2.15 demonstrates an example in which the nestable condition was broken into two steps. Focus on the featuresOK variable, which acts as an intermediary result for hasHood || hasZipper.

```
// Given facts
var sizeOK = true
var hasHood = true
var hasZipper = false
```

```
// Purchase decision
var featuresOK = hasHood || hasZipper
var willPurchase = sizeOK && featuresOK
```

Listing 2.15 Nested Condition Broken Down

There is no harm in nesting conditions in simple cases, but breaking things down helps in complex cases. This concludes our introduction to Boolean variables. Now, let's move forward to numbers.

2.3 Numbers

We already took a sneak peek at variables storing numbers in Section 2.1. Now, we'll broaden your understanding and cover some numeric types that Swift has to offer, contrasting their differences.

2.3.1 Numeric Types

Nearly all applications need to deal with numbers in some form or shape; therefore, it is natural for any programming language to have strong support for numbers. Swift is no exception in that regard. But *number* is a vague term. In a strongly typed language like Swift, the compiler needs to know some things about your numeric variable, such as the following:

- Will you store whole numbers only, or do you need decimals?
- If you need decimals, how precise do they need to be?

Swift needs to know because simpler numeric types require less memory and perform faster in most cases, while complex numeric types require more memory and perform a little slower—but pay off with higher precision, which might be crucial in certain domains, like finance and statistics.

For that reason, Swift features different numeric types for you to choose from. One suggested approach is to pick the simplest numeric type you can live with.

Why do different number types require more memory? Remember Booleans, which could only be represented as one or zero in memory? Numbers follow a similar logic. Because computer memory cells are only able to represent the values 1 (charged) or 0 (discharged), you need to use a binary number system to represent numbers.

For example, as an eight-bit variable, the number 5 would be represented as 00000101 in memory—because that's what it corresponds to in the binary number system. That's nice and easy for the computer; such numbers occupy only eight memory cells.

But if you have a more complex data type (such as double or float), then the computer needs to reserve more memory cells to accurately represent the number. For instance,

Table 2.2 demonstrates the memory representation of the pi number (~3.14) as a 32-bit float—with decimal places and all.

Sign (1 Bit)	Exponent (8 Bits)	Mantissa (23 Bits)
0	10000001	10011001100110011001100

Table 2.2 Representation of Pi in Memory

As you see, it costs 32 bits (memory cells) to represent the complex pi number, in contrast to the representation of the number 5, which merely cost 8 bits of memory. In a large application, such memory-consumption differences may stack up and end up being the deciding factor between a fast and lightweight application or a slow and heavyweight application.

> **Too Technical?**
>
> If the hardware aspect was a bit too technical for you, no worries! The main takeaway is as follows: Pick the simplest or smallest number type possible, and you're good to go! Don't use a number type with decimals if you will be storing whole numbers only, for instance.
>
> Because computers have powerful processors and plenty of memory nowadays, you may get away with picking bloated data types in small, local applications. But in big applications, especially with multiuser scenarios, such memory extravagance may not scale well—so it's best to gain the good habits early on.

Now we'll introduce the different numeric types so that you can pick the best one for your purposes. We'll move from the simplest to the most complex.

Integer

An *integer* can be considered the simplest number type in Swift. It can store whole numbers only; no decimals are allowed. However, both negative and positive numbers are supported. Student counts, page numbers, and ages are some examples in which an integer type would come in handy.

The basic keyword to represent integers in Swift is Int. Check Listing 2.16 for some sample integer definitions.

```
var age: Int = 25
var count: Int = -100
var population: Int = 7_000_000  // Underscore for readability
```

Listing 2.16 Sample Integer Definitions in Swift

You may be wondering about the range of numbers that would fit into an integer. That's a complicated question with multiple answers.

When you use the data type Int (as we did in Listing 2.16), Swift automatically picks Int32 or Int64, depending on your device's architecture (32-bit or 64-bit). Check Table 2.3 for a detailed breakdown.

Platform	Int Autotype	Range
32-bit	Int32	-2,147,483,648 to 2,147,483,647
64-bit	Int64	-9,223,372,036,854,775,808 to 9,223,372,036,854,775,807

Table 2.3 Number Ranges of Int, Depending on Architecture

So on a 64-bit system (which is the standard nowadays), Listing 2.16 and Listing 2.17 would mean the same thing. But as you can't ensure the architecture of each and every target system, it may be a good idea to use the Int type and let the system decide how to interpret it.

```
var age: Int64 = 25
var count: Int64 = -100
var population: Int64 = 7_000_000  // Underscore for readability
```

Listing 2.17 Sample Integer Definitions with Int64

Even with Int32, integers may be taking up a lot more memory space than you need, though. Check Table 2.3: You don't need 10 digits all the time, right? What if you merely need a simpler whole-number variable to represent, say, a day of the week or the age of a cat? Luckily, Swift provides further integer types for a higher level of granularity. Table 2.4 showcases useful alternatives in that regard.

Type	Bits	Sign	Range
Int8	8 bits	Yes	-128 to 127
Int16	16 bits	Yes	-32,768 to 32,767
Int32	32 bits	Yes	-2,147,483,648 to 2,147,483,647
Int64	64 bits	Yes	-9,223,372,036,854,775,808 to 9,223,372,036,854,775,807
UInt8	8 bits	No	0 to 255
UInt16	16 bits	No	0 to 65,535
UInt32	32 bits	No	0 to 4,294,967,295
UInt64	64 bits	No	0 to 18,446,744,073,709,551,615

Table 2.4 Alternative Integer Data Types and Number Ranges

It's important to know about signed and unsigned integer types in advance. Types starting with Int are *signed numbers* and may contain negative numbers, while types starting with UInt are *unsigned numbers* and may not contain negative numbers. Because the sign itself also takes up one bit or memory space, unsigned numbers of the same size can represent larger numbers than their signed counterparts.

Let's work through some thought exercises for picking the correct integer type:

- To represent the day of the week or the age of a cat, you can comfortably use the UInt8 type because the numbers are always positive and won't exceed the upper limit in any case.

- To represent the stock count of an item in a warehouse, you'd better pick a larger number like UInt16 or UInt32 because a large warehouse may contain very large quantities of small items.

- To represent the quantities of incoming/outgoing goods in a store, you should pick a signed number because you'd need positive numbers for incoming goods and negative numbers for outgoing goods. Int16 should be more than enough for a midsize shop.

Those exercises hopefully trained your thoughts to pick the smallest sensible integer type. But there is no harm in being extra careful: When in doubt, you can always pick a larger type to be on the safe side.

Now that you're familiar with integers, let's move to the next level by introducing floats.

Float

A *float* is a mid-tier number type. It differs from integers because it can store decimals of up to six or seven places. This is an ideal number type to pick if you need decimals without a high grade of granularity. Calculations of student grade averages is a good example.

Listing 2.18 showcases some float declaration examples.

```
var pi: Float = 3.1415927
var temperature: Float = -13.15
```

Listing 2.18 Float Variable Examples

Some further use cases for float include the following:

- Speed/height calculations in games
- Transparency/scale levels in a UI
- Sensor data in embedded systems

For the highest degree of granularity, you always have the option to upgrade to a double.

Double

A *double* is the top-tier number type. It is similar to a float but can store up to 15 or 16 decimal places. This is an ideal number type to pick if you need decimals with the highest grade of granularity possible. Financial calculations or statistic applications are typical use cases.

Listing 2.19 showcases some double declaration examples. Note that the pi number can be represented with a higher granularity than its float counterpart in Listing 2.18.

```
var pi: Double = 3.141592653589793
var speed: Double = 299_792_458.0
```

Listing 2.19 Double Variable Examples

Some further use cases for double include the following:

- Scientific/engineering calculations
- Exact GPS coordinates
- Cryptography/security algorithms

Number Conversions

Although Swift is a statically typed language, it offers the option to convert values between different numeric data types. The self-explanatory syntax of type conversion is shown in Listing 2.20.

```
let intNumber: Int = 10
let floatNumber: Float = Float(intNumber)
let doubleNumber: Double = Double(intNumber)
let floatToDouble: Double = Double(floatNumber)
```

Listing 2.20 Simple Conversions Between Numeric Data Types

As a result, each variable will store the initial value of 10, as shown in Figure 2.10.

```
1  let intNumber: Int = 10                              10
2  let floatNumber: Float = Float(intNumber)            10
3  let doubleNumber: Double = Double(intNumber)         10
4  let floatToDouble: Double = Double(floatNumber)      10
```

Figure 2.10 Same Number Converted to Different Types

If only life were so simple! There are some gotchas in type conversions. First, when you cast a number type from one type to another, you may lose decimal points. In Listing 2.21, you are casting from float to integer.

```
let f1: Float = 1.23
let i1: Int = Int(f1)
```

```
let f2: Float = 1.99
let i2: Int = Int(f2)
```

Listing 2.21 Down-casting from Float to Integer

The result of that operation may or may not be desirable for you—but it's important to know the effect. Figure 2.11 shows the result of such an operation. As you can see, the decimal part of the float value was ignored as being converted to an integer, simply because integers can't represent decimals.

```
1  let f1: Float = 1.23      1,23
2  let i1: Int = Int(f1)     1
3
4  let f2: Float = 1.99      1,99
5  let i2: Int = Int(f2)     1
```

Figure 2.11 Result of Float to Integer Conversion

A similar effect is to be expected when down-casting from double to float too, for instance. Check Listing 2.22 for such an operation.

```
let d: Double = 9.12345678901234
let f: Float = Float(d)
```

Listing 2.22 Down-casting from Double to Float

As a result, you'll lose some decimals, simply because float is incapable of representing such high decimal granularities. Check Figure 2.12 to see the exact numbers.

```
1  let d: Double = 9.12345678901234    9,12345678901234
2  let f: Float = Float(d)             9,123457
```

Figure 2.12 Result of Double to Float Conversion

Next, you'll learn about a number-related topic that's often needed in the real world of programming: generation of random numbers.

Random Numbers

Generating random numbers is an essential part of everyday programming. Some sample cases include the following:

- Rolling dice or shuffling cards in a game
- Generating one-time passwords
- Randomizing AI training data
- Performing test automation with random inputs

Swift provides built-in mechanisms to generate random numbers efficiently. You can easily generate random integers, floats, or doubles within a range that you provide.

Listing 2.23 demonstrates a sample statement to generate a random integer between 1 and 100. In this statement, both 1 and 100 are included in the possible outcomes.

```
let luckyNumber = Int.random(in: 1...100)
```

Listing 2.23 Generation of Integer Between 1 and 100

Due to the nature of randomization, this code snippet will generate a different number every time you execute it. Figure 2.13 shows one possible outcome.

```
1  let luckyNumber = Int.random(in: 1...100)    98
```

Figure 2.13 Generated Random Whole Number

Following a similar syntax, you can also generate random floats and doubles. Listing 2.24 demonstrates sample statements to do so.

```
let randomFloat = Float.random(in: 0..<1)
let randomDouble = Double.random(in: 1.5...5.5)
```

Listing 2.24 Generation of Random Float and Double Numbers

Note that the first line ended with <1. This means that you want the upper limit to be less than 1, instead of including the possible outcome of 1. Figure 2.14 shows one possible outcome.

```
1  let randomFloat = Float.random(in: 0..<1)         0,6353782
2  let randomDouble = Double.random(in: 1.5...5.5)   3,176805994773629
```

Figure 2.14 Generated Random Numbers with Decimals

Note that the float outcome has fewer decimals than its double counterpart—just as expected.

How Random Is Random?

Computers often generate random numbers using algorithms, making them not truly random but *pseudorandom*. This means that if you knew the randomization algorithm of the system, you could potentially guess the next random number—a difficult task, but not impossible. This is not a big deal in many cases; the generated random numbers are random enough for daily use if absolute security is not a concern.

However, if you are developing a system targeting the highest grade of security, then you may have to rely on stronger measures. There are, for instance, hardware-based random number generators that derive values from physical processes like atmospheric noise or electrical circuit fluctuations—which is nearly impossible to predict deterministically.

Now, let's move forward to arithmetic operators, which will enable you to make math calculations.

2.3.2   Arithmetic Operators

As math is a fundamental part of any programming language, it is natural to expect support for arithmetic operations. Swift obviously fulfills this exception. Let's inspect what it has to offer in this regard.

Four Standard Arithmetic Operators

The standard arithmetic operations are addition, subtraction, multiplication, and division, as we all know. Table 2.5 shows the Swift operators corresponding to these operations.

Operation	Operator	Example	Result
Addition	+	100 + 70	170
Subtraction	-	100 − 70	30
Multiplication	*	100 * 0.95	95
Division	/	170 / 2	85

Table 2.5 Arithmetic Operators in Swift

To see these operators in action, check Listing 2.25, in which each operator has been used once.

```
let myMoney: Double = 100
let yourMoney: Double = 70

let totalMoney = myMoney + yourMoney
let moneyDifference = myMoney - yourMoney
let myMoneyAsEuro = myMoney * 0.95
let averageMoney = totalMoney / 2
```

Listing 2.25 Arithmetic Operators in Action

Figure 2.15 showcases the results of those math statements.

> **Division by Zero**
>
> Like most programming languages, Swift is prone to errors if you divide a value by zero. If you are making a division between two variables, as in x / y, then some defensive programming is recommended to ensure that the denominator (y) is not zero.

```
1  let myMoney: Double = 100                   100
2  let yourMoney: Double = 70                   70
3
4  let totalMoney = myMoney + yourMoney        170
5  let moneyDifference = myMoney - yourMoney    30
6  let myMoneyAsEuro = myMoney * 0.95           95
7  let averageMoney = totalMoney / 2            85
```

Figure 2.15 Results of Basic Arithmetic Operations

You can express more complex formulas with the help of parenthesis as well—just like complex formulas in math. Listing 2.26 demonstrates an example of this.

```
let price1 = 100
let price2 = 400
let averagePrice = (price1 + price2) / 2
```

Listing 2.26 Usage of Parenthesis in Math

Just as in math, the formulas in the parentheses have priority. Therefore, (price1 + price2) will be executed before the following / 2 calculation, as shown in Figure 2.16.

```
1  let price1 = 100                            100
2  let price2 = 400                            400
3  let averagePrice = (price1 + price2) / 2    250
```

Figure 2.16 Results of Parentheses-Supported Calculation

If you have multiple sets of parentheses, then the innermost set will have the highest priority, while the outer set has a lower priority. In Listing 2.27, (100 + 10) would be executed first because it is within the innermost set of parentheses. That would be followed by the calculation of * 4 because it is within the outer-level parentheses. Finally, - 66 would be executed last because it is not in parentheses at all.

```
let result = ( (100 + 10) * 4 ) - 66
```

Listing 2.27 Swift Statement with Multiple Parenthesis

Always Use Parentheses

Math has a natural order of operation. However, not all programmers may remember those rules correctly. Even if you are the best math wizard around, you should consider the fact that the person reading your code might not be aware of the rules of priority.

Therefore, in the case of a complex math formula, it's recommended to use parentheses to express the priority of operations without leaving any room for doubt. This ensures that other programmers will read and interpret your code correctly.

Besides, this approach can prevent a possible human error on your end as well.

Now that you've learned about basic math operations, we're ready to move forward to more advanced ones.

Remainder

In Swift, the % operator returns the remainder of a division when placed between two numbers. Listing 2.28 demonstrates some remainder operations—including ones with negative numbers.

```
let rem1 = 23 % 3
let rem2 = 10 % 2
let rem3 = -23 % 3
let rem4 = 23 % -3
let rem5 = -23 % -3
```

Listing 2.28 Sample Remainder Operations in Swift

The results of those operations are shown in Figure 2.17. Negative numbers may deserve special attention, but they're all self-explanatory and expected results.

```
1  let rem1 = 23 % 3      2
2  let rem2 = 10 % 2      0
3  let rem3 = -23 % 3     -2
4  let rem4 = 23 % -3     2
5  let rem5 = -23 % -3    -2
```

Figure 2.17 Results of Remainder Operations

Odd and Even Numbers

It's common practice to use the % 2 operation to determine if a number is odd or even. If the remainder is 0 (as in 10 % 2), then the number (10) would be even; otherwise, it would be odd.

Basic Unary Operators

Because you're learning about numbers in Swift, this is a good time to discuss unary operators. A *unary operator* is an operator working on a single operand.

Unary plus (+) and minus (-) are the basic instances, which we used in some previous examples. Listing 2.29 demonstrates an explicit example.

```
let positiveValue = +10 // + is optional
let negativeValue = -10 // Negates the value
```

Listing 2.29 Unary Plus and Minus Example

The minus operator is often used in front of variables as well to mark their value as negative. In Listing 2.30, lines 2 and 3 produce the exact same result of -10, though line 2 is more compact.

```
let positiveNumber = 10
let negativeNumber1 = -positiveNumber
let negativeNumber2 = positiveNumber * -1 // Same as above
```

Listing 2.30 Unary Minus Example with Variable

Logical NOT (!) is used to invert a Boolean value:

- If the Boolean value is true, ! returns false.
- If the Boolean value is false, ! returns true.

Check Listing 2.31 to see it in action.

```
var loggedIn = false
var loggedOut = !loggedIn
```

Listing 2.31 Logical NOT Example

The resulting values of this operation are shown in Figure 2.18. As you can see, !loggedIn returned the opposite value of loggedIn, which is true.

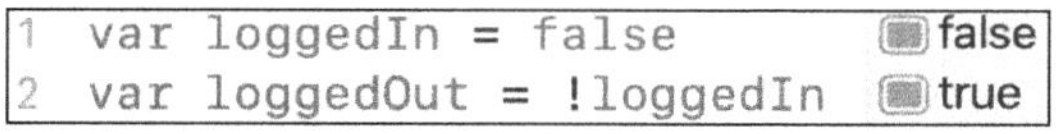

```
1  var loggedIn = false        false
2  var loggedOut = !loggedIn   true
```

Figure 2.18 Results of Logical NOT Example

Further Unary Operators

In due course, you'll learn about other unary operators as well, such as forced unwrapping (!) and optional chaining (?). Because we didn't go over the context for those operators here, their explanation is postponed.

Arithmetic Compound Assignment Operators

Swift features some syntactic sugar to improve readability, and compound assignment operators belong to that category. You can write any program without ever using them if you want, but they are widely used by Swift programmers due to their practicality. Table 2.6 explains those operators.

Operator	Example	Equivalent To
+=	x += 50	x = x + 50
-=	x -= 5	x = x - 5
*=	x *= 3	x = x * 3
/=	x /= 4	x = x / 4
%=	x %= 2	x = x % 2

Table 2.6 Arithmetic Compound Assignment Operators

Listing 2.32 features a Swift code snippet containing all the operators in a row.

```
var x: Int = 42
x += 50
x -= 5
x *= 3
x /= 4
x %= 2
```

Listing 2.32 Arithmetic Compound Assignment Operators in Swift Code

The results of those operations are shown in Figure 2.19.

```
1  var x: Int = 42       42
2  x += 50               92
3  x -= 5                87
4  x *= 3                261
5  x /= 4                65
6  x %= 2                1  ·
```

Figure 2.19 Results of Arithmetic Compound Operations

Revisiting Data Types

To revisit the topic of data types, pay special attention to line 5: The accurate result of 261 / 4 was supposed to be 65.25, but it was calculated as 65. Why? The type of x is integer, which doesn't support decimal places. For more accurate results, you would have to declare x as a float or double.

That concludes our content on arithmetic operators. Now, let's move forward with the equally important neighboring topic of logical operators.

2.3.3 Comparison Operators

Nearly all programs contain code blocks with value comparisons, and many programs use such blocks extensively. To ease the lives of programmers and enable a more elegant coding style, Swift features special operators for that purpose.

These operators compare two values and return a Boolean result—either true or false. You can find them in Table 2.7.

Operator	Meaning	Example
==	Equal to	x == y
!=	Not equal to	x != y
>	Greater than	x > y
<	Less than	x < y
>=	Greater than or equal to	x >= y
<=	Less than or equal to	x <= y

Table 2.7 Swift Comparison Operators

You can run through those operators via Listing 2.33.

```
let x = 10
let y = 5

let isEqualTo = x == y
let isNotEqualTo = x != y
let isGreaterThan = x > y
let isLessThan = x < y
let isGreaterThanOrEqualTo = x >= y
let isLessThanOrEqualTo = x <= y
```

Listing 2.33 Run-through with Swift Comparison Operators

The result of this run-through is shown in Figure 2.20. Note that each comparison returned a Boolean, as expected.

```
1  let x = 10                                      10
2  let y = 5                                       5
3
4  let isEqualTo = x == y                          false
5  let isNotEqualTo = x != y                       true
6  let isGreaterThan = x > y                       true
7  let isLessThan = x < y                          false
8  let isGreaterThanOrEqualTo = x >= y             true
9  let isLessThanOrEqualTo = x <= y                false
```

Figure 2.20 Result of Swift Comparison Operator Run-Through

Those operators will be especially useful when we're dealing with conditions in Chapter 4.

2.3.4   Numeric Literals

In Swift, *numeric literals* represent fixed number values in code. You've already used number literals many times in earlier examples. Now, let's take look at the different numeric formats supported by Swift.

Decimal Numbers

Decimal numbers are the everyday numbers in the basic math we all know (the decimal system or base-10 system). They contain the digits 0–9 and are the core part of all numeric types we covered in Section 2.3.1. To keep this section self-contained, Listing 2.34 features a couple of decimal number examples—with nothing different than you've seen so far!

```
let decimalNumber = 42
let largeNumber = 1_000_000
```

Listing 2.34 Decimal Number Examples

Binary Numbers

Binary numbers represent the core ones and zeros in the binary number system (or base-2 system).

We discussed the fact that under the hood, computers use a binary number system simply because memory cells can represent only two states: 1 (charged) or 0 (discharged). Any number stored in the memory is actually a binary number. For instance, the value we see as 46 is actually 101110 in the memory. In turn, there are *charged, discharged, charged, charged, charged, discharged* memory cells.

This mechanism will probably be challenged and outlived someday by a more advanced technology like quantum computing, but binary numbers are here to stay for now.

Swift lets you express binary values directly. You need to use the 0b prefix, followed by the binary value of ones and zeros. Listing 2.35 contains a code snippet for that in which you express the binary value of 46.

```
let binaryNumber = 0b101110
let intNumber = Int(binaryNumber)
```

Listing 2.35 Binary Number Example in Swift

You can see the result of those expressions in Figure 2.21. Swift successfully recognizes the value 0b101110 as binary, even translating it into the human-readable value of 46!

```
1  let binaryNumber = 0b101110          ■46
2  let intNumber = Int(binaryNumber)    ■46
```

Figure 2.21 Binary Value Results

Octal Numbers

Octal numbers are based on a number system using digits between 0 and 7 (the base-8 system). In the computing world, octal numbers have particular use cases, such as file permissions in Linux (755, 644, etc.) and legacy IPv4 network subnet masking (777000 for 255.255.0.0).

Using octal numbers is similar to using binary numbers, but they need to be expressed with the 0o prefix. Check Listing 2.36 for an example, in which you express the number 46 in its octal form.

```
let octalNumber = 0o56
let intNumber = Int(octalNumber)
```

Listing 2.36 Octal Number Example in Swift

The output of this code snippet is shown in Figure 2.22.

```
1  let octalNumber = 0o56              ■46
2  let intNumber = Int(octalNumber)    ■46
```

Figure 2.22 Octal Value Results

Hexadecimal Numbers

Hexadecimal (hex) numbers are based on a number system using an extended value set: the digits 0–9 and letters A–F (the base-16 system: A = 10, B = 11, ..., F-15).

Hexadecimal is widely used in computer science because it's compact and efficiently represents binary data. For instance, the binary value 0101101010110101 can be represented as the hex value 5AB5—which is much easier to read. The same value would be 23221 in our everyday decimal system.

Some typical use-cases for hexadecimal values are shown in Table 2.8.

Domain	Example Value
Memory addresses	7FFE1234ABCD
Colors in HTML	FF0000 (red)
Network device mac addresses	00:1A:2B:3C:4D:5E
Unique GUID values	0b460769-e0c5-4361-a264-df579446b4e1

Table 2.8 Some Use Cases for Hex Values

Hexadecimal values in Swift need to be expressed with the 0x prefix. Check Listing 2.37 for an example, in which you express the number 46 in its hexadecimal form.

```
let hexNumber = 0x2E
let intNumber = Int(hexNumber)
```

Listing 2.37 Hexadecimal Number Examples in Swift

Scientific Numbers

Swift supports the scientific expression of numbers as well, which has possible use cases in multiple domains: physics, astronomy, finance, chemistry, medicine, and so on. Swift offers a way of writing very large or small numbers in a compact form.

For example, let's convert the value 46,000 to the scientific notation in a multistep approach. To do so, follow through Table 2.9.

Step	Description	Result
1	Initial value	46000
2	Move the decimal left until only one digit remains to the left of it	4.6000
3	Count how many places the decimal moved	Four places
4	Write this in scientific notation	4.6×10^4
5	Express in Swift	4.6e4

Table 2.9 Quick Guide to Scientific Notation

Through this notation style, you can express a very large number like 900000000000 as 9e11, or a very small number like 0.00000000003 as 3e-11. Although both expressions represent the same value, scientific notation is much more readable for humans.

Now, it's Swift time! Listing 2.38 contains an example in which the number 273,6485 is expressed in the scientific form.

```
let scientificNumber = 2.736485e2
let doubleNumber = Double(scientificNumber)
```

Listing 2.38 Floating-Point Number Examples in Swift

The output of this code snippet is shown in Figure 2.23.

```
1  let scientificNumber = 2.736485e2          273,6485
2  let doubleNumber = Double(scientificNumber)  273,6485
```

Figure 2.23 Scientific Value Results

That concludes our section on numbers in Swift. So far, you have (hopefully) witnessed firsthand that Swift is a powerful and expressive language with a simple syntax, which minimizes boilerplate code. Moving forward to the topic of text handling, you'll see the same quality in string functions as well.

2.4 Text

We took a sneak peek at text handling in Swift earlier by assigning string values to variables. In this section, we'll give you a proper introduction to text in Swift, followed by a look at useful Swift mechanisms for string handling.

2.4.1 About Strings

Let's start by getting to know strings a little better. A *string* is a sequence of characters used to represent text in programming. Those characters can be letters, numbers, or symbols. Listing 2.39 demonstrates some basic string declarations in Swift.

```
let greeting: String = "Hello!"
let greeting2: String = "Hello 2!"
let name = "Kerem" // Auto-recognized as String
```

Listing 2.39 Some String Declarations in Swift

So far, so good; there's nothing too different from our former variable examples. But we should go a little deeper and understand how strings work under the hood. After all, we learned that computers store all data as binary values (1s and 0s). If that's the case, how are strings stored in the memory as ones and zeros?

Strings work with the help of a mechanism called *character encoding*, in which each character corresponds to a predefined, fixed number. For example, when you store the value Hi to a string variable, each letter is stored as a number. Although you see the letters H and i on the screen (in the user interface), what you have in memory is the numeric representation of those letters—as shown in Table 2.10.

Character	Decimal Representation	Binary Representation
H	72	01001000
i	105	01101001

Table 2.10 Storage of Letters "H" and "i" in Memory (ASCII)

When you assign the value Hi to a string variable, you are actually assigning the numbers 72 and 105 sequentially to that variable, like an array.

So do all computers know that 72 corresponds to H and 105 corresponds to i?

Almost. There are different character-encoding schemes in the software industry. ASCII is arguably the oldest and most basic encoding scheme, which contains English letters and basic symbols. Table 2.11 showcases a snippet of the standard ASCII encoding scheme.

Character	Decimal	Binary	Hexadecimal
A	65	01000001	0x41
B	66	01000010	0x42
...			
Y	89	01011001	0x59
Z	90	01011010	0x5A

Table 2.11 ASCII Encoding Scheme Snippet

When working with ASCII text, the computer will follow the ASCII translation table and show the value 65 as A, 66 as B, and so on. So in a sense, all computers know that 72 corresponds to H and 105 corresponds to i—so long as they are instructed to use the ASCII table to convert those numbers to text and vice versa.

However, ASCII is not the only encoding scheme in the industry; there are many others. Theoretically speaking, the number 72 may correspond to the letter H in one scheme, while the same number may correspond to X in another scheme. Therefore, the computer always needs to know what encoding scheme to use so that it can interpret the numbers correctly as letters.

Although ASCII may work fine for certain tasks, the golden standard of our age is arguably UTF-8, which supports all known characters, including special characters in non-English languages. Table 2.12 shows a snippet of the UTF-8 encoding scheme.

Character	Decimal	Binary	Hexadecimal
A	65	01000001	0x41
B	66	01000010	0x42
€	226 130 172	11100010 10000010 10101100	0xE2 0x82 0xAC
©	194 169	11000010 10101001	0xC2 0xA9
☺	249 159 152 128	11110000 10011111 10011000 10000000	0xF0 0x9F 0x98 0x80

Table 2.12 UTF-8 Encoding Scheme Snippet

As you can see, basic UTF-8 characters share the same decimal values with ASCII characters; that is, UTF-8 is mostly backward compatible with ASCII. But UTF-8 is an extended character set, containing many more characters.

You will be pleased to know that Swift uses UTF-8 as its default. Although text and its encoding scheme should always be in tandem, assuming the global standard of UTF-8 frees the programmer from incompatibility worries in many cases nowadays—especially in greenfield projects.

But this is no guarantee! You may encounter text files with different encoding schemes during your programming adventures, typically on legacy systems. In such a case, your Swift code should handle the file with the correct encoding scheme, which we'll look at in Chapter 14. For now, you can check Table 2.13 for some encoding schemes used in the industry to get an idea of what's out there.

Encoding	Description
ASCII	Basic Latin characters
ISO-8859-1	ASCII plus Western European characters
ISO-8859-9	Extended ASCII plus Turkish characters
UTF-8	Global standard; contains all languages and symbols
UTF-16	Fixed- or variable-length encoding (two to four bytes per character)
UTF-32	Fixed-length encoding (exactly four bytes per character)
Base64	Encodes binary data into ASCII characters

Table 2.13 Some Common Encoding Schemes

Weird Characters in Editors

This is the reason that you occasionally see weird characters when you open a text file. If the editor application makes an incorrect assumption about the encoding scheme of the file, then it can't display the correct characters. Instead, it displays incorrect characters from another encoding scheme.

For example, the decimal number 163 corresponds to the £ symbol in UTF-8 encoding, whereas the same number corresponds to the Ŀ symbol in ISO-8859-9 encoding. If you save a file as UTF-8 and forcefully reopen it as ISO-8859-9, you will see the incorrect Ŀ character instead of the intended £ character.

Now that have a better understanding of how strings work under the hood, let's advance to how to handle and manipulate text values in Swift.

2.4.2 Basic String Literals

In Swift, basic string literals are enclosed in double quotes; that's the basic syntax rule you need to follow. Although we covered many basic string examples before, Listing 2.40 showcases a basic string example to keep the section self-contained.

```
let myText = "I love Swift!"
let myText2 = "I love programming!"
```

Listing 2.40 Basic String Literals in Swift

2.4.3 Multiline String Literals

You have the option to declare multiline string literals as well, and quite easily! All you need to do is to use triple " tokens instead of single ones. Listing 2.41 showcases a clear example of this.

```
let multiLineString = """
How much wood
would a woodchuck chuck
if a woodchuck could chuck wood?
"""
```

Listing 2.41 Multiline String Literal in Swift

Note that Swift also takes line breaks into consideration, which becomes evident when the content is printed to the screen. Figure 2.24 demonstrates the output.

```
1  let multiLineString = """
2  How much wood
3  would a woodchuck chuck
4  if a woodchuck could chuck wood?
5  """
6  print(multiLineString)
7
■ ▶
How much wood
would a woodchuck chuck
if a woodchuck could chuck wood?
```

Figure 2.24 Playground Output of Multiline String

2.4.4 Escape Sequences

Swift supports certain escape sequences with special purposes for string literals, which are listed in Table 2.14.

Escape Sequence	Purpose
\n	New line / enter
\t	Tab
\\	Backslash
\"	Double quote
\'	Single quote

Table 2.14 Common Escape Characters in Swift

When you place one of those sequences within a string literal, the special purpose of the sequence is fulfilled instead of making the sequence part of the string. For instance, "Hello\tWorld" would result in Hello and World to be separated by a tab character. The \t expression won't be printed as part of the string.

All of this will be crystal clear with some examples! Check Listing 2.42, in which an escape sequence is planted into each string literal.

```
print("Line 1\nLine 2")
print("Hello\tWorld")
print("A\\backslash")
print("Double quote \"What she said\"")
print("Single quote \'What he said\'")
```

Listing 2.42 Swift Escape Sequence Example

The output of that example is shown in Figure 2.25, where the effect of each sequence becomes evident:

```
1  print("Line 1\nLine 2")
2  print("Hello\tWorld")
3  print("A\\backslash")
4  print("Double quote \"What she said\"")
5  print("Single quote \'What he said\'")

Line 1
Line 2
Hello    World
A\backslash
Double quote "What she said"
Single quote 'What he said'
```

Figure 2.25 Escape Sequence Output

- *Statement 1:* Words were broken down into separate lines.
- *Statement 2:* Words were separated by a tab character.

- *Statement 3:* A backslash was included.
- *Statement 4:* A double quote was included.
- *Statement 5:* A single quote was included.

Now what if you literally want to put the characters \n into the string? That seems impossible as Swift would interpret \n as an escape sequence, right? The same question is valid for all other escape sequences as well. Thankfully, there are special methods to work around this!

The "official" method is to put the string literal between #" "# instead of the usual " " characters. As a result, all characters are evaluated individually, and any (coincidental) escape sequences are ignored.

The "unofficial" workaround is to simply use the backslash escape character \\ where you want \ to appear, hiding any (coincidental) escape sequences from the compiler.

Both approaches are shown in Listing 2.43. The code is simpler than the explanation sometimes, right?

```
print(#"Hello\nWorld"#)
print("Hello\\nWorld")
```

Listing 2.43 String Literals with Ignored Escape Sequences

The output of this code is shown in Figure 2.26. Note that both statements have produced the same result. You can use the official method when you want all escape sequences in a string to be ignored and the unofficial workaround when only some escape sequences need to be ignored.

```
1  print(#"Hello\nWorld"#)
2  print("Hello\\nWorld")
```

```
Hello\nWorld
Hello\nWorld
```

Figure 2.26 Ignored Escape Sequence Output

2.4.5   Concatenation

Sometimes, you may encounter the requirement to combine two strings together to build a new string. For example, you may have to combine the name and surname of a student to build up their full name as a singular joint value. This operation is called *concatenation*.

In Swift, you can simply combine strings using the + operator. Although this is a known math operator to sum numbers, it can be used for concatenation as well. An example is shown in Listing 2.44.

```
let name = "Liam"
let middleName = "James"
let surname = "Carter"
let fullName = name + " " + middleName + " " + surname
```

Listing 2.44 Concatenation of Strings Using +

The compound assignment operator += can be used with strings as well! Check Listing 2.45, in which you build fullName step by step, concatenating a new word sequentially.

```
var fullName = "Liam"
fullName += " " + "James"
fullName += " " + "Carter"
```

Listing 2.45 Concatenation of Strings Using +=

The output of that code snippet is shown in Figure 2.27. Note that the fullName variable grew with new values on every new line with help of the += operator.

```
1  var fullName = "Liam"          "Liam"
2  fullName += " " + "James"      "Liam James"
3  fullName += " " + "Carter"     "Liam James Carter"
```

Figure 2.27 Concatenation Output, Demonstrating Sequential Growth of String

2.4.6 String Interpolation

Although concatenation with operators is a solid way to build strings from substrings, there is an alternative approach called *string interpolation*, in which you can plant variables into string literals. Listing 2.46 demonstrates this technique. Note how variable values were planted into string literals using the \(variable) pattern—that is, \(name) and \(age).

```
let name = "Alice"
let age = 30
let msg = "My name is \(name) and I'm \(age) years old."
```

Listing 2.46 String Interpolation Demonstration

As shown in Figure 2.28, when the msg string is built, \(name) is replaced with its value, Alice, and \(age) is replaced with its value, 30.

```
1  let name = "Alice"                                          "Alice"
2  let age = 30                                                30
3  let msg = "My name is \(name) and I'm \(age) years old."    "My name is Alice and I'm 30 years old."
```

Figure 2.28 String Interpolation Output

You could have reached the same result by concatenating strings, as demonstrated in Listing 2.47. Although the result is the same, string interpolation improves readability dramatically. Plus, it handles basic number-to-string conversions automatically, which you had to do manually with `String(age)` during concatenation (the main topic of Section 2.4.11).

```
let name = "Alice"
let age = 30
let msg = "My name is " + name + " and I'm " + String(age) + " years old."
```

Listing 2.47 Same Result Without String Interpolation

Concatenation or Interpolation?

Although both can be used interchangeably in many cases, you can use concatenation when you want to combine string variables and interpolation when you want to include variables in a fixed text literal.

2.4.7   Checking String Contents

In real-world programming, text-oriented requirements go way beyond joining strings. From this section on, you'll learn about using string functions in Swift for various purposes.

Let's start with functions to check string contents. Table 2.15 showcases some common functions for that purpose.

Function	Description
count	Returns the number of characters
isEmpty	Checks if the string is empty
contains(text)	Checks if the string contains a substring
hasPrefix(text)	Checks if the string starts with a substring
hasSuffix(text)	Checks if the string ends with a substring

Table 2.15 Functions to Check String Contents

The names of those functions are already very intuitive, don't you think? Invoking them is just as easy! Check out Listing 2.48 for a demonstration.

```
let intro = "In a hole in the ground there lived a hobbit."

let introCharCount = intro.count
let introIsEmpty = intro.isEmpty

let introHasThe = intro.contains("the")
let introHasDragon = intro.contains("dragon")

let introStartsWithIn = intro.hasPrefix("In")
let introEndsWithGround = intro.hasSuffix("ground")
```

Listing 2.48 Demonstration of String Content Functions

As you can see, a string function is invoked by placing it at the end of the string variable, separated by a dot. For instance, intro.count would return the number of characters in the variable intro. If you had a second variable called outro, you would need to type outro.count to get the number of characters in that variable. This logic applies to all functions discussed earlier.

Class Functions

You'll learn much more about functions in Chapter 5 and classes in Chapter 9. For now, it's enough to understand that the syntax object.variable is used to access a variable of an object, as in intro.count, which contains the character count of the intro string.

Likewise, the syntax object.function() invokes a function/subroutine of that object, as in intro.contains("the"), which tells whether the intro string contains the substring "the".

The syntax and variable/function names are mostly intuitive anyway; any English speaker can follow them!

To see how the functions are performed, check the output shown in Figure 2.29.

```
 1  let intro = "In a hole in the ground there lived a hobbit."      ▣ "In a▐
 2
 3  let introCharCount = intro.count                                 ▣ 45
 4  let introIsEmpty = intro.isEmpty                                 ▣ false
 5
 6  let introHasThe = intro.contains("the")                          ▣ true
 7  let introHasDragon = intro.contains("dragon")                    ▣ false
 8
 9  let introStartsWithIn = intro.hasPrefix("In")                    ▣ true
10  let introEndsWithGround = intro.hasSuffix("ground")              ▣ false
```

Figure 2.29 Output of String Content Functions

Everything seems to have worked as expected! For the sake of extra clarification, Table 2.16 contains a detailed breakdown of how each function operated in the example.

Statement	What It Did	Result
`intro.count`	Counted the number of characters in `intro`	45
`intro.isEmpty`	Checked if `intro` is empty	false, because we have a sentence in the variable
`intro.contains("the")`	Checked if the substring the is contained in `intro`	true, because it does
`intro.contains("dragon")`	Checked if the substring dragon is contained in `intro`	false, because it doesn't
`intro.hasPrefix("In")`	Checked if `intro` starts with the substring In	true, because it does
`intro.hasSuffix("ground")`	Checked if `intro` ends with the substring ground	false, because it doesn't

Table 2.16 Breakdown of Function Operations

Function Return Types
Note that each function returns a value of the appropriate type. For instance, you get a Boolean output for `isEmpty` and a numeric output for `count`.

Swift allows you to use those functions with string literals as well, as demonstrated in Listing 2.49.

```
let nameLength = "Ethan Caldwell".count
```

Listing 2.49 String Function Applied to Literal

One important point of consideration is case sensitivity. Know that `contains`, `hasPrefix`, and `hasSuffix` are case-sensitive functions. In Listing 2.50, you run two different searches: one with `Future` and one with `future`, differing by whether the first character is upper- or lower-case.

```
let movieName = "Back to the Future"
movieName.contains("Future")
movieName.contains("future")
```

Listing 2.50 Capital Letter–Emphasized String Search

As shown in Figure 2.30, the search with `Future` (capital F) did succeed, because it matched the substring in the original string completely. The search with `future` did not.

```
1  let movieName = "Back to the Future"    "Back to the Future"
2  movieName.contains("Future")             true
3  movieName.contains("future")             false
```

Figure 2.30 Result of Case-Sensitive Search

If you want to run a case-insensitive search, you can make use of the `.lowercased` string function, which returns a lowercase version of the string. When you transform both the content and search term to lowercase, a case-insensitive match is expected to be found. Listing 2.51 demonstrates this approach in a dramatic fashion.

```
let movieName = "BaCk tO the FUtUrE"
let searchTerm = "fuTuRe"

let lowMovieName = movieName.lowercased()
let lowSearchTerm = searchTerm.lowercased()

lowMovieName.contains(lowSearchTerm)
```

Listing 2.51 Case-Insensitive String Search

Figure 2.31 illustrates success! After converting both strings to lowercase, a match was found—which wouldn't be the case otherwise. Note that this technique can be used with other string-comparison functions as well.

```
1  let movieName = "BaCk tO the FUtUrE"          "BaCk tO the FUtUrE"
2  let searchTerm = "fuTuRe"                      "fuTuRe"
3
4  let lowMovieName = movieName.lowercased()      "back to the future"
5  let lowSearchTerm = searchTerm.lowercased()    "future"
6
7  lowMovieName.contains(lowSearchTerm)           true
```

Figure 2.31 Case-Insensitive String Search Result

2.4.8 Substrings

When you need to slice and dice strings into smaller substrings, you can resort to corresponding Swift functions. The common ones are listed in Table 2.17.

Function	Description
prefix(n)	Returns the first n characters
suffix(n)	Returns the last n characters

Table 2.17 Common Substring Functions in Swift

Function	Description
dropFirst(n)	Removes the first n characters
dropLast(n)	Removes the last n characters
[startIndex..<endIndex]	Extracts a substring with ranges

Table 2.17 Common Substring Functions in Swift (Cont.)

Now that you're used to invoking string functions, these new ones will be easy to apply! The first four functions are relatively straightforward, so let's handle them quickly in Listing 2.52.

```
let sentence = "My favorite number is 42, folks!"

sentence.prefix(5)       // First 5 characters
sentence.suffix(5)       // Last 5 characters
sentence.dropFirst(5)    // Removing first 5 characters
sentence.dropLast(5)     // Removing last 5 characters


sentence
```

Listing 2.52 Basic Substring Functions

Although the output is not hard to guess, it's shown in Figure 2.32. One point worth noting is the last line: Despite all substring operations, the original value of sentence was never changed! Substring operations don't mutate (change) the original value; they simply return new values built from the original value.

```
1  let sentence = "My favorite number is 42, folks!"        ▣ "My favorite number is 42, folks!"
2
3  sentence.prefix(5)       // First 5 characters            ▣ "My fa"
4  sentence.suffix(5)       // Last 5 characters             ▣ "olks!"
5  sentence.dropFirst(5)    // Removing first 5 characters   ▣ "vorite number is 42, folks!"
6  sentence.dropLast(5)     // Removing last 5 characters    ▣ "My favorite number is 42, f"
7
8  sentence                                                  ▣ "My favorite number is 42, folks!"
```

Figure 2.32 Output of Basic Substring Functions

Now, those functions were dealing with the first or last part of a string. What if you want to access a substring in the middle of the string? That's a bit trickier, but still easy once you understand the template code demonstrated in Listing 2.53.

```
let sentence = "My favorite number is 42, folks!"
let startPos = sentence.index(sentence.startIndex, offsetBy: 6)
let endPos = sentence.index(sentence.startIndex, offsetBy: 10)
sentence[startPos..<endPos] // Result: "orit"
```

Listing 2.53 Substring in Middle of String

First, you declare startPos, which represents the 6th character index within sentence. Next, declare endPos, which represents the 10th character index within sentence. Finally, sentence[startPos..<endPos] finds the characters between the 6th and 10th indexes, which are "orit".

One important aspect to understand is that indexes start with 0 in Swift, as in many other languages. When counting the 6th character in the sentence, Swift would start counting with 0 instead of 1 and find the character "o", as simulated in Table 2.18.

Character Number	Character
0	M
1	y
2	(space)
3	f
4	a
5	v
6	o

Table 2.18 Character Index Simulation

That's why character range [6..<10] produces "orit". You'll encounter the same logic in other collections in due course; this was an early heads-up!

Defensive Programming

When dealing with substring operations in a real-life application, taking defensive measures if essential. If you have a 10-character string and try to access the 15th character, you will get an error because you are requesting a nonexisting character.

As a rule of thumb, always check your indices against the string length (.count) before slicing and dicing the string. Or you can make use of error-handling mechanisms — which is the subject of Chapter 13. The main idea is to prevent an application crash due to a simple string error. This warning is valid for upcoming string functions too.

Defensive programming should be part of any real-world code you will ever write — not just string operations.

2.4.9 Modifying Strings

Now that you're warmed up with string processing, let's continue with string modification. As you would expect, Swift features many useful functions in that regard; the common ones are listed in Table 2.19.

Function	Description	Mutating
replacing(text, with: newText)	Find and replace operation	No
reversed()	Reverses the string	No
uppercased()	Converts the string to capitals	No
lowercased()	Converts the strings to lowercase	No
trimPrefix(text)	Removes the initial part of the string	Yes
insert(text, at: index)	Inserts a substring into the original string	Yes
removeSubrange(range)	Removes any part of the string	Yes

Table 2.19 Common String Modification Functions in Swift

Functions marked as *mutating* modify the original string directly. Others don't modify the original string and return a new modified string instead, which you can observe in Listing 2.54.

```
var quote = "May The Force be with you, always."

quote.replacing("always", with: "forever")
String(quote.reversed())
quote.uppercased()
quote.lowercased()
quote.trimPrefix("May ")

quote.insert(contentsOf:"Shall ", at: quote.startIndex)
let midIndex = quote.index(quote.startIndex, offsetBy: quote.count/2)
quote.insert(contentsOf:" strong", at: midIndex)

let range = quote.index(quote.endIndex, offsetBy: -9)..<quote.endIndex
quote.removeSubrange(range)
```

Listing 2.54 String Modification Functions in Action

This is a relatively long code snippet, but it is merely a demonstration of simple string functions, so stay tuned. Check the output in Figure 2.33, which we'll break down line by line.

```
 1  var quote = "May The Force be with you, always."                              ▣ "May The Force be with you, always."
 2
 3  quote.replacing("always", with: "forever")                                    ▣ "May The Force be with you, forever."
 4  String(quote.reversed())                                                      ▣ ".syawla ,uoy htiw eb ecroF ehT yaM"
 5  quote.uppercased()                                                            ▣ "MAY THE FORCE BE WITH YOU, ALWAYS."
 6  quote.lowercased()                                                            ▣ "may the force be with you, always."
 7  quote.trimPrefix("May ")                                                      ▣ "The Force be with you, always."
 8
 9  quote.insert(contentsOf:"Shall ", at: quote.startIndex)                       ▣ "Shall The Force be with you, always."
10  let midIndex = quote.index(quote.startIndex, offsetBy: quote.count/2)         ▣ Index
11  quote.insert(contentsOf:" strong", at: midIndex)                             ▣ "Shall The Force be strong with you, always."
12
13  let range = quote.index(quote.endIndex, offsetBy: -9)..<quote.endIndex        ▣ Range(Swift.String.Index(_rawBits: 2228487).
14  quote.removeSubrange(range)                                                   ▣ "Shall The Force be strong with you"
```

Figure 2.33 Output of String-Modification Functions

On the first line, we defined the original string as quote. So far, so good.

On line 3, we used the .replacing function to replace the value "always" with "forever". This did not mutate the original quote but returned a new string instead.

On lines 4, 5, and 6, we reversed, uppercased, and lowercased quote using the corresponding functions. None of this mutated the original value.

On line 7, we instructed the computer to remove the "May " prefix from the quote variable. This is a mutating operation, modifying quote on the spot. A heads-up, though: If quote did not start with "May ", then this statement would have no effect. In plain English, quote.trimPrefix("May ") means: "If quote starts with "May ", then remove this prefix; otherwise, do nothing."

On line 9, we inserted the value "Shall " at the beginning of the string using the .insert function. This could be done using simple string concatenation as well, but this alternative method prepares for the more interesting stuff right after.

On lines 10–11, we fulfilled the intention of implanting the value " strong" into the middle of quote. First we determined the middle index of quote with the help of the .index function. Then we have implanted " strong" right at that index, mutating quote in the process.

Finally, on lines 13–14, we removed the last nine characters of quote. Similar to the previous step, we determined the starting index of the removal using the .index function and removed the last few characters via .removeSubrange.

It's simple and straightforward, right?

2.4.10   Comparison Operators

When it comes to comparing strings, Swift features the same operators you learned about for numbers in Section 2.3.3. This is good news: We programmers can reuse our knowledge on the subject! As a memory refresher, you can refer to Table 2.20 to review those operators.

Operator	Meaning	Example
==	Equal to	x == y
!=	Not equal to	x != y
>	Greater than	x > y
<	Less than	x < y
>=	Greater than or equal to	x >= y
<=	Less than or equal to	x <= y

Table 2.20 String Comparison Operators in Swift

To see the operators in action, check Listing 2.55; note that they are used exactly the same way as they are for numbers.

```swift
let word1 = "Piano"
let word2 = "Guitar"

word1 == word2
word1 != word2
word1 > word2
word1 < word2
word1 >= word2
word1 <= word2
```

Listing 2.55 String Comparison Operator Demo

Check the results of those comparisons in Figure 2.34, and see if you find anything you didn't expect.

```
1  let word1 = "Piano"     "Piano"
2  let word2 = "Guitar"    "Guitar"
3
4  word1 == word2          false
5  word1 != word2          true
6  word1 > word2           true
7  word1 < word2           false
8  word1 >= word2          true
9  word1 <= word2          false
```

Figure 2.34 Result of String Comparison Operations

In those results, the == and != operators are intuitive; they simply check if two strings are the same or not. Remember, those operators are case-sensitive. "Guitar" == "guitar" would return false due to the case difference of the first letter.

But how do the rest of the operators work? With numbers, they worked intuitively; 5 > 3 would return true because 5 is greater than 3, for instance. How can "Piano" be greater than "Guitar"?

When we are comparing two strings, Swift compares the sequence of the letters in the corresponding encoding scheme, like ASCII or UTF-8. In ASCII, P (for *piano*) is the 80th character of the scheme, and G (for *guitar*) is the 71st letter of the scheme. Because 80 is greater than 71, "Piano" is considered greater than "Guitar".

In a shallower approach at the cost of accuracy, you can consider Swift to be comparing alphabetical sequences of the letters as numbers. Checking the first letters of those strings, P is the 16th letter of the alphabet and G is the 7th letter of the alphabet. Because 16 is greater than 7, "Piano" is considered greater than "Guitar".

If the first letters were equal, Swift would check the 2nd letters instead. When comparing "Piano" with "Penguin", i would be compared against e, for instance. And if the 2nd letters were the same too, the 3rd letters would be compared, and so on. A detailed demo in that regard is provided in Listing 2.56, followed by the results in Figure 2.35.

```
"Piano" > "Penguin"
"Piano" > "Pizza"
"Piano" > "Pia"
```

Listing 2.56 Some Further String Comparisons

```
1  "Piano" > "Penguin"    true
2  "Piano" > "Pizza"      false
3  "Piano" > "Pia"        true
```

Figure 2.35 Results of Further String Comparisons

The third comparison is worth extra attention. If one string is a prefix of the other, then the longer string is considered greater. That's why "Piano" is considered greater than "Pia".

Naturally, string comparison operators can be combined with Boolean operators too! Listing 2.57 demonstrates such a code snippet in form of *(Comparison 1) AND (Comparison 2)*. This should return true because both comparisons are true.

```
("Piano" == "Piano") && ("Piano" != "Guitar")
```

Listing 2.57 Combination of String and Boolean Operators

2.4.11 String/Number Conversion

In some cases, you have a string that should be interpreted as a number. A typical case would be opening a text file that contains numbers with Swift. Initially, the contents of the file would be returned as a string; it's our job to convert this to numbers before doing math calculations.

> **Handling Files in Swift**
>
> In Chapter 14, we will focus on dealing with files in Swift. If you're curious, you can take a sneak peek now!

Fortunately, Swift features a handful of alternative methods for string-to-number conversion and vice versa, which we'll cover in the following sections.

Simple Conversion

The first method is to simply wrap the string value with the target number type and two parentheses. Check Listing 2.58 for a demonstration, followed by the results in Figure 2.36.

```
let intNumber = Int("123")
let floatNumber = Float("123.45")
let doubleNumber = Double("123.45")

let intNumberNoDec = Int("123.45")
let wrongNumber = Int("123abc")
```

Listing 2.58 Simple String-to-Number Conversions Using Target Type

```
1  let intNumber = Int("123")                  123
2  let floatNumber = Float("123.45")           123,45
3  let doubleNumber = Double("123.45")         123,45
4
5  let intNumberNoDec = Int("123.45")          nil
6  let wrongNumber = Int("123abc")             nil
```

Figure 2.36 Results of String-to-Number Conversions Using Target Type

In this result set, the first three lines are intuitive. Because the strings are neatly formatted as if they were numbers, they can accurately be converted to the target type of integer, float, or double.

Line 5 and 6 deserve special attention. Because integers don't support decimal places, converting "123.45" to an integer spits out a nil value—which means that intNumberNoDec doesn't have a value at all. In simple terms, no value conversion took place. Likewise, "123abc" is obviously not a number in the first place; it is an alphanumeric sequence. Therefore, wrongNumber got a nil value too.

In case of an invalid number, you might want to assign a default value instead of nil. Swift enables this option via the ?? operator. Although this operator will be explained in more detail in Chapter 6, let's take a sneak peek in Listing 2.59. The extension ?? 0 means that if the conversion fails, you want the target variable to take up the value 0 instead of nil.

```
let goodNumber = Int("123") ?? 0
let badNumber = Int("abc") ?? 0
```

Listing 2.59 Using ?? Operator for Default Values

Check the results shown in Figure 2.37. The first line features a correct number; therefore, goodNumber got the value 123 instead of the fallback value of 0. The second line features an incorrect number; therefore, badNumber got the fallback value of 0 instead of nil.

```
1  let goodNumber = Int("123") ?? 0    ▣ 123
2  let badNumber = Int("abc") ?? 0     ▣ 0
```

Figure 2.37 Result of Default Value Assignments

Using the wrap method, it's easy to convert numbers to strings too because it uses the exact same approach! Check Listing 2.60 for a demonstration, followed by the output in Figure 2.38.

```
let doubleVal: Double = 12345.67
let stringVal = String(doubleVal)
```

Listing 2.60 Converting Numbers to Strings Using Wrap Method

```
1  let doubleVal: Double = 12345.67    ▣ 12.345,67
2  let stringVal = String(doubleVal)   ▣ "12345.67"
```

Figure 2.38 Number Converted to String

Number Formatter

If you want to go beyond simple conversions, you have the option to use the NumberFormatter class provided by Swift. Although classes will be discussed in Chapter 9, we will touch the concept briefly here; this section should be easy to understand without an in-depth understanding of classes.

Let's start by converting a number to a string using NumberFormatter. Listing 2.61 showcases a demonstration, followed by its output in Figure 2.39.

```
import Cocoa
let formatter = NumberFormatter()
formatter.numberStyle = .decimal
formatter.number(from: "1500222,12345")!
```

Listing 2.61 Number-to-String Conversion Using NumberFormatter with Default Settings

```
1  import Cocoa
2
3  let formatter = NumberFormatter()          <NSNumberFormat
4  formatter.numberStyle = .decimal           <NSNumberFormat
5  formatter.number(from: "1500222,12345")!   1.500.222,12345
```

Figure 2.39 Number Converted to String Using NumberFormatter with Default Settings

The first two lines are present to prepare a `NumberFormatter` object; you can think of those lines as boilerplate code for now. On the third line, you inform `formatter` that you want to interpret a `decimal` type of number. You have other options—currency, ordinal, percent, and so on—available via the autocomplete feature of the code editor, as shown in Figure 2.40.

```
formatter.numberStyle = .currency
formatter.number(from:   K currency
                         K currencyAccounting
                         K currencyISOCode
                         K currencyPlural
                         K decimal
                         K none
                         K ordinal
                         K percent
```

Figure 2.40 Number Style Autocomplete

Finally, on the fourth line, you've successfully converted the given value of "1500222,12345" to a number. An important aspect to note: A comma was used instead of a dot as the decimal separator, and `NumberFormatter` was still able to recognize the number! How did that magic happen?

On the Mac where this example was executed, the number settings were set up according to the Turkish locale, in which the dot is the grouping separator and the comma is the decimal separator—as shown in Figure 2.41. `NumberFormatter` is clever enough to use those settings as default when converting strings to numbers.

Date format	19.08.2025 ⟳
Number format	1.234.567,89 ⟳
List sort order	Turkish ⟳

Figure 2.41 Number Format Settings on the Mac

But you can override the defaults if you need to! Listing 2.62 provides a code snippet in which we instructed `NumberFormatter` to use custom `groupingSeparator` and `decimalSeparator` values when parsing the string. The successful result is shown in Figure 2.42.

```
import Cocoa

let formatter = NumberFormatter()
formatter.numberStyle = .decimal
formatter.groupingSeparator = ";"
formatter.decimalSeparator = "-"

formatter.number(from: "1;500;222-12345")!
```

Listing 2.62 String Conversion Using Custom Separators

```
1  import Cocoa
2
3  let formatter = NumberFormatter()          ▣ <NSNumberFormatt
4  formatter.numberStyle = .decimal           ▣ <NSNumberFormatt
5  formatter.groupingSeparator = ";"          ▣ <NSNumberFormatt
6  formatter.decimalSeparator = "−"           ▣ <NSNumberFormatt
7
8  formatter.number(from: "1;500;222−12345")! ▣ 1.500.222,12345
```

Figure 2.42 Result of String Conversion Using Custom Separators

Optionals in Swift

Careful readers may have noticed that the `formatter.number(…)!` statements end with an exclamation mark. In upcoming examples, you'll encounter other function calls ending with `!` as well.

This operator is part of the topic of optionals in Swift, which will be explained thoroughly in Chapter 6. Until then, you can simply accept the exclamation mark as an occasional necessity and ignore its effect.

For impatient readers, here is a summary: There are functions that optionally return a value. When you call such a function with an exclamation mark, you basically tell the computer that you're definitely expecting a result from the function. If the exclamation-marked function call doesn't return a value, then Swift will generate an error.

`NumberFormatter` can also be used to convert a number to a string via custom formatting rules. Listing 2.63 contains such an example.

```
import Cocoa

let formatter = NumberFormatter()
formatter.numberStyle = .decimal
formatter.groupingSeparator = ","
formatter.decimalSeparator = "."
```

```
formatter.maximumFractionDigits = 2

formatter.string(from: 1500.12345)!
```

Listing 2.63 Formatting Numbers with Custom Rules

We provided custom groupingSeparator and decimalSeparator values, as well as a maximumFractionDigits value of 2, which indicates that the string output should contain 2 decimals at most, even if the source number has more decimals. Figure 2.43 showcases the string output, formatted exactly as we wanted!

```
1  import Cocoa
2
3  let formatter = NumberFormatter()         ▦ <NSNumber
4  formatter.numberStyle = .decimal          ▦ <NSNumber
5  formatter.groupingSeparator = ","         ▦ <NSNumber
6  formatter.decimalSeparator = "."          ▦ <NSNumber
7  formatter.maximumFractionDigits = 2       ▦ <NSNumber
8
9  formatter.string(from: 1500.12345)!       ▦ "1,500.12"
```

Figure 2.43 Output of Custom Formatted Number

This concludes our intro to string-to-number conversions, which are possible via alternative methods too. You can use the direct conversion in simple cases and NumberFormatter when you need more control over the conversion format.

We'll continue our text adventures next with a topic that can sometimes be a bit scary for even the most experienced programmers.

2.4.12   Regular Expressions

A *regular expression (regex)* is basically a string template, containing certain rules. It is used for matching, finding, and manipulating text. Some typical use cases are as follows:

- Validating user input, such as email addresses and phone numbers
- Ensuring compliance, such as with password safety rules
- Searching and replacing text

For example, the expression ^[a-zA-Z0-9._%+-]+@[a-zA-Z0-9.-]+\.[a-zA-Z]{2,}$ is a sample regex pattern for email address validation. This regex surely looks complex, as most regex patterns do! The main point is that the string joe@gmail.com would match this regex, while the string joe;gmail,com would not. Therefore, using this regex pattern, you can check the validity of a user's email address with a minimum amount of code.

By using further regex patterns, you can implant many similar text-comparison and validation mechanisms into your code easily.

> **Regular Expressions Patterns**
>
> In the software industry, regex is a universal concept, which appears in most programming languages. It is not a Swift-specific feature. Because this is a book on Swift, we won't go into the details of understanding and building regex patterns. We'll merely cover how regex pattern features are invoked in Swift.
>
> There are countless online resources to learn how to build regex patterns, as well as browser-based playgrounds where you can test your regex pattern with various text inputs, allowing for easy unit tests.

Swift features multiple use cases for regex patterns.

String Validation

The basic functionality is to compare a string against a regex pattern, looking for a match or mismatch. Listing 2.64 showcases such a code snippet, covering our introductory email validation example.

```
let emailRegex = /^[a-zA-Z0-9._%+-]+@[a-zA-Z0-9.-]+\.[a-zA-Z]{2,}$/
let validEmail = "joe@gmail.com"
let invalidEmail = "joe:gmail,com"

validEmail.wholeMatch(of: emailRegex)
invalidEmail.wholeMatch(of: emailRegex)
```

Listing 2.64 String Comparison Against Regex Pattern

After defining the regex, we invoke the `wholeMatch` function of each string we have; that's the magic button for simple regex comparison. Check the result in Figure 2.44 and note that we got a nil value for the mismatching string "joe:gmail,com". That's how you spot mismatching strings—such as email inputs with typing errors.

```
1  let emailRegex = /^[a-zA-Z0-9._%+-]+@[a-zA-Z0-9.-]+\.[a-zA-Z]{2,}$/   Regex<Substring>
2  let validEmail = "joe@gmail.com"                                       "joe@gmail.com"
3  let invalidEmail = "joe:gmail,com"                                     "joe:gmail,com"
4
5  validEmail.wholeMatch(of: emailRegex)                                  ["joe@gmail.com"]
6  invalidEmail.wholeMatch(of: emailRegex)                                nil
```

Figure 2.44 Result of Regex Comparison with wholeMatch

Substring Detection

Another use case is to detect if a string contains a certain pattern of substring.

First, finding if a string contains an exact instance of a substring is an easy task, fulfilled with the `contains` function. Check the demonstration in Listing 2.65, followed by its intuitive output in Figure 2.45.

```
let mainText = "Billie Jean is not my lover"
mainText.contains("Jean")
```

Listing 2.65 Exact Substring Search

```
1  let mainText = "Billie Jean is not my lover"    "Billie Jean is not my lover"
2  mainText.contains("Jean")                       true
```

Figure 2.45 Output of Exact Substring Search

But sometimes you need to spot a text *pattern* within your main text, instead of an exact match. For example, you may want to check if the text contains any three digits in a row—like 123, 847, or 225. Instead of running a `contains` function for every three-digit combination possible, you can achieve this goal by applying the correct regex pattern. In Listing 2.66, the regex pattern /\d{3}/ has been used to look for three digits in a row.

```
let threeNumberRegex = /\d{3}/
let missingAddress = "I live in Sunset Way"
let completeAddress = "I live in 176 Sunset Way"


missingAddress.matches(of: threeNumberRegex)
completeAddress.matches(of: threeNumberRegex)
```

Listing 2.66 Three Digits in a Row Detection via Regex

The result of this code snippet is shown in Figure 2.46. Note that the nonmatching value in `missingAddress` returned an empty array, while the matching value in `completeAddress` returned a full array. We will learn about arrays in Chapter 3 – that's where you'll see how to evaluate those results. At this time, it is enough to understand that Swift was able to detect subsequent digits.

```
1  let threeNumberRegex = /\d{3}/                    Regex<Substring>
2  let missingAddress = "I live in Sunset Way"       "I live in Sunset Way"
3  let completeAddress = "I live in 176 Sunset Way"  "I live in 176 Sunset Way"
4
5  missingAddress.matches(of: threeNumberRegex)      Array of 0 Regex<Substring>
6  completeAddress.matches(of: threeNumberRegex)     Array of 1 Regex<Substring>
```

Figure 2.46 Result of Three Digits in a Row Detection

Substring Replacement

Finally, you can replace substring patterns with alternative text using regex. Following the previous pattern, Listing 2.67 showcases a code snippet in which any three subsequent digits are replaced with "***", masking those numbers—perhaps to address privacy concerns.

```swift
let text = "My phone number is 123-456-7890"
let numberPattern = /\d{3}/
let hiddenText = text.replacing(numberPattern, with: "***")
```

Listing 2.67 Substring Pattern Replacement Using Regex

The result of this operation is shown in Figure 2.47. Swift has successfully replaced 123 and 456 with ***, as intended. But when it encountered the value 7890, it replaced only the first three digits because that was the first match to the pattern. The resulting value was ***-***-***0.

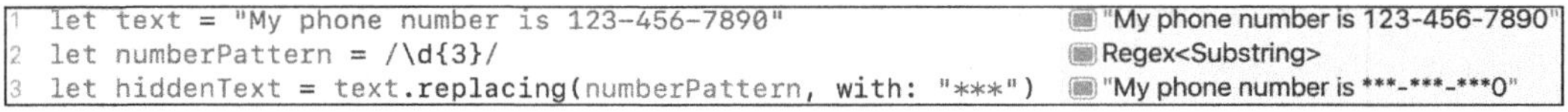

```
1  let text = "My phone number is 123-456-7890"              "My phone number is 123-456-7890"
2  let numberPattern = /\d{3}/                               Regex<Substring>
3  let hiddenText = text.replacing(numberPattern, with: "***")   "My phone number is ***-***-***0"
```

Figure 2.47 Result of Substring Pattern Replacement

This concludes our journey through regular expressions, as well as text processing in Swift. The functionality covered in this section should be enough for many typical string operations in the practical life of a programmer. Now, let's continue with another topic to be encountered equally often: dates and times.

2.5   Dates and Times

In this section, you'll learn how to deal with date/time values in Swift. Let's not name names here, but some programming languages make even the simplest date/time operations notoriously difficult and confusing. Luckily, Swift is not one of them. Quite the contrary: Such operations are quite intuitive and simple to execute, as you will see for yourself.

Let's start with date/time data types and advance from there.

2.5.1   Basic Data Types

When dealing with a date or time, Date is the first class you should learn about.

Classes in Swift

Classes are part of object-oriented programming in Swift. We'll cover this topic thoroughly in Chapter 9. However, in our current scope, the core syntax of date/time classes is very intuitive and easy. As we go through them, you can accept the syntax as it is, with deeper understanding to come in Chapter 9.

Date represents a specific moment, containing both the date and the time—so don't let the name fool you! Listing 2.68 showcases a code snippet to capture the current date/time in a variable, followed by its execution output in Figure 2.48.

```
import Foundation
let rightNow = Date()
```

Listing 2.68 Capturing System Date/Time in Swift

```
1  import Foundation
2  let rightNow = Date()     ▦ "20 Mar 2025 at 12:59"
```

Figure 2.48 System Date/Time Output

What if you need to represent a custom moment, instead of the system date? In that case, you can make use of the DateComponents class, which enables you to set custom values for the year, month, day, and so on. Then you can run those components through a Calendar object, building up the final date. This may sound a bit confusing if you don't have experience with object-oriented programming (yet), but the code example in Listing 2.69 should still be very intuitive.

```
import Foundation

var customDateComp = DateComponents()
customDateComp.year = 2019
customDateComp.month = 1
customDateComp.day = 1
customDateComp.hour = 10
customDateComp.minute = 23

let customDate = Calendar.current.date(from: customDateComp)!
```

Listing 2.69 Building Custom Date Object

All we did here was set custom values for the year, month, day, and so on, and glue them all together with the help of Calendar. Check the playground output in Figure 2.49, where the building up of the custom value can be followed in distinct steps.

```
1  import Foundation
2
3  var customDateComp = DateComponents()
4  customDateComp.year = 2019                                  year: 2019
5  customDateComp.month = 1                                    year: 2019 month: 1
6  customDateComp.day = 1                                      year: 2019 month: 1 day: 1
7  customDateComp.hour = 10                                    year: 2019 month: 1 day: 1 hour: 10
8  customDateComp.minute = 23                                  year: 2019 month: 1 day: 1 hour: 10 minute: 23
9
10 let customDate = Calendar.current.date(from: customDateComp)!   "1 Jan 2019 at 10:23"
```

Figure 2.49 Step-by-Step Execution of Custom Date Buildup

Note that we could have left out day, hour, and minute if we wanted to represent the date alone (without the time); you can try this out as an exercise.

Now that you know the long way to build a custom date, you're due for a shortcut! In Listing 2.70, you can see how to provide literal values for year, month, day, and so on directly during the initial declaration of customDateComp. That leads to shorter and more readable code in many cases.

```
import Foundation

let customDateComp = DateComponents(year: 2019, month: 1, day: 1, hour: 10,
minute: 23)

let customDate = Calendar.current.date(from: customDateComp)!
```

Listing 2.70 Shorter Approach to Custom Date Definition

If you go one step further and nest those statements, then you can build up a custom Date using a single statement—as demonstrated in Listing 2.71. The entire expression is almost readable as plain English, don't you think? That's the beauty of Swift: Even a non-programmer can read and understand many statements.

```
import Foundation

let customDate = Calendar.current.date(from: DateComponents(year: 2019, month:
1, day: 1, hour: 10, minute: 23))!
```

Listing 2.71 Even Shorter Approach to Custom Date Definition

Now that you're familiar with building date/time values, we can advance to date/time operations.

2.5.2 Arithmetic Operations

Swift turns date/time calculations into an easy task, thanks to the ready-to-use features of its underlying framework. We'll walk through these components in this section.

Addition

To add minutes, hours, days, and so on to an existing Date value, all you need to do is invoke an intuitive function. Check the code snippet in Listing 2.72, where we subsequently add minutes, hours, days, weeks, months, and years to the current date.

```
import Foundation

let today = Date()

let nextMinute = Calendar.current.date(byAdding: .minute, value: 1, to: today)!
let nextHour = Calendar.current.date(byAdding: .hour, value: 1, to: today)!
let nextDay = Calendar.current.date(byAdding: .day, value: 1, to: today)!
let nextWeek = Calendar.current.date(byAdding: .weekOfYear, value: 1, to: today)!
let nextMonth = Calendar.current.date(byAdding: .month, value: 1, to: today)!
let nextYear = Calendar.current.date(byAdding: .year, value: 1, to: today)!
```

Listing 2.72 Adding Values to Current Date

When invoking Calendar.current.date for that purpose, you need to provide three parameters—as explained in Table 2.21.

Parameter	Description
byAdding	Time unit of measure (.minute, .hour, etc.).
value	Value to add on top; .minute and 1 means to add 1 minute.
to	The base Date; in this example, we used the current date/time.

Table 2.21 Parameters for Addition Operation

Now that the parameters are clear, check the output in Figure 2.50. Each line added a different time unit on top of the base Date and returned a new Date value—without mutating the original value.

```
1  import Foundation
2
3  let today = Date()                                                              "20 Mar 2025 at 13:00"
4
5  let nextMinute = Calendar.current.date(byAdding: .minute, value: 1, to: today)!  "20 Mar 2025 at 13:01"
6  let nextHour = Calendar.current.date(byAdding: .hour, value: 1, to: today)!      "20 Mar 2025 at 14:00"
7  let nextDay = Calendar.current.date(byAdding: .day, value: 1, to: today)!        "21 Mar 2025 at 13:00"
8  let nextWeek = Calendar.current.date(byAdding: .weekOfYear, value: 1, to: today)! "27 Mar 2025 at 13:00"
9  let nextMonth = Calendar.current.date(byAdding: .month, value: 1, to: today)!    "20 Apr 2025 at 13:00"
10 let nextYear = Calendar.current.date(byAdding: .year, value: 1, to: today)!      "20 Mar 2026 at 13:00"
```

Figure 2.50 Results of Addition Operations

If the purpose is to add a duration to a custom Date (instead of the current date/time), you can do that by simply replacing the initial Date definition, as shown in Listing 2.73.

```
import Foundation

let baseDate = Calendar.current.date(from: DateComponents(year: 2019, month: 1,
day: 1, hour: 10, minute: 23))!

let nextMinute = Calendar.current.date(byAdding: .minute, value: 1, to:
baseDate)!
let nextHour = Calendar.current.date(byAdding: .hour, value: 1, to: baseDate)!
let nextDay = Calendar.current.date(byAdding: .day, value: 1, to: baseDate)!
let nextWeek = Calendar.current.date(byAdding: .weekOfYear, value: 1, to:
baseDate)!
let nextMonth = Calendar.current.date(byAdding: .month, value: 1, to:
baseDate)!
let nextYear = Calendar.current.date(byAdding: .year, value: 1, to: baseDate)!
```

Listing 2.73 Adding Values to Custom Date

If you are targeting a higher level of granularity, you may like the alternative method in Listing 2.74. Here, we are simply invoking the .addingTimeInterval function of the base Date object and passing the seconds we want to add. Because .addingTimeInterval accepts a double value, we can add very small subsecond values by making use of decimals.

```
import Foundation

let today = Date()

let nextMicroSecond = today.addingTimeInterval(0.000001)
let nextMiliSecond = today.addingTimeInterval(0.001)
let nextSecond = today.addingTimeInterval(1)
let nextMinute = today.addingTimeInterval(60)
let nextHour = today.addingTimeInterval(3600)
```

Listing 2.74 Adding Small Values to Current Date

Cool, right? Depending on your target level of granularity, you can follow any method.

Subtraction

If you want to subtract time, you don't need to follow a different path! All you need to do is to provide negative temporal values instead of positive values. Listing 2.75 features a subtraction example for each method.

```
import Foundation

let today = Date()
let prevHour = Calendar.current.date(byAdding: .minute, value: -1, to: today)!
let prevMiliSecond = today.addingTimeInterval(-0.001)
```

Listing 2.75 Subtracting Values from Current Date

2.5.3   Comparison

The Date type is *comparable*, meaning that you can directly compare two Date values using the comparison operators you learned about for numbers and strings! To keep this section self-contained, Table 2.22 provides a reminder of those operators.

Operator	Meaning	Example
==	Equal to	x == y
!=	Not equal to	x != y
>	Greater than	x > y
<	Less than	x < y
>=	Greater than or equal to	x >= y
<=	Less than or equal to	x <= y

Table 2.22 Date Comparison Operators

Listing 2.76 showcases a demonstration of date-comparison operations, followed by the obvious output in Figure 2.51. You see: It's not really any different than a number or string comparison!

```
import Foundation

let calendar = Calendar.current

let date1 = calendar.date(from: DateComponents(year: 2025, month: 1, day: 10))!
let date2 = calendar.date(from: DateComponents(year: 2025, month: 4, day: 3))!
let date3 = calendar.date(from: DateComponents(year: 2025, month: 1, day: 10))!

date1 == date3
date1 != date2
date1 < date2
date1 > date2
```

Listing 2.76 Demonstration of Date Comparison Operators

Although those operators are useful, they compare the entirety of given `Date` values—including their years, months, days, hours, minutes, and seconds. But sometimes you may want to find out if, for example, two dates are on the same day—ignoring their hours, minutes, and seconds.

```
1  import Foundation
2
3  let calendar = Calendar.current                                                              gregorian (gregorian) loca
4
5  let date1 = calendar.date(from: DateComponents(year: 2025, month: 1, day: 10))!   "10 Jan 2025 at 00:00"
6  let date2 = calendar.date(from: DateComponents(year: 2025, month: 4, day: 3))!    "3 Apr 2025 at 00:00"
7  let date3 = calendar.date(from: DateComponents(year: 2025, month: 1, day: 10))!   "10 Jan 2025 at 00:00"
8
9  date1 == date3                                                                    true
10 date1 != date2                                                                    true
11 date1 < date2                                                                     true
12 date1 > date2                                                                     false
```

Figure 2.51 Output of Date-Comparison Demonstration

Swift has a direct solution to that requirement. In Listing 2.77, there are three `Date` values on the same day, but at different times: $07{:}45$, $08{:}44$, and $08{:}46$. Using `Calendar`'s `isDate` function, you can check if those dates are on the same day (ignoring the hour, minute, and second) or on the same day and at the same hour (ignoring the minute and second).

```
import Foundation

let calendar = Calendar.current

let date1 = calendar.date(from: DateComponents(year: 2025, month: 4, day: 3,
hour: 7, minute: 45))!
let date2 = calendar.date(from: DateComponents(year: 2025, month: 4, day: 3,
hour: 8, minute: 44))!
let date3 = calendar.date(from: DateComponents(year: 2025, month: 4, day: 3,
hour: 8, minute: 46))!

calendar.isDate(date1, equalTo: date2, toGranularity: .day)
calendar.isDate(date1, equalTo: date2, toGranularity: .hour)
calendar.isDate(date2, equalTo: date3, toGranularity: .hour)
```

Listing 2.77 Comparing Dates with Special Granularity

Further similar granularities are naturally also supported, but those examples are enough to grasp the logic. If you check the output in Figure 2.52, you'll see that Swift has conducted the comparisons with the specified granularity levels.

```
1  import Foundation
2
3  let calendar = Calendar.current                                          gregorian (gregorian) lo
4
5  let date1 = calendar.date(from: DateComponents(year: 2025, month: 4, day: 3, hour: 7, minute: 45))!   "3 Apr 2025 at 07:45"
6  let date2 = calendar.date(from: DateComponents(year: 2025, month: 4, day: 3, hour: 8, minute: 44))!   "3 Apr 2025 at 08:44"
7  let date3 = calendar.date(from: DateComponents(year: 2025, month: 4, day: 3, hour: 8, minute: 46))!   "3 Apr 2025 at 08:46"
8
9  calendar.isDate(date1, equalTo: date2, toGranularity: .day)              true
10 calendar.isDate(date1, equalTo: date2, toGranularity: .hour)             false
11 calendar.isDate(date2, equalTo: date3, toGranularity: .hour)             true
```

Figure 2.52 Output of Date Comparison with Special Granularity

Finally, Swift features a shortcut functionality for the commonly required question of whether a given Date is yesterday, today, or tomorrow—as demonstrated in Listing 2.78. The output of those statements will depend on the day you execute the code and will be either true or false, as expected.

```
import Foundation

let calendar = Calendar.current

let date1 = calendar.date(from: DateComponents(year: 2025, month: 2, day: 23,
hour: 7, minute: 45))!

calendar.isDateInYesterday(date1)
calendar.isDateInToday(date1)
calendar.isDateInTomorrow(date1)
```

Listing 2.78 Check If Date Is Yesterday, Today, or Tomorrow

2.5.4 Measuring Durations

In programming, it is a common requirement to calculate the duration between two Date values. From games in which you must calculate the level completion duration to accounting applications in which you must detect overdue invoices, this requirement has correspondence all over the industry.

With this awareness, Swift empowers you with multiple options for duration calculation.

For a high granularity in seconds, minutes, or hours, you can make use of the timeIntervalSince function. This function initially calculates the duration in seconds, but you can convert this value to minutes or hours using simple math. Listing 2.79 features an example of this intuitive function, followed by its expected output in Figure 2.53.

```
import Foundation

let calendar = Calendar.current
```

```
let date1 = calendar.date(from: DateComponents(year: 2025, month: 4, day: 3,
hour: 7, minute: 45))!
let date2 = calendar.date(from: DateComponents(year: 2025, month: 4, day: 4,
hour: 8, minute: 45))!

let durationInSeconds = date2.timeIntervalSince(date1)
let durationInMinutes = durationInSeconds / 60
let durationInHours = durationInMinutes / 60
```

Listing 2.79 Duration Calculation in Seconds, Minutes, and Hours

```
1  import Foundation
2
3  let calendar = Calendar.current                                                    gregorian (gregorian) lo
4
5  let date1 = calendar.date(from: DateComponents(year: 2025, month: 4, day: 3, hour: 7, minute: 45))!    "3 Apr 2025 at 07:45"
6  let date2 = calendar.date(from: DateComponents(year: 2025, month: 4, day: 4, hour: 8, minute: 45))!    "4 Apr 2025 at 08:45"
7
8  let durationInSeconds = date2.timeIntervalSince(date1)                             90.000
9  let durationInMinutes = durationInSeconds / 60                                     1.500
10 let durationInHours = durationInMinutes / 60                                       25
```

Figure 2.53 Calculated Durations via timeIntervalSince

If you are targeting a lower granularity in days, months, or even years, then you can make use of the dateComponents class. This approach is easier understood in a demo than an explanation, so check out Listing 2.80.

```
import Foundation

let calendar = Calendar.current

let date1 = calendar.date(from: DateComponents(year: 2024, month: 5, day: 3,
hour: 7, minute: 45))!
let date2 = calendar.date(from: DateComponents(year: 2025, month: 4, day: 4,
hour: 5, minute: 32))!

let dayDiff = calendar.dateComponents([.day], from: date1, to: date2)
let monthDiff = calendar.dateComponents([.month], from: date1, to: date2)
let yearDiff = calendar.dateComponents([.year], from: date1, to: date2)
```

Listing 2.80 Duration Calculation in Days, Months, and Years

In this approach, we invoked dateComponents by providing the enumeration value .day, .month, or .year, declaring our intended level of granularity. Figure 2.54 showcases the calculated results.

```
1  import Foundation
2
3  let calendar = Calendar.current                                                      gregorian (gregorian) I...
4
5  let date1 = calendar.date(from: DateComponents(year: 2024, month: 5, day: 3, hour: 7, minute: 45))!   "3 May 2024 at 07:45"
6  let date2 = calendar.date(from: DateComponents(year: 2025, month: 4, day: 4, hour: 5, minute: 32))!   "4 Apr 2025 at 05:32"
7
8  let dayDiff = calendar.dateComponents([.day], from: date1, to: date2)                 day: 335
9  let monthDiff = calendar.dateComponents([.month], from: date1, to: date2)             month: 11
10 let yearDiff = calendar.dateComponents([.year], from: date1, to: date2)               year: 0
```

Figure 2.54 Calculated Durations in Days, Months, and Years

2.5.5   String/Date Conversion

Converting a date to a string is a common programming requirement. Typical use cases include displaying the date in the UI or printing labels containing dates. It is also a common requirement to convert a string to a date. Typical use cases include parsing the date input of a user or reading a date text from an uploaded text file.

For such requirements, you can make use of Swift's DateFormatter class, which will do much of the heavy lifting on your behalf. We'll explore this in the following sections.

Date to String

Let's start with date-to-string conversions, as shown in Listing 2.81. In this example, we will convert the current date/time to strings in various formats.

```
import Foundation

let date = Date()
let formatter = DateFormatter()

formatter.dateFormat = "yyyy-MM-dd HH:mm:ss"
formatter.string(from: date)

formatter.dateFormat = "yyyy-MM-dd"
formatter.string(from: date)

formatter.dateFormat = "yyyy (mm/ss)"
formatter.string(from: date)
```

Listing 2.81 Date-to-String Conversion in Swift

The code snippet starts by getting the current date/time in the date variable; there's nothing unusual here. It continues by initializing the formatter variable as a DateFormatter object. Following that, you can use formatter to format any date as a string using a custom template of your choice. Check Figure 2.55 to see how each template was used to build a different string output. For instance, "yyyy-MM-dd" resulted in "2025-02-25", containing the year, month, and day separated by hyphens.

```
 1  import Foundation
 2
 3  let date = Date()                                      "20 Mar 2025 at 13:07"
 4  let formatter = DateFormatter()                        <NSDateFormatter: 0x60
 5
 6  formatter.dateFormat = "yyyy–MM–dd HH:mm:ss"           <NSDateFormatter: 0x60
 7  formatter.string(from: date)                           "2025-03-20 13:07:27"
 8
 9  formatter.dateFormat = "yyyy–MM–dd"                    <NSDateFormatter: 0x60
10  formatter.string(from: date)                           "2025-03-20"
11
12  formatter.dateFormat = "yyyy (mm/ss)"                  <NSDateFormatter: 0x60
13  formatter.string(from: date)                           "2025 (07/27)"
```

Figure 2.55 Date Output Using Various Formats

As you can see, different string templates like "yyyy-MM-dd HH:mm:ss" determine how the output should appear. You have the flexibility to include custom characters too, such as hyphens, parentheses, and so on. Note that components like yyyy and dd are placeholders for date parts. Table 2.23 explains the components of date templates.

Template Component	Description
yyyy	Year
MM	Month
dd	Day
HH	Hour
mm	Minute
ss	Second

Table 2.23 String Template Components

This is as easy as string interpolations, right? Just provide your template and let Swift do the conversion for you!

String to Date

Converting a string to a date is equally easy! Let's go through the code example in Listing 2.82 for a comprehensive demonstration, followed by its output in Figure 2.56.

```
import Foundation

let formatter = DateFormatter()

formatter.dateFormat =  "yyyy-MM-dd HH:mm:ss"
let date1 = formatter.date(from: "1978-12-27 21:30:00")
```

```
formatter.dateFormat = "dd/MM/yyyy"
let date2 = formatter.date(from: "27/12/1978")

formatter.dateFormat = "HH:mm:ss"
let date3 = formatter.date(from: "14:30:45")

formatter.dateFormat = "yyyy-MM-dd HH:mm:ss"
let date4 = formatter.date(from: "Dummy text")
```

Listing 2.82 String-to-Date Conversion in Swift

Once again, we're making use of the DateFormatter class.

```
 1  import Foundation
 2
 3  let formatter = DateFormatter()                             <NSDateFormatter: 0x60
 4
 5  formatter.dateFormat = "yyyy-MM-dd HH:mm:ss"                <NSDateFormatter: 0x60
 6  let date1 = formatter.date(from: "1978-12-27 21:30:00")     "27 Dec 1978 at 21:30"
 7
 8  formatter.dateFormat = "dd/MM/yyyy"                          <NSDateFormatter: 0x60
 9  let date2 = formatter.date(from: "27/12/1978")              "27 Dec 1978 at 00:00"
10
11  formatter.dateFormat = "HH:mm:ss"                           <NSDateFormatter: 0x60
12  let date3 = formatter.date(from: "14:30:45")                "1 Jan 2000 at 14:30"
13
14  formatter.dateFormat = "yyyy-MM-dd HH:mm:ss"                <NSDateFormatter: 0x60
15  let date4 = formatter.date(from: "Dummy text")              nil
```

Figure 2.56 String-to-Date Conversion Result

In the first case, we have the string "1978-12-27 21:30:00" at hand, which is formatted as "yyyy-MM-dd HH:mm:ss". So long as the string and template match, DateFormatter will be able to convert it to a date—as shown on line 6 in the output.

In the second case, we have a string formatted as "dd/MM/yyyy"—just a date without time values. In that case, DateFormatter has set the day/month/year in date2 but left the hour/minute as 00:00. This is fine in many cases! For example, if you're reading the expiration date of a product, that's how you probably would store it; the exact time is not important.

In the third case, we have a string formatted as "HH:mm:ss", just a time without date values. In that case, DateFormatter has set the hour/minute/second in date3 but left the day/month/year set to their initial values. This is also fine in many cases! For example, if you're reading the daily alarm time of a simple digital clock, you would only be interested in the hour and minute; the date would be irrelevant.

Finally, in the fourth case, we have a string input, "Dummy text", which is not a date value at all. DateFormatter naturally can't parse this value as a date, but it doesn't raise an exception either. The target variable date4 is simply left as nil, which means that the variable is empty and has not been initialized at all.

Defensive Programming

Here's another heads-up for defensive programming! When dealing with string inputs, you can never trust that the string is in the exact format you hope for. A string that is supposed to contain a date-like value may occasionally produce unexpected values, like "Testing 1-2-3", or misformatted dates, like "20.24-12/25 ((xx".

The reasons behind such occurrences may vary. Programming errors on a remote system, a bug in your own system, or mischievous users trying to enforce a crash are some possibilities. To be on the safe side, always assume that there is a potential for unexpected input and take the appropriate security measures in your code.

In the preceding example, after each string-parsing operation, you could check if the target variable (date1, date2, etc.) is nil or not. If the value is nil, you could raise an exception, indicating that the input is incorrect.

Error raising and handling will be explained in Chapter 13. This was an early heads-up to implant good habits early on.

Quick Formatting

If you don't need custom formatting options and are simply looking for a quick and dirty solution, you can use the dateStyle and timeStyle properties of DateFormatter. Check the code snippet in Listing 2.83, followed by its output in Figure 2.57.

```
import Foundation

let formatter = DateFormatter()
formatter.dateStyle = .medium
formatter.timeStyle = .none

let quickText = formatter.string(from: Date())
```

Listing 2.83 Using Default Format Options of DateFormatter

```
1  import Foundation
2
3  let formatter = DateFormatter()              <NSDateFormat
4  formatter.dateStyle = .medium               <NSDateFormat
5  formatter.timeStyle = .none                 <NSDateFormat
6
7  let quickText = formatter.string(from: Date())  "20 Mar 2025"
```

Figure 2.57 Output of Default Format Options

Both the dateStyle and timeStyle properties accept the values .none, .short, .medium, .long, and .full. Table 2.24 showcases some sample outputs using different styles. As an example, you can try combining different dateStyle and timeStyle values to build up various outputs.

Style	Example Output
.short	2/12/25, 2:30 PM
.medium	Feb 12, 2025 at 2:30 PM
.long	February 12, 2025 at 2:30 PM
.full	Wednesday, February 12, 2025 at 2:30 PM GMT

Table 2.24 Example Outputs of Different Styles

2.5.6 Internationalization

By default, Swift will use the system defaults for date tasks, but this may not be desirable in a large-scale app with users from all over the world. You may have to deal with dates in the locale or time zone of each individual user. As you probably expect, Swift empowers us with multiple tools for that scenario.

Let's start with time zones. Listing 2.84 demonstrates the usage of TimeZone object. Here we have converted the local time to UTC and New York time zone strings.

```
import Foundation

// Preparation
let rightNow = Date()
let formatter = DateFormatter()
formatter.dateStyle = .medium
formatter.timeStyle = .medium

// Local time (default)
formatter.string(from: rightNow)

// UTC time
formatter.timeZone = TimeZone(abbreviation: "UTC")
formatter.string(from: rightNow)

// New York time
formatter.timeZone = TimeZone(identifier: "America/New_York")
formatter.string(from: rightNow)
```

Listing 2.84 Formatting Dates for Given Time Zone

Check the output in Figure 2.58. On line 10, you can see the regular output in the local time zone. Lines 14 and 18 have instead published the output in the UTC and New York time zones; notice the time differences. That's how you would show the date/time for an online event to different users living in different time zones. Neat, eh?

```
1   import Foundation
2
3   // Preparation
4   let rightNow = Date()                                          "20 Mar 2025 at 13:11"
5   let formatter = DateFormatter()                               <NSDateFormatter: 0x60000
6   formatter.dateStyle = .medium                                 <NSDateFormatter: 0x60000
7   formatter.timeStyle = .medium                                 <NSDateFormatter: 0x60000
8
9   // Local time (default)
10  formatter.string(from: rightNow)                              "20 Mar 2025 at 13:11:16"
11
12  // UTC time
13  formatter.timeZone = TimeZone(abbreviation: "UTC")            <NSDateFormatter: 0x60000
14  formatter.string(from: rightNow)                              "20 Mar 2025 at 10:11:16"
15
16  // New York time
17  formatter.timeZone = TimeZone(identifier: "America/New_York") <NSDateFormatter: 0x60000
18  formatter.string(from: rightNow)                              "20 Mar 2025 at 06:11:16"
```

Figure 2.58 Time Zone Output of Local Date/Time

Another cool feature is the RelativeDateTimeFormatter class, which helps you build relative date strings in any given language. Here we're talking about human-readable string outputs, like "three days ago". Check the demonstration in Listing 2.85.

```
import Foundation

let relativeFormatter = RelativeDateTimeFormatter()
relativeFormatter.locale = Locale.init(identifier: "de")

let pastDate = Calendar.current.date(byAdding: .day, value: -3, to: Date())!
relativeFormatter.localizedString(for: pastDate, relativeTo: Date())
```

Listing 2.85 Demonstration of RelativeDateFormatter

After initializing relativeFormatter, we declared our target language to be "de", which stands for German (Deutsch). We could have used other abbreviations, like "fr" (French) or "en" (English), or simply used the expression Locale.current to capture the language of the user's device.

The rest of the code should be intuitive. Check the output in Figure 2.59, where the German output "vor 3 Tagen" was produced—which translates to "three days ago." Cool, right? That's how you can display the "last seen online" value of a user in a messaging app easily. There's no need for any manual internalization effort!

```
1  import Foundation
2
3  let relativeFormatter = RelativeDateTimeFormatter()        <NSRelativeDateTimeForr
4  relativeFormatter.locale = Locale.init(identifier: "de")   <NSRelativeDateTimeForr
5
6  let pastDate = Calendar.current.date(byAdding: .day, value: -3, to: Date())!  "17 Mar 2025 at 13:12"
7  relativeFormatter.localizedString(for: pastDate, relativeTo: Date())    "vor 3 Tagen"
```

Figure 2.59 Output of RelativeDateFormatter

That concludes our content on date/time data types in Swift. As you have seen, Swift contains many standardized solutions for common date/time calculations and formatting requirements, which can be invoked with a minimal amount of code.

Our journey will continue with unit data types, which will prove to be equally easy to figure out.

2.6   Quantities

A *quantity* is a measurable property of an object. Length and weight are common examples; for example, a television may be two meters wide and weigh five kilograms, highlighting two quantities related to a TV.

As evident in the last example, a quantity consists of two parts:

- A numerical value, such as 2 or 5
- A unit of measurement, such as meters or kilograms

Because the value and unit always go hand in hand, Swift empowers us with standardized data types to manage those in tandem, as we'll discuss in this section. Let's start with the core concept of units.

2.6.1   Units

A *unit* represents the type of measurement, such as meters, grams, Celsius, bars, degrees, and so on. Swift provides several predefined Unit subclasses to represent different types of units, which can be seen in Table 2.25

Category	Unit Type	Example Units
Angle	UnitAngle	degrees radians
Energy	UnitEnergy	joules calories kilowattHours
Length	UnitLength	meters kilometers miles feet
Mass	UnitMass	grams kilograms pounds

Table 2.25 Some Unit Subclasses in Swift

Category	Unit Type	Example Units
Pressure	UnitPressure	bars pascals atmospheres
Speed	UnitSpeed	metersPerSecond kilometersPerHour milesPerHour
Temperature	UnitTemperature	celsius fahrenheit kelvin
Time	UnitDuration	seconds minutes hours

Table 2.25 Some Unit Subclasses in Swift (Cont.)

Declaring a Unit variable is not different from declaring any other data type. Check the code sample in Listing 2.86, where various variables are declared for different measurement types. This will clarify how to make use of the unit types shown in Table 2.25 and more.

```
import Foundation

let unit1 = UnitLength.meters
let unit2 = UnitPower.watts
let unit3 = UnitSpeed.kilometersPerHour
let unit4 = UnitVolume.liters
let unit5 = UnitMass.kilograms

var unit6: Unit
unit6 = UnitTemperature.celsius

var unit7: Unit = UnitDuration.minutes
```

Listing 2.86 Unit Variable Declaration Examples

Unit Is an Abstract Class

A tip for readers proficient in object-oriented programming: Unit is an abstract class, and UnitX classes like UnitLength, UnitMass, and so on are concrete classes derived from it. That's why we can cast a subtype like UnitTemparature.celcius to a Unit variable like unit6. In this way, you can declare your own custom units if the necessity arises.

If this explanation was confusing for you, no worries! Chapter 9 will introduce you to object-oriented Swift, and these concepts will become clear then.

Now that you have units under control, we can move to measurements, in which you will combine quantity values with units.

2.6.2   Measurement

In Swift, Measurement represents a complete quantity with a value and Unit. Measurements are very straightforward to declare, as demonstrated in Listing 2.87.

```
import Foundation

let distance = Measurement(value: 5, unit: UnitLength.kilometers)
let weight = Measurement(value: 70, unit: UnitMass.kilograms)
```

Listing 2.87 Declaration of Measurements in Swift

Note that we have provided a value (like 5 or 70) and Unit picked from the previous section. This combination builds up a proper quantity, which is stored in a Measurement type of variable—called distance or weight in the example. Figure 2.60 showcases the output of those variables, with nothing unexpected happening.

```
1  import Foundation
2
3  let distance = Measurement(value: 5, unit: UnitLength.kilometers)  ▣ 5.0 km
4  let weight = Measurement(value: 70, unit: UnitMass.kilograms)      ▣ 70.0 kg
```

Figure 2.60 Output of Measurement Variables

To beautify the output of measurement variables into a human-readable format, you can use the MeasurementFormatter class. Listing 2.88 contains a demonstration in which the string output of a Measurement variable is produced in .long, .medium, and .short formats.

```
import Foundation

let distance = Measurement(value: 5.7, unit: UnitLength.kilometers)
let formatter = MeasurementFormatter()

formatter.unitStyle = .long
let longOutput = formatter.string(from: distance)

formatter.unitStyle = .medium
let mediumOutput = formatter.string(from: distance)

formatter.unitStyle = .short
let shortOutput = formatter.string(from: distance)
```

Listing 2.88 Producing Different Measurement Outputs Using MeasurementFormatter

Check the outputs in Figure 2.61, where a different output has been produced for each given format.

```
 1  import Foundation                                                      ▣ 5.7 km
 2
 3  let distance = Measurement(value: 5.7, unit: UnitLength.kilometers)     ▣ 5.7 km
 4  let formatter = MeasurementFormatter()                                  ▣ <NSMeasuremen
 5
 6  formatter.unitStyle = .long                                            ▣ <NSMeasuremen
 7  let longOutput = formatter.string(from: distance)                      ▣ "5,7 kilometers"
 8
 9  formatter.unitStyle = .medium                                          ▣ <NSMeasuremen
10  let mediumOutput = formatter.string(from: distance)                    ▣ "5,7 km"
11
12  formatter.unitStyle = .short                                           ▣ <NSMeasuremen
13  let shortOutput = formatter.string(from: distance)                     ▣ "5,7km"
```

Figure 2.61 Alternative Measurement Outputs

In .long format, the long text is "5,7 kilometers"; whereas in .short format, the short text is "5,7km". The decimal separator is picked automatically from the user's device settings—very convenient!

If you want to override the locale settings of the user, you can simply set a new Locale in formatter.locale—the same way we did in Section 2.5.6.

2.6.3   Unit Conversions

Converting quantities to different units of measure is such a common daily requirement that most of us do it mentally. Kilometers to miles or kilograms to tons are some examples. To do the same this programmatically in an app, Swift suggests using the converted function of Measurement objects. Check the example in Listing 2.89.

```
import Foundation

let kmDistance = Measurement(value: 5, unit: UnitLength.kilometers)
let milesDistance = kmDistance.converted(to: .miles)

let kgWeight = Measurement(value: 100, unit: UnitMass.kilograms)
let tonWeight = kgWeight.converted(to: .metricTons)
```

Listing 2.89 Demonstration of Unit Conversions in Swift

Once you've initialized the kmDistance variable as usual, you can invoke the kmDistance.converted function to convert the kilometers value to another unit of measure. In this example, we passed the parameter to: .miles to execute the kilometers-to-miles conversion. We applied the same approach to a kilograms to tons conversion, too.

While coding, the autocomplete feature will suggest possible target units, as shown in Figure 2.62. You can pick the appropriate value here; there's no need to memorize them at all.

```
(to: .astronomicalUnits)
        M astronomicalUnits
nit  M centimeters
met  M decameters
        M decimeters
        M fathoms
        M feet
        M furlongs
        M hectometers
```

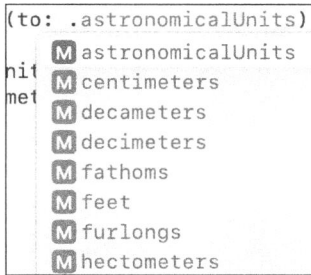

Figure 2.62 Autocomplete for Convertible Units

If you attempt to make an illogical conversion, such as kilometers to kilograms, Swift will politely warn you as shown in Figure 2.63, asking you to correct the mistake.

```
let milesDistance = kmDistance.converted(to: UnitMass.kilograms)   ⊗  Cannot convert
```

Figure 2.63 Swift Error for Invalid Unit of Measure

2.6.4 Arithmetic Operations

In Swift, you can add or subtract two Measurement variables directly, as though they were numeric values! There is a prerequisite, though: All Measurement variables in the equation must have the same unit *category*—like length or mass. For instance, you can do a *kilometers + kilometers* or *kilometers + miles* operation, but you can't do a *kilometers + kilograms* operation—which makes sense, right?

Listing 2.90 demonstrates two addition operations. In the first addition, we are summing two kilometer values, which should produce the expected output of *5 + 4.4 = 9.4 km*. The second addition is more interesting though: We attempted to add kilometers and miles. Any guesses on how Swift will behave?

```
import Foundation

// Summing kilometer values
let km1 = Measurement(value: 5, unit: UnitLength.kilometers)
let km2 = Measurement(value: 4.4, unit: UnitLength.kilometers)
let sum1 = km1 + km2

// Summing kilometers and miles
let miles1 = Measurement(value: 3.2, unit: UnitLength.miles)
let sum2 = km1 + miles1
```

Listing 2.90 Demonstration of Length Addition

When summing up different units in the same category (like kilometers and miles), Swift automatically calculates the result in a common sensible unit. You can see the result in Figure 2.64: Swift assumed *meters* to be a common unit for kilometers and miles; therefore, summing *5 km + 3.2 miles* totals *10149.9008 meters*.

```
1   import Foundation
2
3   // Summing KM values
4   let km1 = Measurement(value: 5, unit: UnitLength.kilometers)      5.0 km
5   let km2 = Measurement(value: 4.4, unit: UnitLength.kilometers)    4.4 km
6   let sum1 = km1 + km2                                              9.4 km
7
8   // Summing KM and miles
9   let miles1 = Measurement(value: 3.2, unit: UnitLength.miles)      3.2 mi
10  let sum2 = km1 + miles1                                           10149.9008 m
```

Figure 2.64 Length Addition Results

If you're targeting a different unit, such as centimeters, you could have converted km1 and miles1 to centimeters before the sum operation. Or you can convert the resulting miles1 value to centimeters after the sum operation.

The previous example featured an addition operation. Keeping the basic logic intact, you can conduct other math operations too! Check the demonstration in Listing 2.91, followed by the calculated results in Figure 2.65.

```
import Foundation

// Declarations
let km1 = Measurement(value: 5, unit: UnitLength.kilometers)
let km2 = Measurement(value: 4.4, unit: UnitLength.kilometers)

// Single measurement operations
let km1Increased = km1 + Measurement(value: 1, unit: UnitLength.kilometers)
let km1Decreased = km1 - Measurement(value: 1, unit: UnitLength.kilometers)
let doubleKm1 = km1 * 2
let halfKm1 = km1 / 2

// Multi measurement operations
let sum = km1 + km2
let difference = km1 - km2
let product = Measurement(value: km1.value * km2.value, unit:
UnitArea.squareKilometers)
let quotient = km1.value / km2.value
```

Listing 2.91 Further Unit Math Operations

```
 1  import Foundation
 2
 3  // Declarations
 4  let km1 = Measurement(value: 5, unit: UnitLength.kilometers)              5.0 km
 5  let km2 = Measurement(value: 4.4, unit: UnitLength.kilometers)            4.4 km
 6
 7  // Single measurement operations
 8  let km1Increased = km1 + Measurement(value: 1, unit: UnitLength.kilometers)   6.0 km
 9  let km1Decreased = km1 - Measurement(value: 1, unit: UnitLength.kilometers)   4.0 km
10  let doubleKm1 = km1 * 2                                                   10.0 km
11  let halfKm1 = km1 / 2                                                     2.5 km
12
13  // Multi measurement operations
14  let sum = km1 + km2                                                       9.4 km
15  let difference = km1 - km2                                                0.5999999999999996 km
16  let product = Measurement(value: km1.value * km2.value, unit: UnitArea.squareKilometers)   22.0 km²
17  let quotient = km1.value / km2.value                                      1,136363636363636
```

Figure 2.65 Unit Math Operation Results

As you can see under **Single measurement operations**, you can multiply or divide a Measurement value directly as if it was a basic number. In such an operation, Swift assumes the unit to remain the same.

However, Swift doesn't let you add or subtract numbers on top of Measurement values, like km1 + 1. Why? Because in a quantity, the value and unit must always go hand in hand! In a statement like km1 + 1, Swift wouldn't want to assume that the value 1 is in kilometers too; it wants you to implicitly state that. That's exactly what we did on line 8: To add 1 km to km1, we created a new Measurement value of 1 km before the addition took place.

Now, let's check the **Multi measurement operations** section. Addition and subtraction between Measurement variables are supported directly, as you've seen before.

Multiplication is a different story though. In math, when you multiply two kilometer values, the result is not kilometers anymore: It's square kilometers. To apply the same logic on line 16, we implicitly created a Measurement variable, which has the calculated value km1.value * km2.value, but the unit squareKilometers. All we're doing here is following basic rules of math, really.

A similar story applies to division. When you divide two kilometer values, the result is not kilometers anymore; it is simply a basic ratio without a unit. Therefore, the division of two Measurement variables (quantities) can't be a Measurement variable (quantity) anymore; it turns into a simple number. That's exactly what happens on line 17: We have to follow basic math and divide km1.value / km2.value, ignoring their units.

Multiplication, Division, and Units

When multiplying or dividing two Measurement variables, the responsibility for unit equality falls to the programmer. For example, if you divide 10 km by 5 minutes, Swift will do the calculation without complaint because you're dividing values only—ignoring their units. If you were to calculate the average kilometers per minute, this calculation makes sense. That said, it's best to check the units beforehand and ensure that they match your intention—and perform conversions if needed.

With this example, we can conclude our content on quantities—as well as singular variables. We'll continue with tuples, which basically represent a group of multiple variables. Take a break if you like, and we'll see you there!

2.7 Tuples

In a nutshell, a *tuple* is a group of values combined together. It's that simple! Instead of having dozens of scattered variables all over the place, tuples let us group related variables together, keeping things tidy and improving code readability—to say the least. Tuples also have other uses, which you'll see in upcoming chapters. In this section, we'll focus on how they work.

2.7.1 Basic Tuples

In the most basic sense, a tuple is a set of variables placed inside two parentheses—nothing more! Check the code in Listing 2.92 to see how easy it is to build a tuple and access the variables within.

```
// Tuple with single type
let cars = ("Mercedes", "Ford", "Volvo")
cars.0
cars.1
cars.2

// Tuple with multiple types
let person = ("John", 30, true)
person.0 // Name
person.1 // Age
person.2 // Married
```

Listing 2.92 Basic Tuple Demonstration

In the first part, we declared the tuple cars, which contains three different strings for car brands. To access different elements of this tuple, we must provide the index of each car: cars.0 will return "Mercedes", cars.1 will return "Ford", and cars.2 will return "Volvo".

All those variables were of the same type: string. But that's not a must with tuples. In the second part, we declared the tuple person, which contains different features of a human: name, age, and married status (string, number, and Boolean). We access those variables with indices, as before.

Check the output of both parts in Figure 2.66.

```
1   // Tuple with single type
2   let cars = ("Mercedes", "Ford", "Volvo")    ("Mercedes", "Ford", "Volvo")
3   cars.0                                       "Mercedes"
4   cars.1                                       "Ford"
5   cars.2                                       "Volvo"
6
7   // Tuple with multiple types
8   let person = ("John", 30, true)             ("John", 30, true)
9   person.0 // Name                            "John"
10  person.1 // Age                             30
11  person.2 // Married                         true
```

Figure 2.66 Output of Basic Tuple Demonstration

As with other types, the mutability of tuples depends on how they were declared. If a tuple was defined as a constant using let, then its values can't be changed after initialization. If it was defined as a variable using var, then its values can be changed later. To keep this section self-contained, Listing 2.93 demonstrates a basic mutable tuple.

```
var person = ("John", 30, true)
person.1 += 10   //40
```

Listing 2.93 Mutating Tuples

The previous examples featured tuples built with literals. However, you can build tuples with existing variables too. Check the code snippet in Listing 2.94, in which we declare the variables name, age, and married before combining them into the tuple person.

```
var name = "John"
var age = 30
var married = true
var person = (name, age, married)
person.1

age += 10
person.1
```

Listing 2.94 Building Tuple Out of Variables

Now the second part of the code sample should be interesting. Do you think that changing the value of age would be mirrored in person.1, or would Swift keep those values isolated? There's only one way to find out!

As shown in Figure 2.67, Swift kept the original variable, age, and the new variable, person.1, as two isolated entities; they don't share their values any longer. Although we changed the value of age to 40, person.1 continued to hold its former value, 30. In more technical terms, the initial tuple assignment was made by value, not by reference.

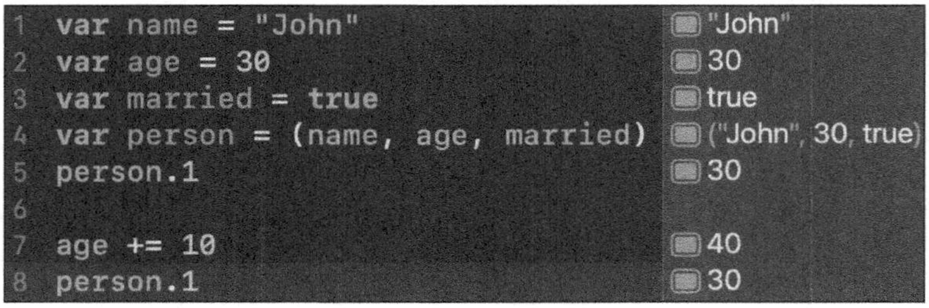

```
1  var name = "John"              ▣ "John"
2  var age = 30                   ▣ 30
3  var married = true             ▣ true
4  var person = (name, age, married)  ▣ ("John", 30, true)
5  person.1                       ▣ 30
6
7  age += 10                      ▣ 40
8  person.1                       ▣ 30
```

Figure 2.67 Variable Change After Tuple Buildup

2.7.2   Named Tuples

In the previous example, we built tuples by dumping variables together and accessed them using indices—like person.1 for age. Although Swift performs fine that way, it makes the code hard to read and maintain, don't you think? Especially in a large tuple with many elements, trying to spot a variable with the correct index is an invitation for errors.

For a cleaner and more understandable syntax, Swift offers an option for *named tuples*. Check the demonstration in Listing 2.95, in which we build the tuple by providing meaningful names for each variable.

```
let person = (name: "John", age: 30, married:true)

person.name      // "John"
person.age       // 30
person.married   // true

person.0         // "John"
person.1         // 30
person.2         // true
```

Listing 2.95 Named Tuple Demonstration

When you declare a tuple that way, you can access the variables in the tuple via meaningful names like person.name, person.age, or person.married. Obviously, the code is much more readable that way. However, the former method of index-based access, such as person.1, can still be used too.

2.7.3   Tuple Type Annotations

You learned about data type annotations in Section 2.1.3. The same approach can be used with tuples as well. It's possible to declare a tuple with designated data types before value assignments take place.

That might have sounded a bit complicated, but it's pretty simple in practice! Check Listing 2.96 for two demonstrations.

```swift
var person1: (name: String, age: Int, married: Bool)
person1 = ("John", 30, true)

var person2: (name: String, age: Int, married: Bool) = ("Jane", 25, false)
```

Listing 2.96 Tuple Type Annotation Demonstration

You have enough experience with Swift variables now to intuitively understand the logic, right? While declaring the tuple, we declared the type of each variable, and that's about it—very similar to what we did with single variables.

2.7.4 Comparison

In previous sections, we went over value comparisons for simple variables like numbers and strings. Following the same syntax, it's possible to compare tuples as well.

Listing 2.97 demonstrates an example of tuple comparison. Because person1 and person2 have the exact same values, person1 == person2 will return true. However, due to their differences, person2 == person3 will return false.

```swift
let person1 = (name: "John", age: 30, married:true)
let person2 = (name: "John", age: 30, married:true)
let person3 = (name: "Jane", age: 25, married:false)

person1 == person2  // true
person2 == person3  // false
```

Listing 2.97 Tuple Value Comparisons

If tuples contain comparable types, you can apply other common operators too. In Listing 2.98, we compared tuples using the > operator.

```swift
let fruit1 = (price: 15, name: "apple")
let fruit2 = (price: 20, name: "banana")
let fruit3 = (price: 20, name: "orange")

fruit1 > fruit2     // false
fruit2 > fruit3     // false
```

Listing 2.98 Application of Alternative Operators

When fruit1 > fruit2 is executed, Swift will go through the tuple components in their natural order and stop at the first difference. In this case, fruit1.price > fruit2.price is false, which is returned as the result.

On the second comparison, `fruit2 > fruit3`, the first components are the same; both `fruit2.price` and `fruit3.price` are 20. In that case, Swift jumps to the next component pair and compares `fruit2.name` and `fruit3.name` as strings—determining the result to be `false`.

That concludes our content on tuples, as well as Swift data types. It surely is a lot to digest, but variables are the very fundamentals of any programming language. Upcoming chapters will be built on top of this knowledge. If you feel like you have some minor knowledge gaps left, don't worry: As you go through variable usage examples more and more, your understanding will solidify.

2.8 Type Aliases

Before you move on to the next chapter, let's address a useful concept: type aliases. A *type alias* allows you to provide a new (and hopefully more meaningful) name for an existing data type. Once a type alias is declared, it can be used instead of the original type name throughout the program. That approach improves the readability of the codebase. In this section, we'll discuss both basic and complex type aliases.

2.8.1 Basic Types

In its basic usage, type aliases can be used as alternative names for basic Swift types. Listing 2.99 demonstrates an example of that approach, in which `MoneyAmount` has been declared as an alternative name for `Float`.

```
typealias MoneyAmount = Float

let budget: MoneyAmount = 100
let price: MoneyAmount = 45
```

Listing 2.99 Type Alias for Float Data Type

This approach has two potential advantages. The first advantage is code readability. Although `budget` and `price` are well-named variables already, having an obvious type alias like `MoneyAmount` improves readability in a more complex codebase.

The second advantage is the centralization of the core data type. Let's imagine that you have hundreds of money-related `Float` variables in your codebase (like `budget` and `price`). Say that due to a management decision, it then becomes necessary to retype them all as `Double`.

Without having a type alias like `MoneyAmount`, you would have to search for all the `Float` variables, spot money-related ones, and replace their type with `Double`—possibly missing some money variables or mistyping some nonmoney variables.

With a type alias, all you have to replace is a single line of code! In Listing 2.100, we have simply changed the type alias of MoneyAmount from Float to Double. That's all we need to replace the type of all money-related variables, like budget and price, automatically. If you had hundreds of money-related variables, the convenience of this feature would be huge.

```
typealias MoneyAmount = Double // replaced Float

let budget: MoneyAmount = 100
let price: MoneyAmount = 45
```

Listing 2.100 Replacing Core Type of Type Alias

You obviously need to be careful about changing data types and take factors like variable length or decimal places into consideration. But that's the case with any data type change, not something specific to type aliases.

2.8.2   Complex Types

The advantages of type aliases can be enjoyed with complex data types as well, such as tuples. Listing 2.101 demonstrates a type alias called Person, which acts as a tuple. Any time we need to declare a variable for a human, we can base it on the type alias Person—as we did for user1 and user2.

```
typealias Person = (name: String, age: Int, married: Bool)

let user1: Person = (name: "John", age: 30, married: true)
let user2: Person = ("Jane", 25, false)
```

Listing 2.101 Type Aliases on Tuples

Other Complex Types
Type aliases can be used for other complex types as well, such as structs and generics. As you progress through the chapters, we'll address those types when they arise.

2.9   Summary

That concludes our chapter on variables! To wrap things up: You learned that Swift is a statically typed language with a strong emphasis on type safety. You must declare the type of a variable, and it can't be changed later. But the option of type inference is also present; if you declare a variable using a value directly, Swift is smart enough to determine the type automatically.

We went over built-in types for Boolean, numeric, text, date/type, and quantity variables, as well as tuples to group variables together. Finally, you learned about type aliases to declare custom names for other types.

This has been the first hands-on chapter on Swift; therefore, extensive details and playground screenshots were shared to get you warmed up and used to the language. Based on that mental framework, upcoming chapters will hopefully be easier to follow through.

Now that you know how to handle individual variables, you'll learn how to handle groups of variables in the next chapter. These groups are called *collections*.

Chapter 3
Collections

Collections are sets of variables glued together as logical units.
This chapter will teach you about collection types in Swift.

Now that you've learned about individual variables, you can advance your Swift journey with collections. In a nutshell, *collections* are data structures that group multiple variables into a single, organized container. Nearly all programming languages use collections—including Swift.

There are three main types of collections, which are the subject of this chapter:

- *Arrays* are ordered variable lists in which each variable has an index.
- *Sets* are unordered variable lists in which each value must be unique.
- *Dictionaries* are key-value pairs in which each key must be unique.

To paraphrase a possible question many of you may have: "But wait a minute—didn't we have groups of variables called *tuples* in the previous chapter? How are collections different than that?"

That's a good question! Collections (arrays, sets, dictionaries) and tuples both hold multiple variables, but they have key differences—which are listed in Table 3.1.

Feature	Tuples	Collections
Size	Fixed size; can't be changed after creation	Can grow or shrink dynamically; you can add/remove elements
Type uniformity	Can group a mix of different types	All elements should have the same type
Element access	Access using position or named properties	Access using indexes (arrays), iteration (sets), or keys (dictionaries)
Use case	Best for grouping related variables of different types	Best for storing a flexible number of variables of the same type

Table 3.1 Tuples Versus Collections

Due to such differences, a tuple is not considered a collection type; it is merely a flexible way of grouping similar variables together. Collections offer more advanced features, which will be highlighted in this chapter.

All clear? Great! Let's start with arrays, then continue with sets and dictionaries.

Screenshots

The previous chapter offered a coding debut to Swift and Xcode. To ensure that you could all get used to Xcode and could follow the examples correctly, screenshots of Xcode outputs were supplied for most of the examples.

Now that everyone is used to how and where Xcode displays outputs, we'll generally show the output of statements as inline comments from this point on. Separate results, such as screenshots or terminal outputs, will be provided only where necessary.

3.1   Arrays

An *array* in Swift is an ordered collection of elements of the same type, allowing you to store multiple values efficiently. For instance, if you are programming a queue system and want to store the customer names ordered by their time of arrival, you could store their names as strings in an array. In this section, you will learn how to create, access, and modify arrays, and learn about convenient features offered by Swift.

3.1.1   Creating Arrays

To begin this example, let's create an array of names for waiting customers, as shown in Listing 3.1.

```
var customers = ["Alice", "Bob", "Charlie"]
```

Listing 3.1 Basic String Array

Check Table 3.2 to see a visual representation of the customers array. Note that indexes begin with 0 as usual.

Index	Value
0	"Alice"
1	"Bob"
2	"Charlie"

Table 3.2 Visual Representation of Customers Array

And there you go: It's that easy! Now the customers array holds three distinct string values, reflecting the names of customers in the queue. As cashiers become available, Alice would be the first customer to be called, followed by Bob and then Charlie. You'll learn how to access those values shortly.

In Chapter 2, you learned about alternative ways of declaring variables using type inference and type interpolation. Likewise, there are alternative ways of declaring arrays using the same methods. Your knowledge of variables will be applicable in that sense. Listing 3.2 showcases the alternatives.

```
// Direct value assignment with type inference
var customers1 = ["Alice", "Bob", "Charlie"]

// Direct value assignment with type annotation
var customers2: [String] = ["Alice", "Bob", "Charlie"]

// Late value assignment with type annotation
var customers3: [String]
customers3 = ["Alice", "Bob", "Charlie"]

// Alternative syntax
var customers4: Array<String>
customers4 = ["Alice", "Bob", "Charlie"]
```

Listing 3.2 Different Methods for Array Creation

3.1.2 Arrays as Constants

In Chapter 2, you learned about the var and let keywords. The var keyword is used to declare a *variable*, which allows value changes later (mutable), whereas let is used to declare a *constant*, which won't allow its initial value to change (immutable).

The same feature applies to arrays too. An array declared with var would be mutable, while an array declared with let would be immutable. It is arguably more common to have mutable arrays, but both states are possible. Listing 3.3 demonstrates both syntaxes for number arrays.

```
var someNumbers = [2, 4, 10, 6, 1, 9]
let lostNumbers = [4, 8, 15, 16, 23, 42]
```

Listing 3.3 Declaration of Mutable and Immutable Arrays

Our examples so far have featured arrays built out of literals. Naturally, you can build mutable or immutable arrays out of variables too—as shown in Listing 3.4, which builds an immutable array out of numbers. The main prerequisite is to have variables of the same type; you can't mix numbers and strings in an array.

```
let n1 = 4
let n2 = 8
let n3 = 15
```

```
let n4 = 16
let n5 = 23
let n6 = 42

let lostNumbers = [n1, n2, n3, n4, n5, n6]
```
Listing 3.4 Building Array Out of Variables

You can be even more adventurous and build arrays out of complex types as well! Listing 3.5 showcases an example, which builds a mutable array out of tuples. This code snippet also highlights the comfort of using type aliases for tuples: Type uniformity for the array is ensured easily and in a human-readable way.

```
typealias Person = (name: String, age: Int, married: Bool)

let user1: Person = (name: "John", age: 30, married: true)
let user2: Person = ("Jane", 25, false)

var people: [Person] = [user1, user2]
```
Listing 3.5 Building Array Out of Tuples

To prevent any confusion, Table 3.3 features a visual representation of the people array.

Index	Value
0	(name: "John", age: 30, married: true)
1	(name: "Jane", age: 25, married: false)

Table 3.3 Visual Representation of People Array

Got it? OK, then! Now, let's go over how to access values in an array.

3.1.3 Accessing Arrays

In this section, you'll learn about accessing arrays in Swift. We'll explore a handful of options in that regard: basic array functions, index-based element access, and first/last elements.

Basic Array Functions

Let's start with array functions. Table 3.4 showcases some basic functions that are used frequently.

Function	Result
isEmpty	true if the array is empty; false otherwise
count	Number of elements in the array
contains(element)	true if the element is in the array; false otherwise

Table 3.4 Basic Array Functions

To see those useful functions in context, check Listing 3.6, which contains the output of each function as a comment. Now that you're familiar with Swift, this code snippet should be intuitive and self-explanatory.

```
let happyCustomers: [String] = ["Alice", "Bob", "Charlie"]
let sadCustomers: [String] = []

happyCustomers.isEmpty              // false
happyCustomers.count               // 3
happyCustomers.contains("Alice")   // true
happyCustomers.contains("Ann")     // false

sadCustomers.isEmpty               // true
sadCustomers.count                 // 0
```

Listing 3.6 Basic Array Properties

Index-Based Element Access

Now that you know about basic functions, let's move forward with element access. As you know, arrays are ordered lists in which each element has an index. Therefore, it's natural to expect the core functionality of being able to access elements via their indexes. As with tuples, indexes start with 0 and increment by one for each element.

Listing 3.7 demonstrates a code snippet for element access by index, in which you extract the first and second person in a bank queue. The intuitive bankQueue[n] expression returns the nth element in the array.

```
var bankQueue = ["Alice", "Bob", "Charlie"]

let firstInLine = bankQueue[0]     // Alice
let secondInLine = bankQueue[1]    // Bob
```

Listing 3.7 Array Element Access by Index

Of course, you can use a variable as an index too! In Listing 3.8, the myIndex variable is used as an array index. Instead of using literal values like 0 or 1, you use the value of myIndex.

```
var bankQueue = ["Alice", "Bob", "Charlie"]

var myIndex = 0                         // 0
var myCustomer = bankQueue[myIndex] // Alice

myIndex += 1                            // 1
myCustomer = bankQueue[myIndex]         // Bob
```

Listing 3.8 Using Variable as Array Index

Check the Index First

If the index value exceeds the number of elements in the array, Swift will naturally generate an error. In Listing 3.8, bankQueue[0] (having the value "Alice") or bankQueue[1] (having the value "Bob") or bankQueue[2] (having the value "Charlie") is fine. However, bankQueue[3] would generate an error because there is no such element.

To prevent such errors, you should always ensure that the index is less than the element count. In this example, the if myIndex < bankQueue.count expression can be placed as a condition before the element access.

Although you will learn much more about if statements in Chapter 4, this heads-up should be a useful detail to have in advance.

First and Last

Access via indexes is cool, but sometimes you simply want to access the first or last element of an array. Swift arrays feature two shortcut functions for that—namely, first and last. These are demonstrated in Listing 3.9.

```
var bankQueue = ["Alice", "Bob", "Charlie"]
let firstInQueue = bankQueue.first!     // Alice
let lastInQueue = bankQueue.last!       // Charlie
```

Listing 3.9 Accessing First and Last Elements of Array

Beyond this basic syntax, first and last also feature a search functionality. For example, if you have a string array, then you can invoke string functions against the elements and look for matches. In Listing 3.10, you run a search in bankQueue to find the first and last customers whose names contain the character l.

```
var bankQueue = ["Alice", "Bob", "Charlie"]

let firstWithL = bankQueue.first(where: { $0.contains("l") })   // Alice
```

```
let lastWithL = bankQueue.last(where: { $0.contains("l") })      // Charlie
let firstWithX = bankQueue.first(where: { $0.contains("x") })    // nil
```

Listing 3.10 Finding First/Last Search Results in String Array

As expected, "Alice" is returned as the first string with l and "Charlie" is returned as the last string with l. When we search for string containing the character x, we get nil as the result simply because there is none.

Closures

{ $0.contains("l") } and similar expressions are *closures*, which are self-contained blocks of code passed as parameters. They are similar to lambda functions in other programming languages. You'll learn more about closures in Chapter 5. For now, you can accept them as common syntax elements and keep your focus on collections.

Naturally, the search functionality can be invoked for other data types too. Listing 3.11 demonstrates a code snippet in which a number search is executed using the < 20 condition.

```
var bingoNumbers = [59, 19, 36, 55, 28]
let firstSmallNumber = bingoNumbers.first(where: {$0 < 20 })     // 19
```

Listing 3.11 Finding First Search Result in Number Array

You can get a little more adventurous and search through an array of complex types as well—such as tuples. In Listing 3.12, there is a patientQueue built out of tuples, in which the name and age of each patient is declared. To find the oldest and youngest patient, you can search through the array using the age property of the tuples.

```
typealias Person = (name: String, age: Int)

var patientQueue: [Person] = [
    (name: "John", age: 30),
    (name: "Jane", age: 15),
    (name: "Jim", age: 80),
    (name: "Jill", age: 20)]

let firstOldPatient = patientQueue.first(where: { $0.age > 65 })     // Jim
let firstYoungPatient = patientQueue.first { $0.age < 18 }           // Jane
```

Listing 3.12 Finding First Search Result in Tuple Array

On the last line of the code snippet, you can also see the shortcut version of running a search; as shown, the parentheses and where: prefix can be omitted if you like.

If you don't want to fetch the resulting element and only want to find the index, you can use firstIndex and lastIndex functions just like you would use *first* and *last*. Listing 3.13 demonstrates using those functions to find the indexes of the first/last elements for the given search conditions.

```
typealias Person = (name: String, age: Int)

var patientQueue: [Person] = [
    (name: "John", age: 30),
    (name: "Jane", age: 15),
    (name: "Jim", age: 80),
    (name: "Jill", age: 20)]

let firstOldIndex = patientQueue.firstIndex { $0.age > 65 }         // 2 (Jim)
let lastJIndex = patientQueue.lastIndex { $0.name.contains("J") }  // 3 (Jill)
```

Listing 3.13 Demonstration of firstIndex and lastIndex Functions

3.1.4 Array Derivation

In Chapter 2, we looked at alternative ways to extract substrings from strings, remember? Swift features similar functions to derive new subarrays from existing arrays. In this section, we'll go through some significant functions for that purpose.

Slicing

The most basic method of array derivation is to *slice* a subarray from an existing array. For example, if you have an array of seven elements, then we can extract the elements between 1 and 3 as a new array. Listing 3.14 demonstrates an example in which workdays are extracted from weekdays using indexes.

```
let weekDays = ["Sun", "Mon", "Tue", "Wed", "Thu", "Fri", "Sat"]
let workDays = Array(weekDays[1...5]) // Mon, Tue, Wed, Thu, Fri
```

Listing 3.14 Slicing Out Subarrays

In the end, workDays becomes an independent array, usable on its own like any other array—just like extracted substrings.

Filter

Another popular technique is *filtering*, in which you filter values of a bigger array to extract values as a smaller array. Listing 3.15 features a basic code snippet in which you filter a number array twice. In this code snippet, $0 represents elements of the main array.

```
let numbers = [1, 2, 3, 4, 5]
let bigNumbers = numbers.filter { $0 > 3 }           // 4, 5
let evenNumbers = numbers.filter { $0 % 2 == 0 }     // 2, 4
```

Listing 3.15 Filtering Numbers

In the end, `bigNumber` contained numbers greater than 3 and `evenNumbers` contained even numbers; both are individual arrays of their own.

The same technique can be used with strings as well. In Listing 3.16, city names are filtered using their character counts, and short city names are extracted into a new array called `shortCities`. Once again, `$0` represents elements of the main array. We could have used any string function, but this example features `$0.count` to filter over string length.

```
let cities = ["New York", "San Francisco", "Los Angeles", "Chicago"]
let shortCities = cities.filter { $0.count < 10 } // New York, Chicago
```

Listing 3.16 Filtering Strings

Why not get a bit more adventurous and filter over tuples, too? The core logic of the filtering won't change—so Listing 3.17 should be pretty easy to follow!

```
typealias Person = (name: String, age: Int)

let people: [Person] = [
    ("Alice", 30),
    ("Bob", 25),
    ("Charlie", 16),
    ("David", 14)
]

let youngPeople = people.filter { $0.age < 18 } // Charlie, David
```

Listing 3.17 Filtering Tuples

Naturally, filters may contain multiple logical conditions too—as demonstrated in Listing 3.18.

```
let numbers = [231, 12, 334, 423, 25]
let smallOddNumbers = numbers.filter { $0 < 100 && $0 % 2 == 1 }     // 25
```

Listing 3.18 Filter with Multiple Conditions

Here we have two distinct conditions, bound together with the && operator. Swift will process those conditions sequentially as demonstrated in Table 3.5 and produce the result shown: an array with a single element.

Condition	Meaning	Result
$0 < 100	Number must be less than 100	[12, 25]
$0 % 2 == 1	Number must be odd	[25]

Table 3.5 Multicondition Filter Process

Map

Mapping is a technique that returns a new array of the same size but with transformed elements. Listing 3.19 features a demonstration that uses the map function to create a new array containing numbers multiplied by 2.

```
let numbers = [1, 2, 3, 4, 5]
let doubleNumbers = numbers.map { $0 * 2 }  // 2, 4, 6, 8, 10
```

Listing 3.19 Mapping Numbers

As usual, $0 represents each element in the array. The map { $0 * 2 } expression declares the intention to multiply each number by 2 and return the results as a new array.

As a second demonstration, Listing 3.20 features a map operation on strings. This time, map { $0.uppercased() } builds a new array out of students, with each name converted to uppercase.

```
let students = ["Alice", "Bob", "Charlie"]
let upperStudens = students.map { $0.uppercased() } // ALICE, BOB, CHARLIE
```

Listing 3.20 Mapping Strings

Reduce

Arrays feature a useful function called reduce, which is called when you want to run a certain calculation over array elements and return a single result.

Listing 3.21 features a basic example that calculates the sum of all elements in an array.

```
let numbers = [1, 2, 3, 4, 5]
let sum = numbers.reduce(0) { $0 + $1 } // 15
```

Listing 3.21 Summing Numbers in Array

Here, (0) is the initial value of the result, and { $0 + $1 } means that you want to add up each number until the array is finished. Table 3.6 showcases the detailed iteration executed by the reduce function.

Iteration	Current Result	Operation	New Result
1	0	+ 1	1
2	1	+ 2	3
3	3	+ 3	6
4	6	+ 4	10
5	10	+ 5	15

Table 3.6 Number Iteration Executed by reduce Function

This useful function can be applied to any array summarization, such as concatenating strings in an array. Listing 3.22 demonstrates this approach through a clean example, followed by its detailed iteration in Table 3.7.

```
let words = ["Hello", "world", "!"]
let sentence = words.reduce("") { $0 + " " + $1 }   // Hello world !
```

Listing 3.22 Concatenating Strings in Array

Iteration	Current Result	Operation	New Result
1	" "	+ " " + "Hello"	" Hello"
2	" Hello"	+ " " + "world"	" Hello world"
3	" Hello world"	+ " " + "!"	" Hello world !"

Table 3.7 String Iteration Executed by reduce Function

Naturally, you also can supply an initial value and use other functions within the reduce clause. In Listing 3.23, we debut the result with the initial value "I say:" and continue by applying the uppercased version of each string in words.

```
let words = ["Hello", "world", "!"]
let sn = words.reduce("I say:") { $0 + " " + $1.uppercased() } // I say: HELLO
WORLD !
```

Listing 3.23 Using String Functions While Reducing

Join

You also have the option of *joining* two existing arrays to build a new one. How cool is that? You can simply use the + operator to combine two arrays, as demonstrated in Listing 3.24. You can imagine this technique as akin to concatenating the arrays.

```
let guestsOnTime = ["Alice", "Bob"]
let guestsLate = ["David", "Eve"]
let allGuests = guestsOnTime + guestsLate // Alice, Bob, David, Eve
```

Listing 3.24 Joining Arrays Using +

The same technique can be applied to arrays with elements of other types too. Just for the fun of it, Listing 3.25 demonstrates the join operation of two arrays containing Measurement objects. In the end, allWeights will contain four Measurement values—which is the joint output of someWeights and otherWeights.

```
import Foundation

let weight1 = Measurement(value: 100, unit: UnitMass.kilograms)
let weight2 = Measurement(value: 1, unit: UnitMass.grams)
let someWeights = [weight1, weight2]

let weight3 = Measurement(value: 44, unit: UnitMass.kilograms)
let weight4 = Measurement(value: 12, unit: UnitMass.grams)
let otherWeights = [weight3, weight4]

let allWeights = someWeights + otherWeights // 4 elements
```

Listing 3.25 Joining Arrays with Measurement Values

This fun example concludes our content on array derivation. Now you know how to create, access, and derive arrays. The next natural step is to modify existing arrays, which will be covered in the next section.

3.1.5 Modifying Arrays

In this section, you will learn how to *modify* arrays. Operations like appending new elements or modifying or deleting existing elements will be covered.

An initial reminder, though: Modifiable arrays must have been declared using a var statement. As you know, let declarations create static/constant arrays that can't be changed later. Therefore, modification examples in this section will inevitably declare arrays using var statements.

Appending Elements

Swift features multiple methods to *append* elements to an array, which you'll discover next. The most basic and straightforward way is to simply invoke the append function of the array object. In Listing 3.26, after the initial declaration of numbers as [1, 2, 3], we execute the append(4) function, extending the array as [1, 2, 3, 4].

```
var numbers = [1, 2, 3]
numbers.append(4) // [1, 2, 3, 4]
```

Listing 3.26 Appending Array Element Using append Function

That's pretty easy, right? It's almost plain English! Now, what if you wanted to append multiple elements instead of just one? Listing 3.27 demonstrates how to do so. For this purpose, you can still use the append function—but instead of providing a single element as the parameter, you provide the contentsOf: [4, 5] parameter, indicating the elements you want to append.

```
var numbers = [1, 2, 3]
numbers.append(contentsOf: [4, 5]) // [1, 2, 3, 4, 5]
```

Listing 3.27 Appending Multiple Elements to Array Using append Function

As a shortcut, you can also make use of the += operator, which you used with strings in Chapter 2. The same logic applies: You can concatenate a subarray to a main array, like concatenating strings. Check the demonstration in Listing 3.28.

```
var numbers = [1, 2, 3]
numbers += [4, 5] // [1, 2, 3, 4, 5]
```

Listing 3.28 Appending Multiple Elements to Array Using += Operator

The examples so far have focused on appending elements to the tail of an array. What if you want to insert elements at a specific index? For that purpose, you should use the insert function of the array object, as demonstrated in Listing 3.29. This function accepts two parameters: the element to insert ("Ann") and the insertion index (2).

```
var people = ["John", "Mary", "Alice"]
people.insert("Ann", at: 2) // John, Mary, Ann, Alice
```

Listing 3.29 Inserting Element at Specific Index

You know by now that indexes begin with 0. That's why the statement at: 2 guided Swift to insert "Ann" after "Mary": Counting 0, 1, 2 makes "Ann" the third element of the array.

To insert *multiple elements* at a specific index, you should modify the parameters of the insert function just as you did with the append function. Check the intuitive demonstration in Listing 3.30, which makes use of the contentsOf parameter once again.

```
var people = ["John", "Mary", "Alice"]
people.insert(contentsOf: ["Ann", "Bob"], at: 2) // John, Mary, Ann, Bob, Alice
```

Listing 3.30 Inserting Multiple Elements at Specific Index

Modifying Elements

Modifying an element of an array is as simple as inserting it. Check Listing 3.31, in which you change the second element of the array from "Banana" to "Blueberry".

```
var fruits = ["Apple", "Banana", "Cherry"]
fruits[1] = "Blueberry" // Apple, Blueberry, Cherry
```

Listing 3.31 Modification of Array Element

To get a bit more adventurous, Listing 3.32 features another example of array element modification. Although it uses the exact same approach, the array contains tuples instead of a basic data type. But if you check the very last line, you will see that the modification syntax doesn't change at all!

```
typealias Instrument = (name: String, price: Double)

var instruments: [Instrument] = []

instruments.append((name: "Guitar", price: 100))   // Guitar
instruments.append((name: "Drums", price: 300))    // Guitar, Drums
instruments.append((name: "Piano", price: 200))    // Guitar, Drums, Piano

instruments[2] = (name: "Keyboard", price: 150)    // Guitar, Drums, Keyboard
```

Listing 3.32 Modification of Array Tuple

Deleting Elements

Deleting elements from an array is equally easy. Array objects feature ready-to-use functions with the remove prefix, which are listed in Table 3.8.

Function	Purpose
removeFirst(n)	Removes the first n elements of the array
removeLast(n)	Removes the last n elements of the array
remove(at: n)	Removes the nth element of the array
removeAll()	Clears the array completely

Table 3.8 Array Functions for Element Removal

A demonstration of those functions is provided in Listing 3.33. The syntax and results are self-explanatory.

```
var fibo = [1, 2, 3, 5, 8, 13, 21, 34, 55]
fibo.removeFirst(3)        // 5, 8, 13, 21, 34, 55
fibo.removeLast(2)         // 5, 8, 13, 21
```

```
fibo.remove(at: 2)          // 5, 8, 21
fibo.removeAll()            // Empty
```

Listing 3.33 Demonstration of Element Deletion from Array

One cool trick is to add a where condition to removeAll. Listing 3.34 demonstrates how to delete elements with a value greater than 10 from the array fibo. The syntax of the where condition is the same as in previous similar examples.

```
var fibo = [1, 2, 3, 5, 8, 13, 21, 34, 55]
fibo.removeAll(where: { $0 > 10 }) // 1, 2, 3, 5, 8
```

Listing 3.34 removeAll Function with where Condition

Sorting Arrays

Finally, we will show you how to sort arrays. And once again, it's very easy: You simply invoke the sort function of the array object. Listing 3.35 demonstrates how to sort an array in ascending and descending order.

```
var numbers = [5, 2, 8, 3, 1]
numbers.sort()       // 1, 2, 3, 5, 8
numbers.sort(by: >) // 8, 5, 3, 2, 1
```

Listing 3.35 Sorting Array in Swift

A similar function is sorted, which does almost the same job as sort. However, instead of mutating the original array, it returns a new sorted array. Check Listing 3.36 for a demonstration.

```
var numbers = [5, 2, 8, 3, 1]
var sortedNumbers = numbers.sorted() // 1, 2, 3, 5, 8
```

Listing 3.36 Creating New Sorted Array

The next step is *array iteration*, in which you loop through the elements of an array for bulk operations.

3.1.6   Iterating Through Arrays

When you have an array at hand, it is a natural expectation to visit each element sequentially. In many cases, this is the reason to build an array in the first place. Imagine an array of phone numbers, in which you have to call each customer in line sequentially. This would require iterating through the numbers in the array, right?

For such cases, Swift offers various methods to iterate through the elements of an array. In this section, we will discuss those iteration methods and their differences.

For Clause

The most fundamental approach to array iteration is to use a for ... in ... statement. Check the demonstration in Listing 3.37, which iterates through phone numbers.

```
var phones = ["123-4567", "890-1234", "543-2109", "234-5678"]

for phone in phones {
    print(phone)
}
```

Listing 3.37 Iteration Through Phone Numbers Using for Statement

The initial part of this code snippet is familiar: The phones array has been declared with some mock values. The next part is the interesting one! Using the for phone in phones statement, you tell Swift to iterate through all elements in phones, assigning a new value to the phone variable on every iteration. The code block between { and } will be executed for each phone sequentially.

That was a mouthful; let's break it down now! Table 3.9 showcases each iteration, including the value assigned to phone and how print(phone) looks.

Iteration	Value in Phone	Print Statement	Output
1	"123-4567"	print("123-4567")	123-4567
2	"890-1234"	print("890-1234")	890-1234
3	"543-2109"	print("543-2109")	543-2109
4	"234-5678"	print("234-5678")	234-5678

Table 3.9 Iteration Broken Down

It should be clear now! For each iteration, phone was assigned a new value, pulled from phones sequentially. The terminal output is shown in Figure 3.1 for even more clarity.

```
1  var phones = ["123-4567", "890-1234", "543-2109", "234-5678"]
2
3  for phone in phones {
4      print(phone)
5  }

123-4567
890-1234
543-2109
234-5678
```

Figure 3.1 Output of for Iteration

Naturally, the code between { and } can be as complex as needed, but the basic idea doesn't change: The code between { and } is executed with every phone value through the iteration. Listing 3.38 demonstrates the same iteration with slightly more complex code, making use of features you learned about before.

```
var phones = ["123-4567", "890-1234", "543-2109", "234-5678"]

for phone in phones {
    let cleanPhone = phone.replacing("-", with: "")
    let operatorText = "You can dial \(cleanPhone) now"
    print(operatorText)
}
```

Listing 3.38 Iteration with Slightly More Complex Code

As evident in the output in Figure 3.2, the iteration ran the same way as before—even if the code between { and } was a little different.

```
1   var phones = ["123-4567", "890-1234", "543-2109", "234-5678"]
2
3   for phone in phones {
4       let cleanPhone = phone.replacing("-", with: "")
5       let operatorText = "You can dial \(cleanPhone) now"

You can dial 1234567 now
You can dial 8901234 now
You can dial 5432109 now
You can dial 2345678 now
```

Figure 3.2 Output for Slightly More Complex Iteration

Using Enumerated

Arrays contain a cool function called enumerated(). When this function is used in a for iteration, you get access to the element and its index simultaneously. The example in Listing 3.39 invokes this functionality: Instead of executing the for iteration against the bankQueue array itself, you execute it against bankQueue.enumerated(). In return, you get access to the so-called queueEntry object, which contains the element index in queueEntry.offset (sequentially, 0, 1, 2) and the element value in queueEntry.element (sequentially, "James", "John", "Robert").

```
var bankQueue = ["James", "John", "Robert"]

for queueEntry in bankQueue.enumerated() {
    print(queueEntry.offset)
    print(queueEntry.element)
}
```

Listing 3.39 Accessing Index and Element Throughout Iteration

A breakdown of this code snippet is provided in Table 3.10, where the operation in each iteration is shown clearly.

Iteration	queueEntry.offset	queueEntry.element	Expected Output
1	0	"James"	0 James
2	1	"John"	1 John
3	2	"Robert"	2 Robert

Table 3.10 Enumerated Iteration Broken Down

It's time to test the code and see if you get the expected result. Check the playground output in Figure 3.3: Things seem to be OK!

```
1  var bankQueue = ["James", "John", "Robert"]
2
3  for queueEntry in bankQueue.enumerated() {
4      print(queueEntry.offset)
5      print(queueEntry.element)
6  }
7

0
James
1
John
2
Robert
```

Figure 3.3 Output of Enumerated Iteration

Iteration Control Flow

It is possible to manipulate the iteration flow using keywords like break or continue, with which you might be familiar from other programming languages. This concept will be covered in Chapter 4, which is focused on control flow.

As stated before, the code between { and } can be as complex as needed. Listing 3.40 features the same iteration with slightly more complex code, in which you prepare a more intuitive cashier text for each customer in bankQueue.

```
var bankQueue = ["James", "John", "Robert"]

for queueEntry in bankQueue.enumerated() {
    let number = queueEntry.offset + 1
    let name = queueEntry.element
    let cashierText = "Call \(number). customer: \(name)"
    print(cashierText)
}
```

Listing 3.40 Slightly More Complex Enumerated Iteration

The output of this code snippet is shown in Figure 3.4. Once again, the core iteration didn't change at all; we merely changed the displayed output.

```
1  var bankQueue = ["James", "John", "Robert"]
2
3  for queueEntry in bankQueue.enumerated() {
4      let number = queueEntry.offset + 1
5      let name = queueEntry.element
6      let cashierText = "Call \(number). customer: \(name)"
7      print(cashierText)
8  }

Call 1. customer: James
Call 2. customer: John
Call 3. customer: Robert
```

Figure 3.4 Output Containing Intuitive Cashier Text

Where Conditions

So far, we've gone through iterations in which we access all elements in an array sequentially. What if we want to access only some of them? There will be cases in which you only want to sequentially access elements matching a certain condition.

In such scenarios, you can add a where clause to the for iteration, containing the desired conditions. It works just like array filters, with a slightly different syntax.

Listing 3.41 demonstrates such an example. In this code snippet, numbers is a regular array. While iterating through numbers, you simply add the condition where number > 100. As a result, only values greater than 100 are processed through the iteration, and thus it only prints the values 435 and 522.

```
let numbers = [1, 25, 38, 435, 522]

for number in numbers where number > 100 {
    print(number)     // 435, 522
}
```

Listing 3.41 Iterating with where Condition

All kinds of logical operators and parenthesis can be used in a where condition. Listing 3.42 demonstrates an example in which a where clause with two conditions is present.

```
let people = ["John", "Ann", "Alice", "Bob"]

for person in people
where person.count > 3 && person.hasPrefix("A") {
    print(person) // Alice
}
```

Listing 3.42 Complex where Condition with Logical Operators

A breakdown of those where conditions is provided in Table 3.11. In the end, Swift is only able to print the value "Alice".

Initial Elements	Condition	Eliminated	Remaining
John Ann Alice Bob	person.count > 3	Ann Bob	John Alice
John Alice	person.hasPrefix("A")	John	Alice

Table 3.11 Breakdown of where Conditions

Such where conditions, and iterations in general, can be used with more complex data types too—like tuples. Listing 3.43 demonstrates an example in which an iteration through a tuple array is executed—including a where condition, too!

```
typealias Person = (name: String, age: Int)

var clients: [Person] = [
    (name: "Alice", age: 30),
    (name: "Bob", age: 25),
    (name: "Charlie", age: 22)]
```

```
for client in clients where client.age >= 25 {
    print("\(client.name) is \(client.age) years old")
}
```

Listing 3.43 Iteration Through Tuple Array

As you can see, the core syntax of the iteration didn't change at all; we simply threw in some tuples as a mental exercise. The output of this iteration is shown in Figure 3.5.

```
1  typealias Person = (name: String, age: Int)
2
3  var clients: [Person] = [
4      (name: "Alice", age: 30),
5      (name: "Bob", age: 25),
6      (name: "Charlie", age: 22)]
7
8  for client in clients where client.age >= 25 {
9      print("\(client.name) is \(client.age) years old")
10 }
```
Alice is 30 years old
Bob is 25 years old

Figure 3.5 Output of Tuple Iteration

Randomization

Swift empowers programmers with options for random access to array elements. Listing 3.44 features an example in which you access a random element in fruits. Every time you execute this code, randomFruit gets a random value, such as "Apple", "Cherry", or "Orange".

```
let fruits = ["Apple", "Banana", "Cherry", "Mango", "Orange"]
let randomFruit = fruits.randomElement()
```

Listing 3.44 Picking Random Array Element

This code could be used, for example, in an app that suggests a random daily fruit to consume.

Another randomization feature lets you shuffle elements in an array. Just like shuffling cards before starting a card game, you can shuffle an array to randomly change the positions of its elements. Listing 3.45 features such an example, in which you use shuffle() to shuffle the names in participants and declare the first three as winners. It's a handy feature for a lottery app, right?

```
// List of participants
var participants = ["Alice", "Bob", "Charlie", "David", "Eve", "Frank", "Grace"]

// Shuffle the array
participants.shuffle()
```

```
// Get the first three as winners
let winners = participants.prefix(3)

// Print the winners
print("Winners:")
for (index, winner) in winners.enumerated() {
    print("\(index + 1). \(winner)")
}
```

Listing 3.45 Array Shuffle Example

Due to the nature of randomization, you will get a different output every time you execute this code. Nevertheless, Figure 3.6 demonstrates a possible output in which three random participants were declared as winners. Apparently, those names were in the first three positions in participants when participants.shuffle() was executed.

```
1  // List of participants
2  var participants = ["Alice", "Bob", "Charlie", "David", "Eve", "Frank", "Grace"]
3
4  // Shuffle the array
5  participants.shuffle()
6
7  // Get the first three as winners
8  let winners = participants.prefix(3)
9
10 // Print the winners
11 print("Winners:")
12 for (index, winner) in winners.enumerated() {
13     print("\(index + 1). \(winner)")
14 }
■ ▶
Winners:
1. Frank
2. Bob
3. Alice
```

Figure 3.6 One Possible Shuffle Output

That final example concluded our content on arrays, one of three major collection types, along with sets and dictionaries. We will continue the journey with sets, which are similar to arrays but bring the uniqueness constraint to the table. Take a break, get some fresh air if you need to, and see you there!

3.2 Sets

In Swift, a *set* is a collection, like an array; they share some common ground. The core difference is that a set is an unordered collection of unique values. Unlike arrays, sets don't allow duplicate elements and don't guarantee a specific order by index—but they promise a very high access speed, which is a significant benefit in large datasets.

Table 3.12 describes the differences between arrays and sets clearly.

Feature	Array	Set
Order	Ordered	Unordered
Duplicate values	Allowed	Not allowed
Access by index	Allowed	Not allowed
Element lookup speed	Slow	Fast

Table 3.12 Comparison Between Arrays and Sets

The selling points of a set are basically element uniqueness and access speed. Here are some general suggestions for working with a set:

■ Use an array when you need ordered elements with possible duplicates.

■ Use a set when you need fast lookup with unique elements.

To strengthen your understanding, Table 3.13 contrasts some use cases in which either an array or set would be preferred as an element container.

Case	Preference	Reason
Shopping cart	Array	Duplicate items should be allowed.
Queue	Array	Order matters and a queue is accessed in entry order.
Member list	Set	Members are unique and access by user name would be fast.
Product categories	Set	Categories are unique and order doesn't matter.

Table 3.13 Use Cases for Arrays and Sets

As you can see, arrays and sets can't fully replace each other. You could attempt to use an array instead of a set, but it won't ensure uniqueness and access speed would be unnecessarily slow. Likewise, you could attempt to use a set instead of an array, but your app will fail on duplicate elements and won't follow the given element order.

Why Are Sets Faster?

Sets are faster than arrays for lookups because they use hashing, while arrays use linear searches.

In more common terms, when checking if an element exists in an array, Swift must scan each item one by one until it finds a correspondence between elements. The runtime cost may be negligible in small datasets, but it would make a significant difference in large datasets.

Sets, meanwhile, use a hash table to help spot elements, which allows for instant lookup. The runtime cost is constant regardless of the set size. When an element is added to a set, Swift (secretly) computes a hash value, which is used as in index. When checking for an element, Swift directly jumps to the hashed index instead of scanning every item.

Now that you have a general notion of sets, we can move forward to hands-on examples to introduce the corresponding syntax and further features.

3.2.1 Creating Sets

Set declarations can be made using a familiar syntax, like that for arrays. No big surprise, right? Listing 3.46 showcases alternative methods for set declarations. Here, set1, set2, and set3 will end up having the exact same elements.

```
// Explicit type declaration
var set1 = Set<Int>()
set1.insert(1)
set1.insert(2)
set1.insert(3)

// Explicit type declaration with values
var set2: Set<Int> = [1, 2, 3]

// Type inference
var set3 = Set([1, 2, 3])
```

Listing 3.46 Alternative Methods of Set Declaration

In the first part, you explicitly declare the type of set1 as Set<Int>() and insert elements afterward. In the second part, you declare the type of set2—but insert the initial elements immediately. Finally, set3 is declared using type inference, which means that you let Swift "guess" the types of elements on your behalf.

Insert Versus Append

You might have caught a syntax difference here. In arrays, new elements are added via the array.append function. In sets, new elements are added via the set.insert function. Although their purposes are similar, the difference is worth noting.

Note that array.append will add the new element to the tail of the array because an array is an ordered collection. However, set.insert will do the hash calculations and add the new element "somewhere"; no particular order is guaranteed.

You can create a set from an array as well! The catch is that you will lose duplicate entries in the process. Whether that's desirable or not depends on the use case. Listing 3.47 features such a conversion.

```
var allStudents = ["Alice", "Bob", "Charlie", "Bob", "Alice"]
var uniqueStudents = Set(allStudents) // Alice, Charlie, Bob
```

Listing 3.47 Array to Set Conversion

3.2.2   Sets as Constants

As with any other data type, sets can be declared as mutable via var or immutable (read-only) via let. To keep this section self-contained, Listing 3.48 demonstrates both declaration types.

```
var changeableSet = Set([1, 2, 3, 4, 5])
let readOnlySet = Set([1, 2, 3, 4, 5])
```

Listing 3.48 Set Declaration as Constant

In this example, you can modify the contents of changeableSet in due course—because it was declared to be mutable using var. On the other hand, readOnlySet can't be modified later—because it was declared to be immutable using let.

3.2.3   Accessing Sets

In this section, you will learn how to access set elements. We are going to cover basic set functions and how to access the first element of a set.

Basic Set Functions

First things first: Basic array functions, which were covered in Section 3.1.3, are available with sets too! To keep this section self-contained, Table 3.14 showcases those functions.

Function	Result
isEmpty	true if the set is empty; false otherwise
count	Number of elements in the set
contains(element)	true if the element is in the set; false otherwise

Table 3.14 Basic Set Functions

To see those functions in context, look at Listing 3.49; the functionality is quite intuitive.

```
var animals: Set<String> = ["dog", "cat", "bird", "elephant"]

animals.isEmpty              // false
animals.count               // 4
animals.contains( "bird" )  // true
animals.contains( "snake" ) // false
```

Listing 3.49 Demonstration of Basic Set Functions

A cool feature is to use a where clause within the contains function. Because you're already familiar with where clauses in Swift, the syntax of the demonstration in Listing 3.50 should be intuitive.

```
var animals: Set<String> = ["dog", "cat", "bird", "elephant"]
animals.contains(where: { $0.count > 5 })        // true due to "elephant"
animals.contains(where: { $0.contains("x") })    // false
```

Listing 3.50 Where Clause Within contains Function

First Function

Because sets are unordered collection types, they don't support index-based access. A supported complementary function is first, which returns the "first" element in the set. Be careful though: Because sets are not ordered like arrays, Swift doesn't guarantee that first returns the initial inserted element; instead, it simply returns an element based on the internal order of the set, which could be any of them.

Having that said, Listing 3.51 demonstrates the usage of the first function. Go ahead and try it: Every time you execute this code snippet, you should get a different element because the calculated hash values will change on each execution—and therefore, so does the "first" element behind the scenes.

```
var numbers: Set = [10, 20, 30, 40, 50]
numbers.first! // Output: Could be any of 10, 20, 30, 40, 50
```

Listing 3.51 Demonstration of first Function for Sets

3.2.4 Set Derivation

Just like arrays, sets feature functions through which you can derive new sets. They are mostly similar to those for array derivation—with the obvious deviation that index-based access would not make sense. In this section, we will walk through the available derivation functions for sets.

Filter

Set filters share the same logic and syntax as array filters. The `filter` function allows you to create a subset of a set based on a condition. Listing 3.52 showcases two examples, in which you extract even numbers and fruits starting with the letter *A*.

```
let numbers: Set = [1, 2, 3, 4, 5, 6, 7, 8, 9, 10]
let evenNumbers = numbers.filter { $0 % 2 == 0 }
print(evenNumbers) // 2, 4, 6, 8, 10; order varies

let fruits: Set = ["Apple", "Banana", "Cherry", "Avocado", "Blueberry"]
let aFruits = fruits.filter { $0.hasPrefix("A") }
print(aFruits) // Apple, Avocado; order varies
```

Listing 3.52 Set Filtering Demonstration

Although this code will work just fine, you can't ensure the element order in `evenNumbers` and `aFruits`; Swift will order them as it pleases, based on hash values.

Map

Like it is in arrays, the `map` function for sets in Swift is used to transform each element in the set into a new value. The output is a new set containing those new values. Listing 3.53 features two mapping examples.

```
let numbers: Set = [1, 2, 3, 4, 5]
let squaredNumbers = numbers.map { $0 * $0 }
print(squaredNumbers) // Output: [1, 4, 9, 16, 25] (order varies)

let fruits: Set = ["Apple", "Banana"]
let uppercasedFruits = fruits.map { $0.uppercased() }
print(uppercasedFruits) // Output: ["APPLE", "BANANA"] (order varies)
```

Listing 3.53 Set Mapping Demonstration

In the first example, you square the elements in `numbers` and collect them into a new set called `squaredNumbers`. In the second example, you convert the elements in `fruits` to uppercase and collect them into a new set called `uppercasedFruits`. As usual, element order is undeterminable in either sample.

Reduce

Like in arrays, the `reduce` function for sets in Swift allows you to combine all elements in a set into a single value. The detailed breakdown for this function was provided in Section 3.1.4 already, so you can jump directly into a couple of examples provided in Listing 3.54.

```
let numbers: Set = [10, 20, 30, 40, 50]
let sum = numbers.reduce(0) { $0 + $1 }
print(sum)  // Output: 150

let words: Set = ["Follow", "the", "white", "rabbit"]
let sentence = words.reduce("") { $0 + " " + $1 }
print(sentence)  // Output: " the Follow rabbit white" (order varies)
```

Listing 3.54 Set reduce Demonstration

In the first part, you have a set of numbers. By invoking numbers.reduce(0) { $0 + $1 }, you tell Swift to sum those numbers, the result of which is 150. Because the element order doesn't affect the result of an addition operation, this part works just like it does for arrays.

In the second part, though, we have a different story. The words set contains a collection of unique strings. The words.reduce("") { $0 + " " + $1 } expression should return a concatenation of all strings in the set, right?

It kind of does—but with a twist: Because words didn't index the strings in the order they were provided, the sentence output will seemingly have shuffled the order of the strings—which is the expected behavior for sets! If you're aiming for control over the order of elements, you should use an array instead.

Union

You already know that it's possible to merge existing arrays; this topic was covered in Section 3.1.4. Likewise, it's possible to merge existing sets, combining all their elements—as shown semantically in Figure 3.7.

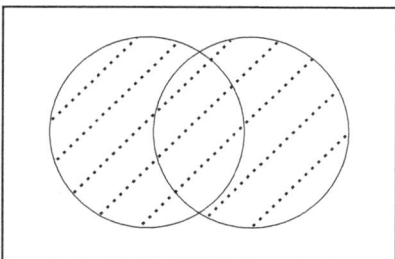

Figure 3.7 Scope of Set Union

You can either build a joint set out of two sets or insert rows of a set into another one to extend it. In this section, we will go over both options.

Listing 3.55 demonstrates the usage of the union function to merge two sets and create a new one. When setA.union(setB) is executed, Swift will merge all elements of setA and setB into the target mergedSet variable. Meanwhile, the original setA and setB sets are not mutated.

```
let setA: Set = [1, 2, 3]
let setB: Set = [3, 4, 5]

let mergedSet = setA.union(setB)
print(mergedSet) // 1, 2, 3, 4, 5 (order varies)
```

Listing 3.55 Joining Sets Using union Function

Note that 3 is a common element in setA and setB. Due to the uniqueness requirement of sets, common elements like 3 are not duplicated in the target set.

What if you had three sets to merge instead of two? That's easy! You can simply chain the union function as shown in Listing 3.56. Because union returns a new set anyway, you can keep chain-executing this function for all sets you need to merge.

```
let setA: Set = [1, 2]
let setB: Set = [2, 3]
let setC: Set = [3, 4]

let mergedSet = setA.union(setB).union(setC)
print(mergedSet) // 1, 2, 3, 4 (order varies)
```

Listing 3.56 Chain-Executing Set Unions

The examples so far focused on producing a new set out of existing sets. But you can also merge a set into another set, extending the target set with new elements. The way to do so is to invoke the formUnion function of the target set. In the demonstration in Listing 3.57, setA was extended with new elements from setB. Naturally, setB is not affected or mutated by this operation.

```
var setA: Set = [10, 20, 30]
let setB: Set = [30, 40, 50]

setA.formUnion(setB)
print(setA) // 10, 20, 30, 40, 50 (order varies)
```

Listing 3.57 Merging Set into Another Set Using formUnion

Intersection

In this section, our purpose is to find common elements of two sets. The semantics of this operation is shown in Figure 3.8.

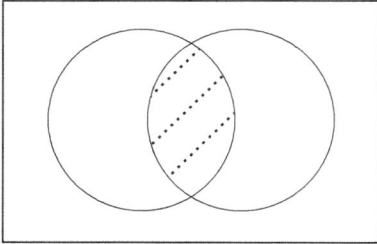

Figure 3.8 Scope of Set Intersection

For this purpose, you'll invoke the `intersection` function of either set. Listing 3.58 demonstrates how to do so; note the syntactic similarity to the former union function.

```
let setA: Set = [1, 2, 3, 4, 5]
let setB: Set = [3, 4, 5, 6, 7]

let commonElements = setA.intersection(setB)
print(commonElements) // 3, 4, 5 (order varies)
```

Listing 3.58 Finding Common Elements of Sets Using intersection Function

The `intersection` function doesn't mutate any of the sets; it simply creates a new set out of their common elements. An alternative is to invoke the `formIntersection` function, which will mutate the target set and reduce its elements to common elements of the two sets.

Listing 3.59 contains a demonstration of this function. After executing `setA.formIntersection(setB)`, `setA` will contain only elements in common with `setB`, replacing its former contents. `setB` is not mutated.

```
var setA: Set = [10, 20, 30, 40, 50]
let setB: Set = [30, 40, 50, 60, 70]

setA.formIntersection(setB)
print(setA) // 30, 40, 50 (order varies)
```

Listing 3.59 Demonstration of formIntersection Function

Subtracting

Now that you know about spotting the common elements of two sets, it's time to learn how to spot the differing elements of two sets! Figure 3.9 showcases the semantics of this operation.

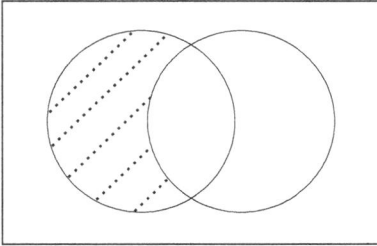

Figure 3.9 Scope of Set Subtraction

For this purpose, you can invoke the `subtracting` function (Listing 3.60), which will build a new set from the differing elements of the given sets.

```
let setA: Set = [1, 2, 3, 4, 5]
let setB: Set = [3, 4, 5, 6, 7]

let diffSet = setA.subtracting(setB)
print(diffSet) // Output: 1, 2 (order varies)
```

Listing 3.60 Finding Differing Elements of Sets Using subtracting Function

As with `union` and `intersection`, you can also mutate the original set with the differing values if you want. For that purpose, you can invoke the `subtract` function. Check Listing 3.61 for a demonstration: The execution of `setA.subtract(setB)` will remove elements from `setA`—but not the ones in common with `setB`. Meanwhile, `setB` is not mutated.

```
var setA: Set = [1, 2, 3, 4, 5]
let setB: Set = [3, 4, 5, 6, 7]

setA.subtract(setB)
print(setA) // Output: 1, 2 (order varies)
```

Listing 3.61 Demonstration of subtract Function

This example concludes our content on set derivation. We went through various functions that will help you produce new sets out of existing ones. Our next topic is a natural follow-up: How to check if sets are supersets or subsets of each other.

3.2.5   Checking for Supersets

For this topic, it might be a good idea to understand the terms *superset* and *subset* before jumping to the playground. Figure 3.10 illustrates a superset and subset.

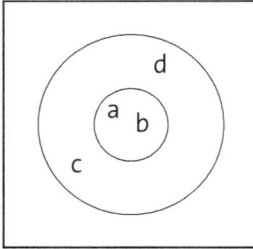

Figure 3.10 Superset and Subset Diagram

In this diagram, the inner circle is a subset and the outer circle is a superset—which contains the entire subset, plus more. The (inner) subset contains the elements *a* and *b*. The (outer) superset automatically contains the subset elements *a* and *b*, as well as the additional elements *c* and *d*. Table 3.15 showcases the element lists in pseudocode format.

Set	Elements
Subset	`["a", "b"]`
Superset	`["a", "b", "c", "d"]`

Table 3.15 Subset and Superset Elements

So far, so good! Moving forward to Swift, you can easily check whether a set is the superset of another set. Naturally, you can also check if a set is the subset of another one.

Listing 3.62 contains a demonstration of such checks. The isSuperset function is invoked to find if a set is a superset of another set, and isSubset is invoked to find out if a set is a subset of another set.

```
let set1: Set = ["a", "b"]
let set2: Set = ["a", "b", "c", "d"]
let set3: Set = ["e", "f"]

set1.isSubset(of: set2)     // true
set2.isSuperset(of: set1)   // true
set3.isSubset(of: set2)     // false
set2.isSuperset(of: set3)   // false
```

Listing 3.62 Checking for Superset and Subset Relations

3.2.6 Modifying Sets

Now that you know about creating, accessing, and deriving sets, it's time to learn about set mutation. In this section, you will learn about modifying sets and changing their contents.

Sets Aren't Indexed

Because sets don't have indexes, set-modification functions will deviate from their array counterparts. Although there is a reasonable overlap, index-based functions are naturally not available for sets.

Inserting Elements

In Section 3.2.1, we went through an example that created a set by using `insert` to insert its initial elements. However, you can invoke the `insert` function to insert elements later too! Listing 3.63 demonstrates this technique by adding new animals to the `animals` set.

```
var animals: Set = ["cat", "dog", "bird"]
animals.insert("fish")
animals.insert("snake")
print(animals) // cat, dog, bird, fish, snake (order varies)
```

Listing 3.63 Inserting Elements into Set

As you know by now, a feature of sets is their promise to contain unique values. Even if you attempt to insert duplicates into a set, as in Listing 3.64, Swift will ignore the duplicate insertion and preserve the uniqueness of the set's elements.

```
var animals: Set = ["cat", "dog", "bird"]
animals.insert("cat")
animals.insert("dog")
print(animals) // cat, dog, bird (order varies)
```

Listing 3.64 Inserting Duplicates into Set

Deleting Elements

Deleting elements from a set is a straightforward operation. Set objects feature ready-to-use functions via the `remove` prefix; these are listed in Table 3.16.

Function	Purpose
remove(element)	Removes the element from the set
removeFirst()	Removes the first element from the set (random)
removeAll()	Clears the set completely

Table 3.16 Set Functions for Element Removal

A demonstration of those functions is given in Listing 3.65. After the initial set definition, `numbers.remove(2)` is executed, which spots and removes the given element. In the second part, `numbers.removeFirst()` is executed. Because you can't be sure of the element order, this statement removes a random element from the set. Finally, `numbers.removeAll()` clears the entire set.

```
var numbers: Set = [1, 2, 3, 4, 5]

numbers.remove(2)       // Removes specific element
print(numbers)          // 1, 3, 4, 5 (order varies)

numbers.removeFirst()   // Removes an arbitrary element
print(numbers)          // Remaining three elements, varies

numbers.removeAll()     // Clears the set
print(numbers)          // (empty)
```

Listing 3.65 Demonstration of Element-Removal Functions

Set Element Modification
Because set elements are uniquely hashed, you can't modify an element in a set like an array. As a workaround, you can emulate element modification by deleting the old value and inserting the new value.

3.2.7   Iterating Through Sets

Although sets are typically preferred for single-element access, it is possible to iterate through set elements too. In fact, the functions to iterate through sets are nearly identical to their array counterparts. In this section, you will discover those functions, as you did for arrays.

As a reminder: Keep in mind that sets don't guarantee an element order, so when you iterate through a set, the access order will be virtually random.

For Clause

The most fundamental approach to set iteration is to use a `for … in …` statement. Check the demonstration in Listing 3.66, which iterates through phone numbers.

```
var phones: Set = ["123-4567", "890-1234", "543-2109", "234-5678"]

for phone in phones {
    print(phone) // Prints each phone, order varies
}
```

Listing 3.66 Iterating Through Set Using for Clause

The logic of for clauses is identical in arrays and sets. To prevent content duplication, we won't dive into those details again here.

Using Enumerated

Likewise, the use of the enumerated function is also identical to its use in arrays. Check the demonstration in Listing 3.67, which iterates through each element of the set.

```
var customers: Set = ["James", "John", "Robert"]

for customer in customers.enumerated() {
    print(customer.offset)
    print(customer.element)
}
```

Listing 3.67 Iterating Through Set Using Enumerated

The output of this code snippet is shown in Figure 3.11, but there are a couple of things to keep in mind:

- Every time you execute this code, you will get a different order because values are hashed, not ordered.
- customer.offset should merely be seen as an index, starting from 0 and increasing with each element. It does not symbolize the (nonexistent) "element order" in the set.

```
1  var customers: Set = ["James", "John", "Robert"]
2
3  for customer in customers.enumerated() {
4      print(customer.offset)
5      print(customer.element)
6  }
0
Robert
1
John
2
James
```

Figure 3.11 Set Enumeration Demonstration Output

Despite those limitations, the ability to browse through set elements is a neat feature to have!

Where Conditions

Just like arrays, sets can be partially iterated with the help of where conditions. Even the syntax is the same! Nevertheless, Listing 3.68 demonstrates an iteration example in which you iterate only through numbers bigger than 100.

```
let numbers: Set = [1, 25, 38, 435, 522]

for number in numbers where number > 100 {
    print(number)    // 435, 522; order varies
}
```

Listing 3.68 Set Iteration Supported by where Condition

You can check Section 3.1.6 for additional where examples, ones used in our discussion of arrays.

Randomization

Finally, you will learn about fetching random elements out of a set—which is, once again, the same as for arrays. Listing 3.69 demonstrates the usage of randomElement in a set.

```
let fruits: Set = ["Apple", "Banana", "Cherry", "Mango", "Orange"]
let randomFruit = fruits.randomElement()
```

Listing 3.69 Picking Random Element from Set

No Shuffle for Sets

Because sets are unordered collections with a virtually random element order, it doesn't make sense to shuffle a set. That's why Swift sets don't have a shuffle function like arrays.

This section has concluded our content on sets. It was probably easy to follow through as sets share a lot of common ground with arrays. By reading about sets, your knowledge of arrays was also solidified. Now we'll move forward to a new collection type, one that's a little different than arrays and sets.

3.3 Dictionaries

The final collection type, dictionaries, is a little different than arrays and sets. Arrays and sets are lists of elements of the same type. Their selling point is their ability to contain multiple elements. A dictionary, meanwhile, is a flat structure containing key-value pairs.

The difference is highlighted in Figure 3.12. On the left side, userNames is an array holding a list of strings, reflecting the names of users in a system. On the right side, currentUser is a dictionary containing details about the current user.

userNames	currentUser	
"Joe"	"name"	"Joe"
"Mary"	"age"	25
"George"	"isAdmin"	true
"Linda"		
"Emma"		

Figure 3.12 Array and Dictionary, Side by Side

In this example, you can imagine the array as a list of elements (user names) and the dictionary as zoomed in on an element (like a user) to show all its details. That's the usual selling point of a dictionary, anyway.

But wait: Didn't we use named tuples for that purpose in Chapter 2? How are dictionaries different? Good point! Their main differences are contrasted in Table 3.17.

Feature	Tuple	Dictionary
Structure	Fixed set of values	Key-value pairs
Access	By index or name	By key
Size	Fixed; set at declaration	Dynamic; can grow or shrink
Mutability	Can't add/remove elements	Can add/remove key-value pairs
Best use case	Grouping related values	Flexible key-value storage

Table 3.17 Tuples Versus Dictionaries

Despite those differences, tuples and dictionaries do have some overlapping functionality. If you are aiming for flexibility, though, dictionaries are the way to go because you can add new key-value pairs as needed.

As we go over some hands-on coding examples, you'll get a better idea of what makes dictionaries unique. Without further ado, let's begin!

3.3.1 Creating Dictionaries

In Chapter 2, you learned about alternative ways of declaring variables using type inference and type interpolation. Likewise, there are alternative ways of declaring dictionaries using the same methods, as we'll discuss in this section.

Dictionaries with a Single Type

Let's start with the basic example in Listing 3.70, featuring type inference, as a relaxed warm-up.

```
var myCar = [
    "make": "Nissan",
    "model": "Qashqai",
    "color": "Black",
    "bodyType": "SUV"
]

print(myCar["make"]!)    // Nissan
print(myCar["color"]!)   // Black
```

Listing 3.70 Dictionary Declaration Using Type Inference

In this example, myCar is a dictionary. It contains various properties as key-value pairs. If you had to list those properties in a table, it would look like Table 3.18.

Key	Value
make	"Nissan"
model	"Qashqai"
color	"Black"
bodyType	"SUV"

Table 3.18 Key-Value Pairs of myCar

That's clear, right? You can add as many properties as necessary. All those key-value pairs are logically properties of myCar.

The last part of Listing 3.70 demonstrates the basic way to access elements in a dictionary: myCar["make"] would return "Nissan", while myCar["color"] would return "Black".

Now that you have seen the basic approach to dictionary declaration, we can move forward to further methods. Listing 3.71 showcases examples of alternative syntax that serve the same purpose. You can pick any alternative that suits your needs.

```
// Type annotation
var myCar: [String: String] = [:]
myCar["make"] = "Nissan"
myCar["model"] = "Qashqai"

// Type annotation - alternative syntax
var herCar = Dictionary<String, String>()
herCar["make"] = "Hyundai"
herCar["model"] = "Accent"
```

```
// Type annotation with initial values
var hisCar: [String: String] = [
    "make": "Toyota",
    "model": "Corolla"
]
```

Listing 3.71 Alternative Dictionary Declaration Methods

Dictionaries with Multiple Types

So far, we have declared dictionaries with a single type, meaning that all values were strings. More often than not, though, a dictionary needs to contain values of various types. Revisiting the introduction to this section, Figure 3.13 features such an example.

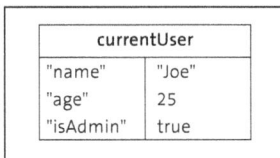

Figure 3.13 Dictionary with Multiple Types

If you look closely, you'll see that currentUser has keys with different data types, which are listed in Table 3.19.

Key	Data Type
name	String
age	Integer
isAdmin	Boolean

Table 3.19 Data Types of currentUser Keys

To declare such dictionaries with flexible/multiple data types, you need to use the Any keyword. An implementation of this is provided in Listing 3.72. Note that we have provided Any as the value data type here, indicating that Swift should behave in a flexible manner and accept any provided value type.

```
var currentUser: [String: Any] = [
    "name": "Joe",
    "age": 25,
    "isAdmin": true
]

print(currentUser["name"]!)     // Joe
```

```
print(currentUser["age"]!)      // 25
print(currentUser["isAdmin"]!)  // true
```

Listing 3.72 Declaration of Dictionary with Flexible/Multiple Data Types

Multidimensional Dictionaries

So far, we've covered *flat dictionaries*, in which each key corresponds to a single value. However, Swift supports *multidimensional dictionaries* too! You can declare a *nested dictionary*, in which an element is a collection instead of a simple variable.

Listing 3.73 demonstrates how to declare an array as a dictionary element. bassGuitar has an element called availableColors, which is a string array.

```
var bassGuitar: [String: Any] = [
    "brand": "Fender",
    "model": "Precision",
    "strings": 5,
    "availableColors": ["Black", "Sage Green", "Maple"]
]
```

Listing 3.73 Declaring a Dictionary Containing an Array

Likewise, a dictionary may contain a set as well—as demonstrated in Listing 3.74.

```
var bassGuitar: [String: Any] = [
    "brand": "Fender",
    "model": "Precision",
    "strings": 5,
    "availableColors": Set(["Black", "Sage Green", "Maple"])
]
```

Listing 3.74 Declaring a Dictionary Containing a Set

You can even include a dictionary inside another dictionary, making it a nested dictionary. In Listing 3.75, specs is a subdictionary of bassGuitar, containing key-value pairs of its own.

```
var bassGuitar: [String: Any] = [
    "brand": "Fender",
    "model": "Precision",
    "strings": 5,
    "specs": [
        "bodyWood": "Alder",
        "neckWood": "Maple",
        "fingerWood": "Rosewood",
        "quarterSawn": true,
```

```
        "scaleLength": 34,
        "frets": 21
    ]
]
```

Listing 3.75 Nested Dictionary Demonstration

As a mental exercise, Listing 3.76 demonstrates a complex dictionary, containing both a subset and a subdictionary as elements.

```
var bassGuitar: [String: Any] = [
    "brand": "Fender",
    "model": "Precision",
    "strings": 5,
    "availableColors": Set(["Black", "Sage Green", "Maple"]),
    "specs": [
        "bodyWood": "Alder",
        "neckWood": "Maple",
        "fingerWood": "Rosewood",
        "quarterSawn": true,
        "scaleLength": 34,
        "frets": 21
    ]
]
```

Listing 3.76 Complex Nested Dictionary Example

JSON Similarity

Readers with JSON experience might have noticed that complex dictionaries start to look like JSON files. That's correct—and you can use that similarity as a mental hook to understand dictionaries a little better.

Zipping Dictionaries

A cool trick in dictionary creation is to use the `zip` keyword. In Listing 3.77, the keys are in the `keys` array and the values in the `values` array.

```
let keys = ["brand", "model", "strings"]
let values = ["Fender", "Precision", "5"]

var bassGuitar = Dictionary(uniqueKeysWithValues: zip(keys, values))
print(bassGuitar) // ["brand": "Fender", "model": "Precision", "strings": "5"]
```

Listing 3.77 Dictionary Creation Using Zip

To build the bassGuitar dictionary, you "zip" keys with values: The first key (brand) gets the first value (Fender); the second key (model) gets the second value (Precision); and so on.

3.3.2   Dictionaries as Constants

As with any other data type, dictionaries can be declared as mutable via var or immutable (read-only) via let. To keep this section self-contained, Listing 3.78 demonstrates a mutable and an immutable dictionary declaration.

```
// Mutable
var currentUser: [String: Any] = [
    "name": "Joe",
    "age": 25,
    "isAdmin": true
]

// Immutable
let previousUser: [String: Any] = [
    "name": "Mary",
    "age": 33,
    "isAdmin": false
]
```

Listing 3.78 Mutable and Immutable Dictionary Declarations

3.3.3   Accessing Dictionaries

Now that you know how to create dictionaries, it's time to access them. After all, why create a dictionary you will never read, right? We'll walk through various methods of accessing dictionaries in this section.

Direct Access Using a Key

Because dictionaries are basically key-value pairs, the first natural expectation would be to give the key and receive the corresponding value, right? Listing 3.79 demonstrates the basic syntax for this. The bassGuitar["brand"] expression returns the value of "brand" within the dictionary—and the same approach applies to "model".

```
var bassGuitar: [String: Any] = [
    "brand": "Fender",
    "model": "Precision",
    "strings": 5,
    "availableColors": ["Black", "Sage Green", "Maple"]
]
```

```
print(bassGuitar["brand"]!)      // Fender
print(bassGuitar["model"]!)      // Precision
```

Listing 3.79 Accessing Simple Dictionary Element

If you query a nonexisting key, Swift will simply return `nil`, as demonstrated in Listing 3.80.

```
var bassGuitar: [String: Any] = [
    "brand": "Fender",
    "model": "Precision",
    "strings": 5,
    "availableColors": ["Black", "Sage Green", "Maple"]
]

bassGuitar["price"] // nil
```

Listing 3.80 Accessing Nonexisting Dictionary Element

Optionals in Dictionary Access

Because it aims to be a type-safe language, Swift is strict about being absolutely sure about the type of any variable. When a dictionary is defined as [String: Any] as in Listing 3.80, this means that you can assign any type of value for a dictionary key. That's a potential area for safety issues because you could mistakenly unwrap an existing element to an incorrect variable type, like assigning the bassGuitar["availableColors"] array to an integer variable.

Or you could basically access a nonexisting element, as in bassGuitar["price"], which may potentially lead to an app crash.

Due to such concerns, Swift always assumes that accessing a dictionary element "optionally" returns a value of the desired type. The dictionary may contain another type of data, or it may not contain such a key at all.

In Chapter 6, you will learn about optionals in Swift. In that chapter, once you understand the core logic of optionals, we will go over dictionary-specific examples of optional element access. But for now, we'll stick to the basics.

Checking If a Key Exists

Each dictionary comes with a property called `keys`. By invoking `keys.contains`, you can check if a dictionary contains a key or not. Listing 3.81 demonstrates the necessary syntax for that.

```
var bassGuitar: [String: Any] = [
    "brand": "Fender",
```

```
    "model": "Precision",
    "strings": 5,
    "availableColors": ["Black", "Sage Green", "Maple"]
]

bassGuitar.keys.contains("brand")    // true
bassGuitar.keys.contains("price")    // false
```

Listing 3.81 Checking if Key Exists in Dictionary

In this example, bassGuitar.keys.contains("brand") returns true because "brand" is an existing key in the dictionary. However, bassGuitar.keys.contains("price") returns false because "price" isn't a key within the dictionary. That's clear, right?

This feature will be useful when you learn about control flow in Chapter 4. You will be able to make your app behave differently depending on the existence of a key in a dictionary.

3.3.4 Modifying Dictionaries

In this section, you will learn how to make changes to dictionaries after their initial declaration. Adding, changing, and removing key-value pairs will be the focus.

Inserting Elements

Inserting a new key-value pair into a dictionary is as simple as merely providing the key and the value! Listing 3.82 features the required syntax.

```
var mobilePhone: [String: Any] = [
    "brand": "Apple",
    "model": "iPhone 15 Pro",
    "operatingSystem": "iOS 17",
    "screenSize": 6.1,
    "storageOptions": [128, 256, 512, 1024],
    "has5G": true,
    "colors": Set(["Black", "Blue", "White", "Titanium"])
]

mobilePhone["weight"] = 187
mobilePhone["waterResistant"] = true
```

Listing 3.82 Inserting New Key-Value Pairs into Dictionary

On the last two lines, the weight and water-resistance features of the mobile phone were added to the dictionary. It's simple as that!

Modifying Elements

Modifying the value for a key uses the exact same syntax as insertion. Listing 3.83 demonstrates both item modification and insertion to showcase them side-by-side.

```
var mobilePhone: [String: Any] = [
    "brand": "Apple",
    "model": "iPhone 15 Pro",
    "operatingSystem": "iOS 17",
    "screenSize": 6.1,
    "storageOptions": [128, 256, 512, 1024],
    "has5G": true,
    "colors": Set(["Black", "Blue", "White", "Titanium"]),
    "weight": 187
]

print(mobilePhone["weight"]!)          // Output: 187
mobilePhone["weight"] = 186            // Changes to 186
print(mobilePhone["weight"]!)          // Output: 186

mobilePhone["waterResistant"] = true   // Inserts new element
```

Listing 3.83 Dictionary Element Modification Demonstration

In this code snippet, weight was initially declared as 187. The mobilePhone["weight"] = 186 expression then changes this value to 186. Note that this line has the exact same syntax as element insertion. The action here will depend on a couple of factors:

- If the dictionary contains an entry for weight already, Swift will update its value.
- Otherwise, Swift will insert the provided key-value pair, as happens for waterResistant.

Deleting Elements

In Swift, element removal from a dictionary is as straightforward as it gets. Listing 3.84 demonstrates two dictionary functions for this task.

```
var mobilePhone: [String: Any] = [
    "brand": "Apple",
    "model": "iPhone 15 Pro",
    "operatingSystem": "iOS 17",
    "screenSize": 6.1,
    "storageOptions": [128, 256, 512, 1024],
    "has5G": true,
    "colors": Set(["Black", "Blue", "White", "Titanium"]),
    "weight": 187
]
```

```
mobilePhone.removeValue(forKey: "has5G")    // Removes single element
mobilePhone.removeAll()                     // Removes all elements
```

Listing 3.84 Dictionary Element Removal Demonstration

The function removeValue will remove the key-value pair for the provided key, whereas removeAll will remove all key-value pairs from the dictionary, leaving an empty collection behind.

3.3.5   Iterating Through a Dictionary

There might be cases in which you have a dynamically created dictionary without knowing the exact key names. For example, a third-party library might have parsed a JSON file and returned a dictionary. In such a case, you might want to access all keys and/or values in the dictionary sequentially, as an array.

In such cases, you can use a for clause to iterate through key-value pairs, as demonstrated in Listing 3.85.

```
var mobilePhone: [String: Any] = [
    "brand": "Apple",
    "model": "iPhone 15 Pro",
    "operatingSystem": "iOS 17",
    "screenSize": 6.1,
    "storageOptions": [128, 256, 512, 1024],
    "has5G": true,
    "colors": Set(["Black", "Blue", "White", "Titanium"]),
    "weight": 187
]

for (key, value) in mobilePhone {
    print(key, value, separator: " : ")
}
```

Listing 3.85 Iterating Through Keys and Values of Dictionary

In this example, for (key, value) in mobilePhone is the iteration command—which works via the same logic as for arrays and sets. The code between { and } is executed for each key-value pair in the dictionary. On each iteration, Swift assigns the next key to the key variable and the next value to the value variable.

The output is shown in Figure 3.14. In this example, all we did between { and } was to print the keys and values. In upcoming chapters, we will do more interesting things as you learn more about Swift.

```
 1  var mobilePhone: [String: Any] = [
 2      "brand": "Apple",
 3      "model": "iPhone 15 Pro",
 4      "operatingSystem": "iOS 17",
 5      "screenSize": 6.1,
 6      "storageOptions": [128, 256, 512, 1024],
 7      "has5G": true,
 8      "colors": Set(["Black", "Blue", "White", "Titanium"]),
 9      "weight": 187
10  ]
11
12  for (key, value) in mobilePhone {
13      print(key, value, separator: " : ")
14  }
15
```

```
has5G : true
colors : ["White", "Blue", "Black", "Titanium"]
weight : 187
model : iPhone 15 Pro
brand : Apple
operatingSystem : iOS 17
screenSize : 6.1
storageOptions : [128, 256, 512, 1024]
```

Figure 3.14 Output of Dictionary Iteration

And voilà! This final example concludes our content on dictionaries, as well as our entire chapter on collections.

3.4 Summary

In this chapter, you learned about collections in Swift. Basically, collections are data structures that group multiple variables into a single, organized container. Swift offers three main collection types, which are summarized in Table 3.20.

Collection Type	Content	Duplicates	Use Case	Specialty
Array	Indexed/ordered collection of elements	Yes	Queue	Flexibility
Set	Hashed/unordered collection of elements	No	Member list	Fast element access
Dictionary	Hashed/unordered key-value pairs	No	Product properties	JSON-like

Table 3.20 Swift Collection Type Summary

Due to type safety concerns, Swift expects programmers to make use of optionals when accessing collections with flexible/uncertain data types, such as Any. You will learn more about optionals in Chapter 6 and will see corresponding examples there.

At this point, you know about variables and collections in Swift, which are the basic building blocks for any program. Now you can venture one step further and learn about control flow. In the next chapter, you will learn how to "flow" through different code snippets based on conditions.

Chapter 4
Control Flow

*Control flow is the basic steering mechanism in software development.
It dictates in which direction the program should go, which is often
determined by conditions.*

So far, we've gone over variables and collections in Swift. In a nutshell, those are core features to make apps memorize and remember values. Now that those basics are in place, we can move forward to the building block of algorithms: control flow.

In the examples so far, you wrote code snippets in Swift language, and the computer processed them sequentially—line by line—as illustrated in Figure 4.1.

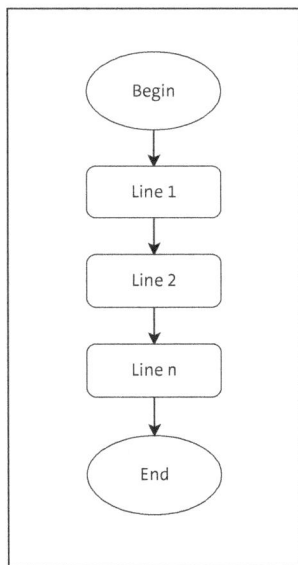

Figure 4.1 Simple Flow, Executed Line by Line

However, such a straightforward flow is a rare occurrence in the real world. In many cases, you want the code to flow differently depending on certain circumstances. Figure 4.2 demonstrates a user login flow, which changes directions depending on the user's credentials:

- If the authentication is successful, the user will be redirected to the main menu.
- Otherwise, the user will see an error message.

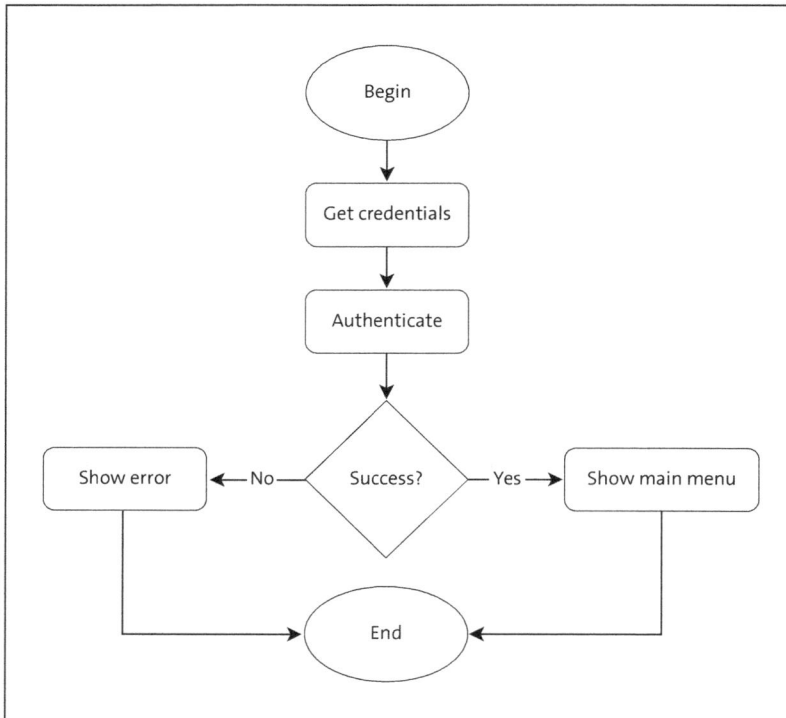

Figure 4.2 Program Flow for Login Process

This is a very simple control flow example, in which the flow will change directions depending on a particular condition. This chapter will start with that: conditional statements. After that, we will continue with different control flow options like loops and control transfers. Without any further ado, let's go!

4.1   Conditional Statements

Conditional statements are programming constructs that allow programmers to guide the flow of the code based on specified conditions. That mechanism lets us enable/ disable different blocks of code, based on the circumstances. This section will introduce several types of conditional statements.

4.1.1   If-Else

The most widely known programming keyword is arguably if. This keyword is so common that many programming languages use it for the same purpose. Chances are that even nonprogrammers know about this intuitive keyword. In this section, you will discover this core conditional statement and its variations.

If

Let's start with the simplest form and move from there. Listing 4.1 demonstrates the basic usage of if. After the if keyword, we provide the condition age >= 18. If this condition is true, then the code snippet between { and } is executed. Otherwise, it's skipped.

```
var age = 33

if age >= 18 {
    print("You are an adult")
}

print("Program finished")
```

Listing 4.1 Basic Usage of If

The expression after the if statement is always a Boolean expression. You can use any Boolean expression, such as age >= 18, after an if statement. You learned about Booleans in Chapter 2; you can revisit that chapter if you need to.

In the example, because we hard-coded the age = 33 value, the code between { and } will be executed (see Figure 4.3).

```
1  var age = 33
2
3  if age >= 18 {
4      print("You are an adult")
5  }
6
7  print("Program finished")

You are an adult
Program finished
```

Figure 4.3 Output of Basic if Example when Age Is 33

Now, what if age is less than 18? Let's find out in Listing 4.2.

```
var age = 15

if age >= 18 {
    print("You are an adult")
}

print("Program finished")
```

Listing 4.2 Basic Usage of if with Age = 15

In this case, because age >= 18 is false, the code between { and } will be ignored; the computer will jump to the statement after the brackets. The output should be as shown in Figure 4.4.

```
1  var age = 15
2
3  if age >= 18 {
4      print("You are an adult")
5  }
6
7  print("Program finished")
■  ▶
Program finished
```

Figure 4.4 Output of Basic if Example when Age Is 15

As a visual representation, the simple algorithm of this code snippet is shown in Figure 4.5.

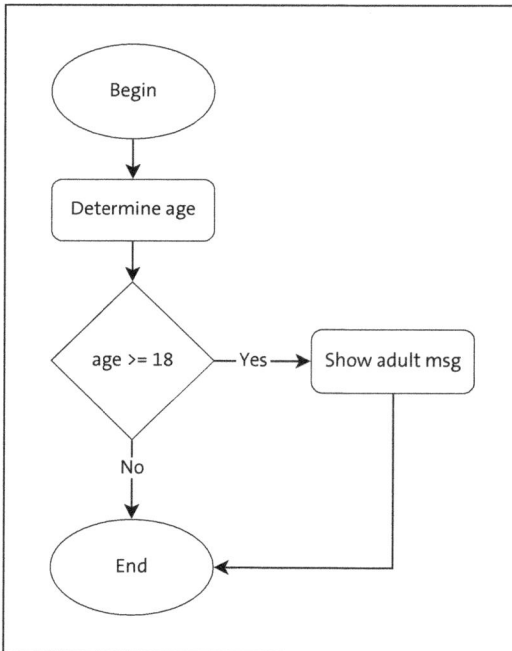

Figure 4.5 Flowchart of Basic if Example

If Else

If statements are often used together with else statements. Check the intuitive example in Listing 4.3. If age >= 18 is true, then the code right after the following { ... } will be executed—just like in the last example. However, if the age condition is false, then the code snippet in else { ... } will be executed.

```
var age = 15

if age >= 18 {
    print("You are an adult")
} else {
    print("You are a minor")
}

print("Program finished")
```

Listing 4.3 Demonstration of if else Statement

In other words, else points toward the alternative code block to be executed if the main condition is not fulfilled. The output of this code snippet is shown in Figure 4.6: Because age is less than 18, the else block has been executed.

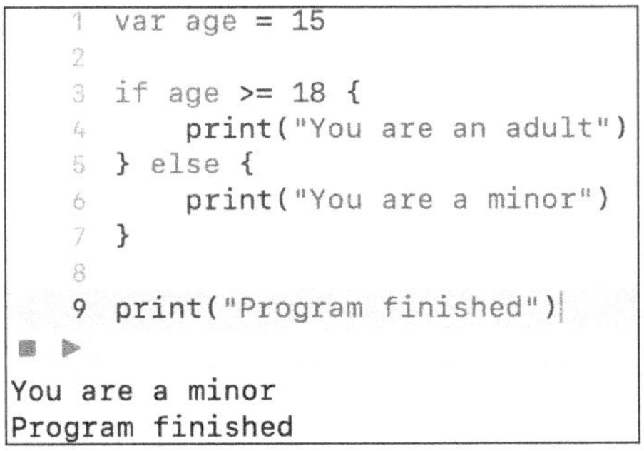

Figure 4.6 Output of else Demonstration

To make this example a little more exciting, let's transform it into a command line app! Create a new command line tool project in Xcode, as explained in Chapter 1, and enter the code snippet in Listing 4.4 into it. Note that the code is mostly the same as before; the only change is that we threw in a couple of lines to get an age input from the user.

```
print("Enter your age: ")
let ageText = readLine()
let age = Int(ageText!)!

if age >= 18 {
    print("You are an adult")
} else {
    print("You are a minor")
}

print("Program finished")
```

Listing 4.4 Command Line Tool for Age Evaluation

Although this code snippet lacks any kind of defensive programming, it should work fine so long as valid integer values are provided. When you execute this program, the computer will ask for your age and categorize you as an adult or a minor. Figure 4.7 shows the output if the user typed "19" for their age.

```
Enter your age:
19
You are an adult
Program finished
```

Figure 4.7 Possible Output of Command Line Tool

Cool, right?

Chain If Else

Swift lets you chain multiple if-else statements too! Check the extended example in Listing 4.5, which offers a more detailed age categorization.

```swift
print("Enter your age: ")
let ageText = readLine()
let age = Int(ageText!)!

if age >= 65 {
    print("You are old")
} else if age >= 30 {
    print("You are an adult")
} else if age >= 18 {
    print("You are a young adult")
} else {
    print("You are a minor")
}

print("Program finished")
```

Listing 4.5 Chaining Multiple if-else Statements

In this code snippet, Swift will start with the initial if statement of age >= 65. If that statement is not fulfilled, the code skips to the second if statement, age >= 30, and continues likewise through all conditions. If none are fulfilled, the code block in the final else statement is executed. Figure 4.8 presents a visual representation of the flow.

As a demonstration, Figure 4.9 showcases a possible outcome of the age condition chain. As an exercise, you can follow the flowchart in Figure 4.8 to find out which path the program took.

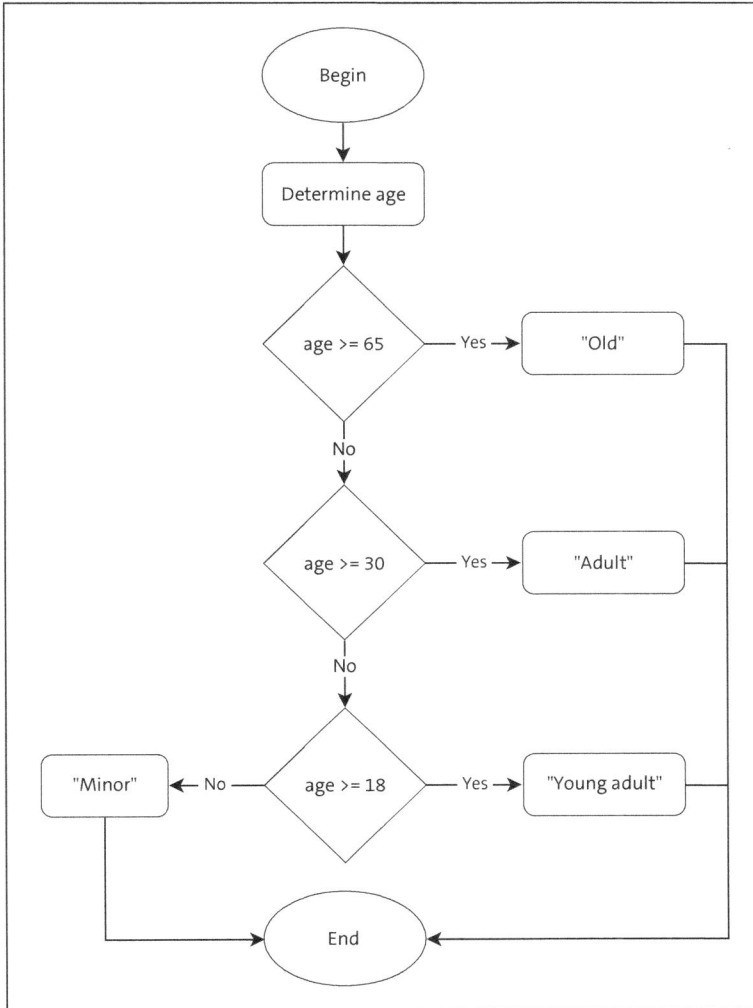

Figure 4.8 Flowchart of Age Condition Chain

```
Enter your age:
44
You are an adult
Program finished
```

Figure 4.9 One Possible Outcome of Age Condition Chain

Naturally, Boolean functions can be used as conditions as well. In Listing 4.6, the contains array function has been used directly as the if condition. Because winningNumbers. contains returns a Boolean anyway, nothing stops us from using it as a condition!

```
var winningNumbers = [12, 22, 23, 32, 33, 40]

let randomNumber = Int.random(in: 1...49)
print("Your number is: \(randomNumber)")

if winningNumbers.contains(randomNumber) {
    print("You won!")
} else {
    print("Sorry, you lose.")
}
```

Listing 4.6 Array Function as if Condition

Every time you execute this code, you will get a different output, depending on the generated random number. If the random number is in the winningNumbers array, then you will win; otherwise, you will lose.

To make this example a little more exciting, let's enhance it as a mini lottery app! Listing 4.7 provides the code snippet to use, which features an if-else chain too.

```
let bigWinningNumbers = Set((1...49).shuffled().prefix(6))
let smallWinningNumbers = Set((1...49).shuffled().prefix(6))

print("Enter your number: ")
let numberText = readLine()
let number = Int(numberText!)!

if bigWinningNumbers.contains(number) {
    print("You win big prize!")
} else if smallWinningNumbers.contains(number) {
    print("You win small prize!")
} else {
    print("You lose!")
}
```

Listing 4.7 Mini Lottery App

The code starts by generating two sets; bigWinningNumbers and smallWinningNumbers contain six random numbers each. Then it asks the user for a number. If their number is in bigWinningNumbers, they get a big prize. Otherwise (else), if their number is in smallWinningNumbers, they get a small prize. Otherwise (else), they lose.

Ignoring Overlaps

In this example, we are purposefully ignoring the case in which bigWinningNumbers and smallWinningNumbers might have overlapping members. Our purpose is to demonstrate

Swift commands, not to write an actual lottery app—so let's not complicate the code more than we should.

Figure 4.10 presents a visual flowchart of this algorithm.

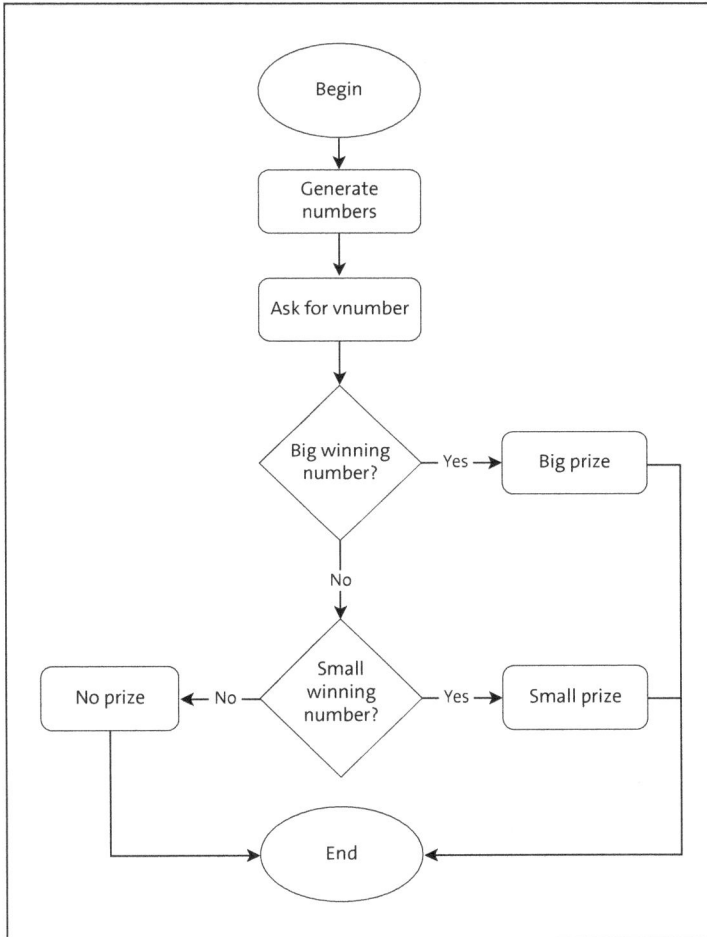

Figure 4.10 Visual Representation of Lottery Flow

Figure 4.11 showcases the program output if the user guesses one of the big winning numbers. Don't ask how many times we had to try to get a winning number!

```
Enter your number:
12
You win big prize!
```

Figure 4.11 Lottery Win Output

Now that you've grasped the basics of if conditions, let's make them more comprehensive by taking advantage of logical operators.

4.1.2   Logical Operators

You saw logical operators in Chapter 2 when you were learning about Boolean variables. Because if conditions also rely on Boolean expressions, they naturally accept logical operators in the same way. This will become clear with the examples we'll walk through in the following sections.

AND Operator

Let's start with a small reminder from Chapter 2: In Swift, the AND operator is &&. With that in mind, inspect Listing 4.8, which demonstrates an if condition with the && operator.

```
print("Username: ")
let user = readLine()!

print("Password: ")
let pass = readLine()!

if user == "admin" && pass == "secret" {
    print("Logged in!")
} else {
    print("Invalid credentials")
}
```

Listing 4.8 AND Condition Demonstration

This example builds a very simple login mechanism. First, it asks the user for their user name and password. Then, it checks if two distinct conditions are true simultaneously: user == "admin" && pass == "secret". This means, "if user is "admin" AND pass is "secret"," you let the user log in. Otherwise, they can't log in.

A possible output for this example is shown in Figure 4.12. Although the user name was given correctly as "admin," the password entered, "abcde," was incorrect—leading to the else condition in which the user isn't allowed to log in.

```
Username:
admin
Password:
abcde
Invalid credentials
```

Figure 4.12 Output of Login Demonstration

Naturally, a real-world login mechanism would be more comprehensive than that, but this simplified example serves its purpose. As you can see, the && operator works exactly the same as it does for Booleans (see Chapter 2)—which is natural as the if condition is a Boolean itself.

OR Operator

You will see that similarity for OR operators too. A reminder from Chapter 2: In Swift, the OR operator is ||. With that in mind, inspect Listing 4.9, which simulates a policy for guest acceptance into a club. Each guest must either be at least 18 years old or have their parents present to enter the club.

```
print("Age?")
let age = Int(readLine()!)!

print("Parents present? (y/n)")
let parent = readLine()!

if age >= 18 || parent == "y" {
    print("May enter")
} else {
    print("May not enter")
}
```

Listing 4.9 OR Condition Demonstration

The dual OR conditions have been coded as if age >= 18 || parent == "y". If either condition is true, Swift will execute print("May enter"); otherwise, it will execute print("May not enter").

Note that each individual condition has a Boolean output, and the condition1 || condition2 combination has a Boolean output too. That's why you can use them directly as if conditions.

A possible output of this code snippet is presented in Figure 4.13. The guest in question is 45 years old, so age >= 18 is true. Their parents are not present, so parent == "y" is false. Because the conditions are combined with an OR operator, age >= 18 || parent == "y", either condition resulting in true is good enough—and therefore this guest may enter the club.

```
Age?
45
Parents present? (y/n)
n
May enter
```

Figure 4.13 Output of Club Entry Demonstration

Nested Conditions

As you saw in Chapter 2, you can build complex nested Boolean conditions using multiple subconditions too. If needed, you can use parentheses for extra clarity as well. To demonstrate such if conditions, Listing 4.10 features an apparel suggestion app.

```
// Get user input
print("Temperature in Celsius?")
let temp = Int(readLine()!)!

print("Raining? (y/n)")
let isRaining = readLine()! == "y"

print("Snowing? (y/n)")
let isSnowing = readLine()! == "y"

// Evaluate and suggest outfit
if temp <= 15 && (isRaining || isSnowing) {
    print("Wear a warm coat and boots")
} else if temp >= 25 && !(isRaining || isSnowing)  {
    print("Wear a shirt and sandals")
} else {
    print("No special suggestions")
}
```

Listing 4.10 Nested Condition Demonstration

The app starts by getting weather conditions from the user. Ideally, you would automatically pull the conditions from a weather service—but let's keep things simple at this point.

Once the app learns the temperature and rain/snow conditions, it evaluates the weather using nested conditions and generates suggestions accordingly. Table 4.1 showcases each condition and its outcome.

Expression	Interpretation	Outcome
temp <= 15 && (isRaining || isSnowing)	Temperature should be less than (or equal to) 15; also, it must be raining or snowing	**Wear a warm coat and boots**
temp >= 25 && !(isRaining || isSnowing)	Temperature should be greater than (or equal to) 25; also, it must be neither raining nor snowing	**Wear a shirt and sandals**
else	Any other case	**No special suggestions**

Table 4.1 Nested Conditions and Their Interpretations

See? There's nothing different here than what you learned in Chapter 2. All we did was glue nested Boolean conditions to an if statement to manipulate the program flow. One possible outcome of the app is shown in Figure 4.14.

```
Temperature in Celsius?
27
Raining? (y/n)
n
Snowing? (y/n)
n
Wear a shirt and sandals
```

Figure 4.14 One Output of Outfit Suggestion App

The answers from the user corresponded to the condition temp >= 25 && !(isRaining || isSnowing); therefore, the **Wear a shirt and sandals** output has been displayed. Simple, right?

Simple But Powerful

Don't let the simplicity of these examples deceive you! Because our focus is the Swift language itself, we are limiting the UI aspect to the terminal for now. When the time comes for you to develop comprehensive applications with advanced user interfaces, RESTful APIs, database access, and so on, the core Swift syntax will remain the same as in these simplified examples. Let's focus on those core muscles today; you'll build further skills on them easily later.

4.1.3   Ternary Operator

Now that you've learned about if conditions, you can reward yourself with some syntactic sugar. Ternary operators offer an optional but welcome syntax for in-line if conditions, which shortens the code and improves readability in many cases.

To understand ternary operators, we will write a short program using an if condition, then transform the code using a ternary operator. Listing 4.11 contains a simple code snippet that evaluates a student's grade to decide if they passed or failed. A grade higher than (or equal to) 50 is enough to pass, while lower grades result in failure.

```
print("Enter your grade")
let grade = Int(readLine()!)!

var status = ""

if grade >= 50 {
    status = "Passed"
} else {
```

```
    status = "Failed"
}
```

```
print("Your status: \(status)")
```

Listing 4.11 Grade Evaluation via if Condition

So far, so good: nothing new. Now let's transform the code using a ternary operator: Check Listing 4.12.

```
print("Enter your grade")
let grade = Int(readLine()!)!
```

```
let status = grade >= 50 ? "Passed" : "Failed"
```

```
print("Your status: \(status)")
```

Listing 4.12 Grade Evaluation via Ternary Operator

The if condition in Listing 4.11 and the ternary operator in Listing 4.12 have the exact same meaning. Figure 4.15 showcases a comparison between those syntaxes; as you can see, they are very similar. Logically, the ? operator replaces the if keyword and the : operator replaces the else keyword. The similarity between those alternative syntaxes is shown in Figure 4.15.

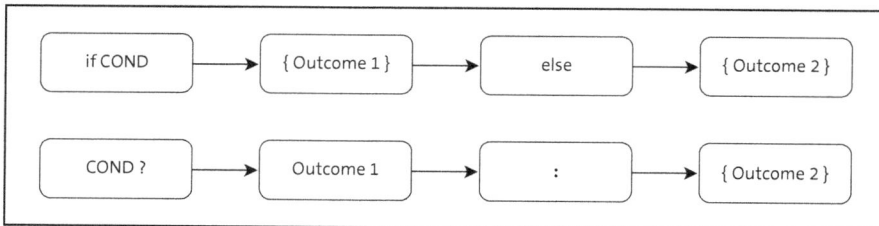

Figure 4.15 Comparison of if Condition and Ternary Operator Syntax

To make the example a little more comprehensive, Listing 4.13 extends the code to apply the ternary operator to a bulk operation.

```
typealias ExamResult = (name: String, grade: Int)
```

```
let examResults: [ExamResult] = [
    ("Alice", 90),
    ("Bob", 85),
    ("Charlie", 42),
    ("David", 88),
    ("Eve", 34)
]
```

```
for examResult in examResults {
    let status = examResult.grade >= 50 ? "Passed" : "Failed"
    print("\(examResult.name): \(status)")
}
```

Listing 4.13 Bulk Grade Evaluation via Ternary Operator

In the first part, we declare `examResults` as a tuple array. In a real-world application, you would pull those results from a database or the like, but for the sake of simplicity, we use hard-coded mock values here.

In the second part, each exam result was evaluated with a ternary operator to determine if each student passed or failed. The `examResult.grade >= 50 ? "Passed" : "Failed"` syntax is nearly the same as in the previous example—a reminder that the core Swift syntax remains the same even if the case at hand is more comprehensive. Finally, the result is printed to the terminal.

The expected output of this example is shown in Figure 4.16, where grades less than 50 led to a failure.

```
Alice: Passed
Bob: Passed
Charlie: Failed
David: Passed
Eve: Failed
```

Figure 4.16 Output of Bulk Grade Evaluation

When to Use the Ternary Operator

Ternary operators are best used for short, simple conditions, as demonstrated in this section. For complex logic containing multiple conditions, `if` statements should be preferred as ternary operators would make code harder to read. You may also prefer other conditional statements you will learn about in this chapter.

You should always remember that code is not done once it is written. Every app has a lifecycle, and throughout time, the code you write today will need to be read, understood, and maintained—by you or by other programmers. Therefore, it is important to ensure that your code is as clear and human-readable as possible.

Computers will understand any code, but your target audience should mostly be humans.

4.1.4 Nil-Coalescing Operator

Another useful bit of syntactic sugar for conditions is the *nil-coalescing operator*, which can be used to declare default fallback values. To ensure a solid understanding, let's begin with a simple example using if conditions and convert it to a version using nil-coalescing operators. Start with the sample code in Listing 4.14.

```
typealias Customer = Dictionary<String, String>

let customers: [Customer] = [
    ["name": "John" , "email": "john@example.com"],
    ["name": "Jane" , "phone": "+1234567890"],
    ["name": "Jim"  , "email": "jim@example.com"],
    ["name": "Jill" , "phone": "+0987654321", "email": "jill@example.com"],
    ["name": "Bob"]
]

for customer in customers {
    var contactInfo: String = ""

    if customer["phone"] != nil {
        contactInfo = customer["phone"]!
    } else if customer["email"] != nil {
        contactInfo = customer["email"]!
    } else {
        contactInfo = "Unknown"
    }

    print("\(customer["name"]!)'s contact: \(contactInfo)")
}
```

Listing 4.14 Prioritized Contact Info Determination with if Conditions

This example begins with the declaration of customers, which is a simple dictionary array. In a real-world application, this dataset would probably come from a database or JSON file, but here we've hand-coded some mock values for the sake of simplicity.

Each entry has the name of a customer and their contact information. However, some customers have an email value, some have a phone value, some have both, and some have none. So when you access a customer record, you aren't sure what kind of contact information is present. That's the nature of real-world datasets anyway: Clean data is hard to come by!

In the following iteration, you need to print valid contact information. Because you aren't sure what kind of contact data is present, a prioritized access sequence is declared—as summarized in Table 4.2.

Priority	Condition	Contact Info
1	`customer["phone"] != nil`	`customer["phone"]`
2	`customer["email"] != nil`	`customer["email"]`
3	—	`"Unknown"`

Table 4.2 Summary of Prioritized Contact Info Determination

The output of this program is shown in Figure 4.17. Note how the prioritized access worked for each customer:

- John doesn't have a phone number, so his email was shown as the secondary priority.
- Jane has a phone number, so her phone was shown as the primary priority.
- Jim doesn't have a phone number, so his email was shown as the secondary priority.
- Jill has a phone number, so her phone was shown as the primary priority.
- Bob has no contact info at all, so `"Unknown"` was shown as a fallback.

```
John's contact: john@example.com
Jane's contact: +1234567890
Jim's contact: jim@example.com
Jill's contact: +0987654321
Bob's contact: Unknown
```

Figure 4.17 Contact Info Output

So far, so good. We took care of this task using an `if-else` chain like you learned about before. Now, let's transform the code to a version that generates the exact same output but uses the nil-coalescing operator instead of an `if-else` chain. Check the new version in Listing 4.15 and see how short and readable the code becomes.

```swift
typealias Customer = Dictionary<String, String>

let customers: [Customer] = [
    ["name": "John" , "email": "john@example.com"],
    ["name": "Jane" , "phone": "+1234567890"],
    ["name": "Jim"  , "email": "jim@example.com"],
    ["name": "Jill" , "phone": "+0987654321", "email": "jill@example.com"],
    ["name": "Bob"]
]

for customer in customers {
    var contactInfo: String = customer["phone"] ?? customer["email"] ??
```

```
"Unknown"
    print("\(customer["name"]!)'s contact: \(contactInfo)")
}
```

Listing 4.15 Prioritized Contact Info Determination with Nil-Coalescing Operator

Instead of the `if-else` chain, we applied the nil-coalescing operator, `??`. This operator is used to chain prioritized values. If the first `customer["phone"]` option is not `nil`, then it is taken as the `contactInfo`. If the first option is `nil`, then Swift will check the second option, `customer["email"]`, and if it is not `nil`, then it is taken as the `contactInfo`. The final fallback option is `"Unknown"`, which is taken if all former values are `nil`.

The similarities among these alternative syntaxes are shown in Figure 4.18.

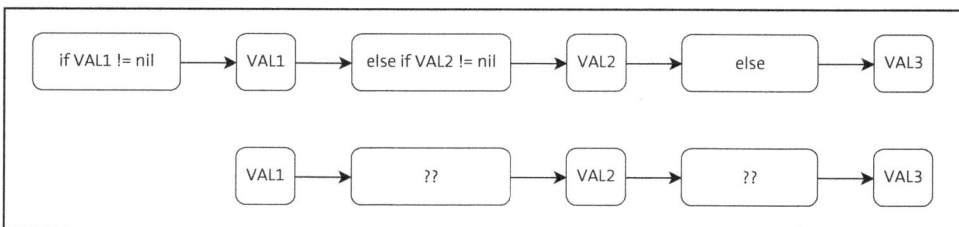

Figure 4.18 Comparison of if Condition and Nil-Coalescing Operator Syntax

4.1.5   Switch

As in many other programming languages, Swift features `switch` statements, which can replace complex `if-else` chains in many cases. Like ternary and nil-coalescing operators, you can code an entire application by ignoring `switch` altogether and use `if-else` statements throughout the program. However, when used appropriately, `switch` will improve the readability of your code, and other programmers will thank you for that — including future you. In this section, you will learn about the syntax alternatives for `switch` statements offered by Swift.

Switch with Literals

Let's start with a simple example, in which we write the code using an `if-else` chain first and transform it into a `switch` after. Listing 4.16 features a code snippet that evaluates the current weekday and prints out a corresponding interpretation.

```
import Foundation

let dateFormatter = DateFormatter()
dateFormatter.dateFormat = "EEEE" // Full weekday name (e.g., Monday)
let currentDay = dateFormatter.string(from: Date())
```

```
if currentDay == "Monday" {
    print("The week is starting.")
} else if currentDay == "Friday" {
    print("Almost the weekend!")
} else if currentDay == "Saturday" || currentDay == "Sunday" {
    print("Weekend!")
} else {
    print("A regular weekday.")
}
```

Listing 4.16 Weekday Evaluation Using if-else Statements

So far, there is nothing fancy about this code. There's a simple if-else chain, which checks the value of currentDay and prints out an interpretation. Figure 4.19 showcases a possible program output, but the actual output will naturally change depending on the day on which the code is executed.

```
A regular weekday.
```

Figure 4.19 Possible Weekday Evaluation Output

Now, let's transform this code using a switch statement to understand its function. Check the transformed code in Listing 4.17.

```
import Foundation

let dateFormatter = DateFormatter()
dateFormatter.dateFormat = "EEEE" // Full weekday name (e.g., Monday)
let currentDay = dateFormatter.string(from: Date())

switch currentDay {
case "Monday": print("The week is starting.")
case "Friday": print("Almost the weekend!")
case "Saturday", "Sunday": print("Weekend!")
default: print("A regular weekday.")
}
```

Listing 4.17 Weekday Evaluation Using Switch Statement

Let's check the differences. The new statement begins with the switch currentDay expression, which indicates that the following { ... } will contain conditional code based on the value of currentDay. In the follow-up, there is a case entry for each possible currentDay value. Finally, the default entry indicates the final fallback condition—just like the final else condition of an if statement.

To help with clarity, Figure 4.20 shows a sketch in which the if-else chain and corresponding switch statement are displayed side by side. Note that they produce the exact same output with a different syntax.

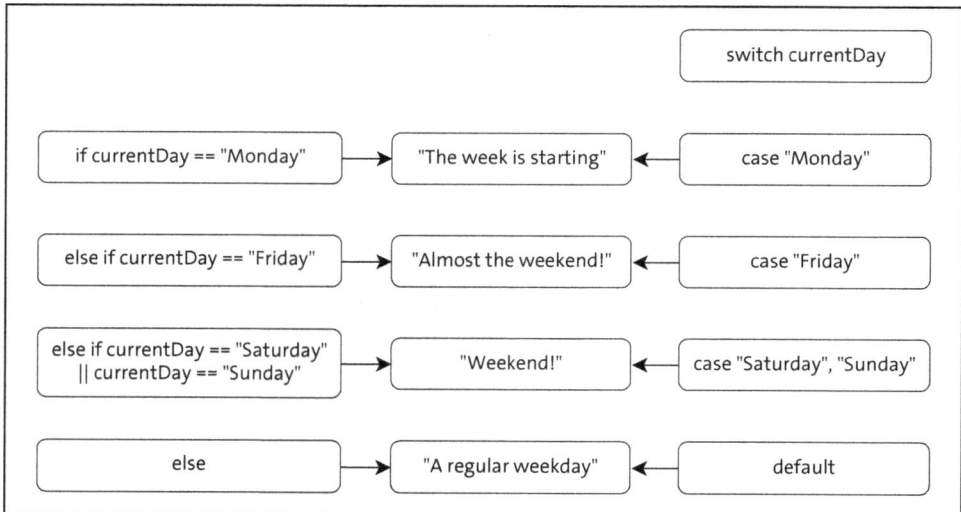

Figure 4.20 Side-by-Side Comparison of if-else Chain and switch Statement

When to Use Switch

The switch and if statements both control program flow conditionally. switch is preferred when the flow is based on the value of a single variable (or condition), such as currentDay, whereas if statements are preferred when more complex conditions for multiple variables are present.

Some developers and clean coding advocates argue that a switch statement can be an antipattern, but this isn't about the switch keyword itself; it's about the way the code is structured. The same antipattern can be hidden behind if statements too. The major objection is that switch statements shouldn't be used for polymorphic behavior because they violate the open-closed principle by requiring modification every time a new case is added.

This is an advanced and controversial topic, which we'll let lie for now—but experienced readers are more than welcome to research further.

Switch with Ranges

Note that switch statements can be used as ranges too, which expands their usability dramatically. Listing 4.18 demonstrates an example that categorizes the scores of the users as lettered grades.

```
print("Enter your score:")
let score = Int(readLine()!)!

switch score {
case 90...100:
    print("Grade: A")
case 80..<90:
    print("Grade: B")
case 70..<80:
    print("Grade: C")
case 60..<70:
    print("Grade: D")
default:
    print("Grade: F")
}
```

Listing 4.18 Range-Based switch Statement Demonstration

Note that the switch … case … default syntax remains the same. The only thing that changed was the expression for each case. Instead of providing string literals, we have provided intuitive number ranges. A possible output of this program is shown in Figure 4.21.

```
Enter your score:
85
Grade: B
```

Figure 4.21 Possible Score Grade Output

You'll be pleased to know that open-ended ranges are also supported! Listing 4.19 demonstrates such an example, in which case ..<0 and case 30... are provided as open-ended ranges.

```
print("Enter temperature:")
let temperature = Int(readLine()!)!

switch temperature {
case ..<0:
    print("Freezing cold")
case 0..<10:
    print("Very cold")
case 10..<20:
    print("Cool")
case 20..<30:
    print("Warm")
case 30...:
```

```
    print("Hot")
default:
    print("Unknown")
}
```

Listing 4.19 Open-Ended Ranges in switch Statements

In this example, any temperature below 0 will be handled by case . . <0 and any temperature above 30 will be handled by case 30. . . . Neat, right? A possible output of this example is shown in Figure 4.22.

```
Enter temperature:
20
Warm
```

Figure 4.22 Possible Output of Open-Ended switch Range Example

Where Clauses

In more advanced scenarios, you can throw where statements into switch statements. This technique enables you to run advanced calculations as case conditions.

As usual, we'll start with an example built with if-else conditions and transform it into a switch statement. Listing 4.20 showcases a password-strength evaluator, which expects a password of at least eight characters in length. The password must also contain a number and a letter.

```
print("Enter password: ")
let password = readLine()!

if password.count < 8 {
    print("Password too short")
} else if !password.contains(where: { $0.isNumber }) {
    print("Password must contain a number")
} else if !password.contains(where: { $0.isLetter }) {
    print("Password must contain a letter")
} else {
    print("Password OK")
}
```

Listing 4.20 Password-Strength Evaluator Using if-else Statements

Although a more realistic application would look for more, such as special symbols and sequential numbers, this will do as a simplified example. Figure 4.23 showcases a possible output of this program, which didn't approve the password "VerySecret" because it doesn't contain a number.

```
Enter password:
VerySecret
Password must contain a number
```

Figure 4.23 Possible Output of Password Evaluator

Now, let's check the alternative syntax in Listing 4.21, which uses a switch … case … where statement.

```
print("Enter password: ")
let password = readLine()!

switch password {
case let pwd where pwd.count < 8:
    print("Password too short")
case let pwd where !pwd.contains(where: { $0.isNumber }):
    print("Password must contain a number")
case let pwd where !pwd.contains(where: { $0.isLetter }):
    print("Password must contain a letter")
default:
    print("Password OK")
}
```

Listing 4.21 Password-Strength Evaluator Using switch-case-where Statements

We started here with the switch password expression, which indicates that the value of password will be evaluated throughout the condition. For each case, we evaluate a different property of password and print the evaluation result accordingly. The if and switch expressions are compared in Table 4.3.

if Expression	switch Expression
password.count < 8	pwd.count < 8
!password.contains(where: { $0.isNumber })	!pwd.contains(where: { $0.isNumber })
!password.contains(where: { $0.isLetter })	!pwd.contains(where: { $0.isLetter })
else	default

Table 4.3 Comparison of if Conditions and switch Conditions

As you can see, the conditions are very similar. However, switch … case … where statements really shine when you need to compare multiple properties of a complex type, such as a tuple or dictionary. Listing 4.22 demonstrates an example in which employee bonus percentages are calculated based on their success rating and number of years of experience.

```
typealias Employee = (name: String, rating: Int, years: Int)

let employees: [Employee] = [
    ("Alice", 9, 10),
    ("Bob", 7, 5),
    ("Charlie", 5, 2),
    ("David", 8, 3)
]

for employee in employees {
    var bonusPerc = 0

    switch employee {
    case let emp where emp.rating >= 9 && emp.years >= 3:
        bonusPerc = 20
    case let emp where emp.rating >= 7 && emp.years >= 5:
        bonusPerc = 15
    case let emp where emp.rating >= 5 && emp.years >= 2:
        bonusPerc = 10
    case let emp where emp.rating < 5:
        bonusPerc = 0
    default:
        bonusPerc = 5
    }

    print("\(employee.name) gets a \(bonusPerc)% bonus")
}
```

Listing 4.22 Employee Bonus Calculation Based on Rating and Years of Experience

If you look closely at the switch statement, you will see that both rating and years were taken into consideration to calculate bonusPerc for each employee. In a real-world application, both employee data and bonus-calculation rules would be pulled from a database, but we're keeping things simple here to focus on the Swift syntax. Figure 4.24 showcases the output of this example.

```
Alice gets a 20% bonus
Bob gets a 15% bonus
Charlie gets a 10% bonus
David gets a 10% bonus
```

Figure 4.24 Calculated Employee Bonus Rates

This example concludes our content on conditional statements. You learned about if statements, which are the core operations of all conditional statements. As shortcut options, you also learned about the ternary operator ?, nil-coalescing operator ??, and `switch` statements.

Now, let's move forward to an entirely different control flow option: loops. Take a break if you need to, and see you there!

4.2 Loops

The good news first: You already know the basics about loops! In Chapter 3, you learned about iterations over collections when we ran a code snippet for each element in a collection (like an array) sequentially. That is the very core logic of loops. In this section, we will build up further knowledge on top of that and see how loops function as a generic control flow mechanism.

4.2.1 For-In Loops

We'll start with `for … in …` loops, which should already be familiar. As a memory refresher, Listing 4.23 features a simple iteration over an array, which prints out each element sequentially.

```
let fruits = ["apple", "banana", "orange"]

for fruit in fruits {
    print(fruit)    // apple, banana, orange
}
```

Listing 4.23 Iteration over Array

The `for fruit in fruits { … }` statement instructs Swift to sequentially do the following:

- Assign each element in `fruits` to the variable `fruit`.
- Execute the code between { and } with the value in `fruit`.

In this example, `print(fruit)` will be executed three times:

- Once with `fruit` = "apple" (will print out "apple")
- Once with `fruit` = "banana" (will print out "banana")
- Once with `fruit` = "orange" (will print out "orange")

You can revisit the iteration subsections in Chapter 3 to see different applications of collection iterations if you need to. Once you've remembered the basics, we'll move forward to alternative applications of `for` loops in the following sections.

Closed Range Operator

With the basic forms of closed range operators, you can instruct Swift to execute a code snippet for each number within a given range. Listing 4.24 demonstrates an example in which each number between one and five is processed in a for loop.

```
for number in 1...5 {
    print(number, terminator: " ") // 1 2 3 4 5
}
```

Listing 4.24 Closed Range Operator Demo with Numbers

The for number in 1...5 statement instructs Swift to execute the code between { and } multiple times—each time with a new value assigned to number sequentially, from 1 to 5. As a result, the output **1 2 3 4 5** is shown in the terminal; all we did was print those numbers. Easy, right? This isn't a far cry from array iterations.

It's also possible to process the number range in reverse order as demonstrated in Listing 4.25.

```
for number in (1...5).reversed() {
    print(number, terminator: " ") // 5 4 3 2 1
}
```

Listing 4.25 Reverse Closed Range Operator Demo

Range operators are not limited to numbers, though; you can also loop through characters with a little help. Listing 4.26 makes use of the UnicodeScalar function, which does the necessary conversion between a character and its unique number in the Unicode encoding scheme. That allows you to loop through characters using their numeric codes.

```
for charCode in UnicodeScalar("a").value...UnicodeScalar("e").value {
    print(UnicodeScalar(charCode)!, terminator: " ")    // a b c d e
}
```

Listing 4.26 Looping Through Characters from "a" to "e"

> **Character Encoding**
> You learned about characters and their numeric values in encoding schemes in Chapter 2 when strings were introduced. If you need a memory refresher, you can peek back at that chapter.

You can also loop through dates, as demonstrated in Listing 4.27. The only new component of this code snippet is the stride keyword.

```
import Foundation

// Prepare date formatter
let dateFormat = DateFormatter()
dateFormat.dateStyle = .medium

// Prepare date range
let calendar = Calendar.current
let startDate = calendar.date(from: DateComponents(year: 2024, month: 3, day: 17))!
let endDate = calendar.date(from: DateComponents(year: 2024, month: 3, day: 21))!

// Loop through the date range
// One day in seconds: 60 * 60 * 24 = 86400
for date in stride(from: startDate, through: endDate, by: 86400) {
    print(dateFormat.string(from: date))
}
```

Listing 4.27 Looping Through Days

You learned about dates in Chapter 2. Using that know-how, you start here by preparing `DateFormatter` and two dates: `startDate` and `endDate`. To loop through these dates, you use the `for date in stride(from: startDate, through: endDate, by: 86400)` expression.

The `stride` function lets you generate a sequence of values from a starting point to an ending point with a given step size. In this example, you tell Swift to stride from `start-Date` until `endDate`, with a step size of `86400` seconds (one day) for each iteration. The result will be as shown Figure 4.25, where each day from the start to the end date is printed out.

```
17 Mar 2024
18 Mar 2024
19 Mar 2024
20 Mar 2024
21 Mar 2024
```

Figure 4.25 Day Loop Output

You could use the same technique to loop through months, hours, years, seconds, and so on by changing the step size. Listing 4.28 demonstrates a modified version of the same example, looping through minutes this time. Its output is shown in Figure 4.26.

```
import Foundation

// Prepare date formatter
let dateFormat = DateFormatter()
dateFormat.dateStyle = .medium
dateFormat.timeStyle = .short
```

```
// Prepare date range
let calendar = Calendar.current
let startDate = calendar.date(from: DateComponents(year: 2024, month: 3,
day: 17, hour: 15, minute: 17))!
let endDate = calendar.date(from: DateComponents(year: 2024, month: 3,
day: 17, hour: 15, minute: 22))!

// Loop through the date range
// One minute in seconds: 60
for date in stride(from: startDate, through: endDate, by: 60) {
    print(dateFormat.string(from: date))
}
```

Listing 4.28 Looping through Minutes

```
17 Mar 2024 at 15:17
17 Mar 2024 at 15:18
17 Mar 2024 at 15:19
17 Mar 2024 at 15:20
17 Mar 2024 at 15:21
17 Mar 2024 at 15:22
```

Figure 4.26 Minute Loop Output

Another typical use case of for loops is to iterate through an array using indexes. Listing 4.29 demonstrates an example that iterates through customers in a queue and prints out their names.

```
var queue = ["John", "Jane", "Mary"]

for no in 0...queue.count-1 {
    let customer = queue[no]
    print("Customer \(no) is \(customer)")
}
```

Listing 4.29 Array Iteration Using Indexes

A point worth remembering from Chapter 3: Array indexes start with 0 (not 1). In this example, customer indexes start with 0 and end with 2, as shown in Table 4.4.

Index	Customer
0	John
1	Jane
2	Mary

Table 4.4 Indexes of Customers in Queue

Therefore, the index iteration should run from 0 to 2 (not 1 to 3). That's why you iterate with the for no in 0...queue.count-1 expression. The output produced should look like Figure 4.27.

```
Customer 0 is John
Customer 1 is Jane
Customer 2 is Mary
```

Figure 4.27 Output of Queue Iteration

If the output had to start with 1 instead of 0, you could easily achieve that by changing the print statement as shown in Listing 4.30, which would produce the output shown in Figure 4.28.

```
var queue = ["John", "Jane", "Mary"]

for no in 0...queue.count-1 {
    let customer = queue[no]
    print("Customer \(no+1) is \(customer)")
}
```

Listing 4.30 Array Iteration with Output Starting from 1

```
Customer 1 is John
Customer 2 is Jane
Customer 3 is Mary
```

Figure 4.28 Output of Queue Iteration, Starting from 1

Half-Open Range Operator

So far, we've walked through closed range operator examples, in which you provide both the start and end values of the range. Swift also features *half-open range operators*, in which you provide a fixed start value and a certain condition as the end value.

Listing 4.31 demonstrates how half-open ranges differentiate from closed ranges. The expression 0..<5 instructs Swift to iterate through numbers, starting from 0 and going as long as the number is less than 5.

```
for i in 0..<5 {
    print(i, terminator: " ") // 0 1 2 3 4
}
```

Listing 4.31 Half-Open Range Operator with Numbers

Half-open ranges occasionally offer a more readable syntax—especially when used in array loops. Listing 4.32 demonstrates such a case with the former queue example; note the readability improvement in 0..<queue.count.

```
var queue = ["John", "Jane", "Mary"]

for no in 0..<queue.count {
    let customer = queue[no]
    print("Customer \(no+1) is \(customer)")
}
```

Listing 4.32 Half-Open Range Operator with Arrays

One-Sided Ranges

In both closed ranges and half-open ranges, you must declare a start value and an end value, between which Swift will operate and execute the code between { and } sequentially. In contrast, one-sided ranges allow you to skip the start or end of a range and iterate through the remaining elements.

Listing 4.33 demonstrates this technique by slicing an array into different pieces.

```
let numbers = [10, 20, 30, 40, 50]

print(numbers[...2])   // [10, 20, 30]
print(numbers[..<2])   // [10, 20]
print(numbers[2...])   // [30, 40, 50]
```

Listing 4.33 Slicing Arrays Using One-Sided Ranges

For a better understanding, Table 4.5 showcases translations of those range expressions.

Expression	Meaning	Result
[...2]	Elements 0, 1, 2	10, 20, 30
[..<2]	Elements 0, 1	10, 20
[2...]	Elements 2, 3, ... to the end	30, 40, 50

Table 4.5 One-Sided Range Expressions and Their Meanings

Naturally, one-sided ranges can be used with for loops too, as demonstrated in Listing 4.34.

```
var carQueue = ["LKP-9482",
                "TZW-1037",
                "BQN-5520",
                "HRX-7889",
                "JVD-4216"]

print("Next three cars in the queue:")
```

```
for car in carQueue[..<3] {
    print(car)
}
```

Listing 4.34 Listing First Three Cars in Queue

In this example, you iterate through cars in carQueue, starting from the first index (0) and continuing so long as the index is less than three. Therefore, cars with index 0, 1, and 2 are processed and their license plates are printed, as shown in Figure 4.29.

```
Next three cars in the queue:
LKP-9482
TZW-1037
BQN-5520
```

Figure 4.29 Car Queue Output

A small reminder about defensive programming: If the size of carQueue was less than 3, this code would generate an error. As mentioned before, the size of an array should always be validated before accessing it via indexes. This suggestion is valid for all index-based array accesses—but why not solidify it now?

Now that you know about conditional statements, you can easily achieve this size-validation goal using a ternary operator! In Listing 4.35, instead of blindly attempting to list the first three cars, you can check the number of cars in the queue first. If the car count is less than 3, you can accept the car count as the outputSize. That minor defense will prevent a program crash if there's a small queue.

```
var carQueue = ["LKP-9482",
                "TZW-1037"]

let outputSize = carQueue.count < 3 ? carQueue.count : 3

print("Next \(outputSize) cars in the queue:")

for car in carQueue[..<outputSize] {
    print(car)
}
```

Listing 4.35 Safe Array Access with Indexes

Defensive Programming Reminder

Because we don't want to overcomplicate the code samples, measures of defensive programming are sometimes left out here. However, this is an integral part of real-world applications. Without defensive programming, your app might crash very quickly.

When writing code, you should always be aware that your assumptions regarding array size, user input, RESTful API response, data format, and so on might be wrong; you can always get unexpected values. Case in point: You can never assume that a car queue has at least three license plates present. On a quiet day, your queue may only have one or two entries, or no entries at all.

Therefore, we did a little defensive programming here and determined the `outputSize` based on the queue size, instead of enforcing an output size of 3.

Long story short, the preceding example should serve as a reminder that defensive programming is an integral part of real-word applications. Don't let the simplified examples make you think otherwise.

Loops Without Variables

In some cases, you might want simply to repeat an operation a few times without assigning a value to the `for` variable. In such cases, you can use the _ (underscore) token as a variable placeholder and conduct the `for` loop as usual. This simple technique is demonstrated in Listing 4.36, followed by its output in Figure 4.30.

```
for _ in 1...4 {
    print("Let it be")
}
```

Listing 4.36 Loop Demonstration Without Variable

```
Let it be
Let it be
Let it be
Let it be
```

Figure 4.30 Output of Loop Demonstration Without Variable

You might find this technique useful for operations that simply need to be repeated, such as the following:

- Playing a sound multiple times
- Repeating a CPU-heavy operation for performance tests
- Generating dummy/random mock datasets

Listing 4.37 demonstrates the last use case by generating five random numbers between 0 and 999. Because we don't need the index value itself, the loop was initiated using the _ token in `for _ in 0..<5`.

```
var randomNumbers: [Int] = []

for _ in 0..<5 {
```

```
    randomNumbers.append(Int.random(in: 0..<1000))
}

print(randomNumbers) // Random output of 5 numbers
```

Listing 4.37 Building Array of Five Random Numbers

And that example concludes our content on for-in loops, which was a partially familiar concept from previous chapters. Now, let's move forward to an unfamiliar but similar technique: while loops.

4.2.2 While Loops

A while loop in Swift is used to repeat a block of code so long as the given condition is true. It's useful for cases in which you don't know in advance how many times you need to loop.

As an initial syntax-revealing example, check Listing 4.38. In this example, you instruct Swift to keep generating random numbers until a generated number is greater than 50.

```
var randomNumber = 0
var attempts = 0

while randomNumber <= 50 {
    attempts += 1
    randomNumber = Int.random(in: 0...100)
    print("Attempt \(attempts): \(randomNumber)")
}
```

Listing 4.38 Demonstration of While Loops

In this code snippet, while randomNumber <= 50 tells Swift to repeat the code between { and } so long as randomNumber is less than (or equal to) 50. Between { and }, we assign a random number between 0 and 100 to randomNumber, so we will eventually exit the while loop when randomNumber produces a value above 50.

Due to the nature of randomization, every execution of this example will generate a different output; one possible output is shown in Figure 4.31.

```
Attempt 1: 19
Attempt 2: 42
Attempt 3: 84
```

Figure 4.31 Possible Output of while Loop Demonstration

As you see, the while loop ended as soon as a number greater than 50 was generated—which is 84 in this case. If you didn't have an algorithm in { … } to eventually increase randomNumber above 50, the while loop would just repeat forever and the app would get stuck—which is definitely something to avoid.

Using your Swift knowledge so far, let's write a simple game with while loops at its heart! In this game, the computer will think of a random number between 1 and 10 and ask the player to guess it. The player will be asked for their guess until the correct number is found. The code for this simple game is provided in Listing 4.39.

```
let correctAnswer = Int.random(in: 1...10)
var userAnswer = 0

while userAnswer != correctAnswer {
    print("Guess my number (1-10):")
    userAnswer = Int(readLine() ?? "0") ?? 0
}

print("Correct!")
```

Listing 4.39 Number-Guessing Game

The core mechanism of this game is the while loop. The player is asked to guess until they eventually get the number right. Figure 4.32 showcases the output of the game played. Here, the computer generated the number 9 and the player guessed the number correctly after a few attempts.

```
Guess my number (1-10):
4
Guess my number (1-10):
5
Guess my number (1-10):
6
Guess my number (1-10):
7
Guess my number (1-10):
8
Guess my number (1-10):
9
Correct!
```

Figure 4.32 Possible Number Guessing Game Output

Note that you couldn't possibly have achieved this functionality in a for loop because for loops don't iterate forever: They only iterate a fixed number of times. In this game, the idea was to keep the user in the loop until they guessed the number, and you can't possibly foresee on which attempt they will get the number right. Table 4.6 contrasts the use cases for both loop types.

Use Case	Suggested Loop Type
Looping a fixed number of times	`for`
Looping until a condition is met	`while`
Looping until a random condition	`while`

Table 4.6 When to Use For Versus While Loops

Avoid Infinite Loops

Infinite loops are the bane of apps. An infinite loop occurs when the program gets stuck in a `while` loop, which never ends because the `while` condition never turns to `false`. Infinite loops typically happen due to poor coding, semantic errors, or wrong assumptions in the { ... } code block.

When an app gets stuck in an infinite loop, it will probably become unresponsive and force the user to quit altogether—which is not a good user experience.

To avoid such situations, it is advised to take defensive programming measures and exit the loop if it's repeated a high number of times. In Section 4.3, you will learn about control transfer statements, which will help break loops early in such cases.

4.2.3   Repeat-While Loops

The sibling of a `while` loop is a `repeat-while` loop; they are nearly identical with one core difference. In `while` loops, conditions are checked before running the { ... } code and after iterations, but in `repeat-while` loops, conditions are checked only after iterations. Therefore, the first execution of the { ... } code is always guaranteed.

Such a loop is demonstrated in Listing 4.40, which asks the user to enter their password.

```
let correctPassword = "abcde"
var enteredPassword = ""

repeat {
    print("Enter your password:")
    enteredPassword = readLine() ?? ""
} while enteredPassword != correctPassword

print("Password correct!")
```

Listing 4.40 Password Request Using a Repeat-While Loop

Because the { ... } code block begins with the `repeat` keyword, Swift will execute the password query at least once—with no conditions attached. After the { ... } code block is completed, Swift will check the `while enteredPassword != correctPassword` condition and repeat the { … } block only if it's true.

When To Prefer repeat-while Loop

If you want the { … } code block to execute at least once, then using repeat-while makes sense. If even the first execution depends on a condition, then using while makes sense.

Figure 4.33 showcases the output of this example, where the user was allegedly able to remember their password on their third attempt.

```
Enter your password:
12345
Enter your password:
xxx
Enter your password:
abcde
Password correct!
```

Figure 4.33 Output of Password Request Example

Naturally, the logical condition after while can be as complex as needed. You can use any logical/nested condition, which you learned about in Chapter 2. So long as your expression returns a single Boolean value, it can be used. In that regard, Listing 4.41 demonstrates an extended version of the previous example. Here, both the user name and the password need to be correct.

```
let correctUsername = "admin"
let correctPassword = "abcde"

var enteredUsername: String
var enteredPassword: String

repeat {
    print("Username:")
    enteredUsername = readLine() ?? ""
    print("Password:")
    enteredPassword = readLine() ?? ""
} while !(enteredUsername == correctUsername && enteredPassword ==
correctPassword)

print("Credentials correct!")
```

Listing 4.41 Credential Request Using repeat-while Loop

If you check the while condition in this example, you'll see a relatively complex expression, but it returns either true or false—and therefore it is perfectly valid and usable. If the expression returns `true`, the code in { ... } will reexecute; otherwise, Swift will break the loop and move to the next statement.

The output of this example is shown in Figure 4.34, where the user allegedly entered correct credentials on the second attempt.

```
Username:
admin
Password:
xxx
Username:
admin
Password:
abcde
Credentials correct!
```

Figure 4.34 Output of Credential Request Example

With this example, we conclude our content on loops. You learned about `for-in`, `while`, and `repeat-while` loops. Before the end of this chapter, there is a final topic of utmost importance: control transfer statements. Without this knowledge, your control over the program flow would be limited.

4.3   Control Transfer Statements

In this section, you'll learn about *control transfer statements*, which are used to manipulate the natural flow of a code block, such as in the following ways:

- Exiting the block earlier than anticipated
- Skipping an iteration
- Enforcing an iteration
- Raising an error
- Performing initial condition checks for execution enablement

Now, let's go over the keywords that lead to such control transfer statements and implement corresponding code examples.

4.3.1   Break

In Swift, the `break` keyword is used to exit a loop early, before it naturally finishes. To see this command in a familiar context, Listing 4.42 demonstrates the former login example—with a small enhancement.

```
let correctPassword = "abcde"
var enteredPassword = ""
var passwordCorrect = false
```

```
for i in 1...3 {
    print("Enter your password (attempt \(i)):")
    enteredPassword = readLine() ?? ""
    passwordCorrect = enteredPassword == correctPassword
    if passwordCorrect { break }
}

print(passwordCorrect ? "Welcome!" : "Access denied.")
```

Listing 4.42 Using break Keyword to Exit for-in Loop Early

This time, you give the user a maximum of three chances for a successful login via the for i in 1...3 statement. At the end of the { ... } code block, if passwordCorrect { break } ensures that the for loop exits early if the user enters the correct password before their third attempt.

Check the output in Figure 4.35. Although the initial loop was planned to run three times, it exited early when the user entered the correct password on their second attempt. The break keyword enabled that early exit.

```
Enter your password (attempt 1):
12345
Enter your password (attempt 2):
abcde
Welcome!
```

Figure 4.35 Correct Login on Second Attempt

Now, let's see the break keyword in the context of a while loop through a familiar example. Listing 4.43 features an enhanced version of the number guessing game created earlier.

```
let correctAnswer = Int.random(in: 1...10)
var userAnswer = 0
var success = false

while userAnswer != correctAnswer {
    print("Guess my number (1-10) or 99 to break:")
    userAnswer = Int(readLine() ?? "0") ?? 0
    if userAnswer == 99 { break }
}

print( userAnswer == correctAnswer ? "Correct!" : "The answer was
\(correctAnswer)")
```

Listing 4.43 Using break Keyword to Exit while Loop Early

This time, you give the user the option to give up by typing "99". This option is realized via the if userAnswer == 99 { break } statement. If the user types "99", then the break command will be executed, leaving the while loop early.

Figure 4.36 showcases an output of one run through this enhanced game, in which the user gives up after the second attempt by typing "99". The break keyword caused the loop to exit early, continuing the flow after the { ... } code block.

```
Guess my number (1-10) or 99 to break:
6
Guess my number (1-10) or 99 to break:
5
Guess my number (1-10) or 99 to break:
99
The answer was 7
```

Figure 4.36 User Giving Up Game

4.3.2 Continue

You've learned that the break keyword is used to exit a loop currently. The continue keyword, meanwhile, is used to skip a single iteration in a loop and move to the next one. Listing 4.44 showcases an extremely simple demonstration to show how continue works.

```
for number in 1...5 {
    if number == 3 {
        continue    // Skip printing 3
    }
    print(number, terminator: " ")   // 1 2 4 5
}
```

Listing 4.44 Basic Demonstration of continue Keyword

In this example, continue is executed when number == 3. As a result, the print command (and any other commands after continue) is skipped and the iteration continues with the next number, which is 4. In the end, you get the output **1 2 4 5**.

As mentioned, continue skips only one iteration—which was number == 3 in this case.

Listing 4.45 features a more realistic example in which you need to list single customers in your company. In this case, you execute the continue command only for married customers, skipping them during the iteration. Thus, the print statement is only executed for single customers.

```
typealias Customer = (name: String, married: Bool)

var customers: [Customer] = [
```

```
        ("Alice", true),
        ("Bob", false),
        ("Charlie", true),
        ("David", false),
        ("Eve", true)
]

for customer in customers {
    if customer.married {
        continue
    }

    print("\(customer.name) is single")
}
```

Listing 4.45 Listing Single Customers

The output of this example is shown in Figure 4.37. Although the iteration was initiated for all customers, the print command was executed only for single customers due to the placement of the continue command.

```
Bob is single
David is single
```

Figure 4.37 Single Customers

Naturally, continue can be used in a while loop as well. To demonstrate that, Listing 4.46 features a simple app that builds a shopping list by collecting a list of missing items from the user.

```
var shoppingList: [String] = []

print("Enter missing items (done to stop):")

while true {
    let userInput = readLine() ?? ""
    if userInput == "" { continue }
    if userInput == "done" { break }
    shoppingList.append(userInput)
}

print("Shopping List:" + shoppingList.joined(separator: " "))
```

Listing 4.46 Using continue in while Loop

This code sample contains two significant points worth noting:

- If the user simply presses `Enter` without typing anything, the app ignores that empty input using `continue`.
- If the user types "done", the app exits the infinite `while` loop altogether, jumping to the final `print` statement.

4

The contrast between `continue` and `break` is more evident now, right? Check the shopping list built by this little app in Figure 4.38 and note how empty entries were skipped via `continue`.

```
Enter missing items (done to stop):
cheese
coffee

milk

bread
done
Shopping List:cheese coffee milk bread
```

Figure 4.38 Shopping List Output

Isn't it exciting that the more you learn about Swift, the more realistic the examples become? You're making good progress: Let's keep it up!

4.3.3 Fallthrough

You already know about `switch` conditions, which were covered in Section 4.1.5. A usable control transfer statement in that context is the `fallthrough` keyword.

As an initial memory refresher, Listing 4.47 represents a simple code snippet containing a `switch` statement.

```
print("Enter your score: ", terminator: "")
let userScore = Int(readLine() ?? "0") ?? 0

switch userScore {
case 90...100:
    print("Amazing!")
case 70...89:
    print("You did well!")
case 50...69:
    print("Not bad.")
default:
    print("You failed.")
}
```

Listing 4.47 Usual Switch Sample, Without Fallback

Due to the nature of switch statements, Swift will process only the first corresponding case and exit the block. As shown in Figure 4.39, when the user entered the value "80", case 70…89 was processed and the output **You did well** was produced. No other case was processed.

```
Enter your score: 80
You did well!
```

Figure 4.39 Score Interpretation for 80

The rigidness of switch statements has a workaround. Let's revise the sample as shown in Listing 4.48 using a fallthrough statement and see how it affects the program flow.

```
print("Enter your score: ", terminator: "")
let userScore = Int(readLine() ?? "0") ?? 0

switch userScore {
case 90...100:
    print("Amazing!")
case 70...89:
    print("You did well!")
    fallthrough
case 50...69:
    print("Not bad.")
default:
    print("You failed.")
}
```

Listing 4.48 Switch Sample with fallthrough Added

Did you spot the fallthrough statement within case 70…89? This statement instructs Swift to process the next immediate case (or default) statement as well. This behavior becomes evident in Figure 4.40. After case 70…89 was processed, the fallthrough statement forced the case 50…69 block to be processed as well—even though the user's score of 80 is not within this range.

```
Enter your score: 80
You did well!
Not bad.
```

Figure 4.40 Score Interpretation for 80 with fallthrough Statement

Naturally, you can have multiple fallback statements in a single switch condition too. To see this case in action, let's work through a membership app that lists the benefits of a member's tier.

In typical membership plans, higher (and more expensive) tiers cover everything that lower tiers offer and add more features on top. As illustrated in Figure 4.41, Tier D is the lowest option. Going one step higher, Tier C covers everything in Tier D and adds more features. On the highest level, Tier A offers every feature possible.

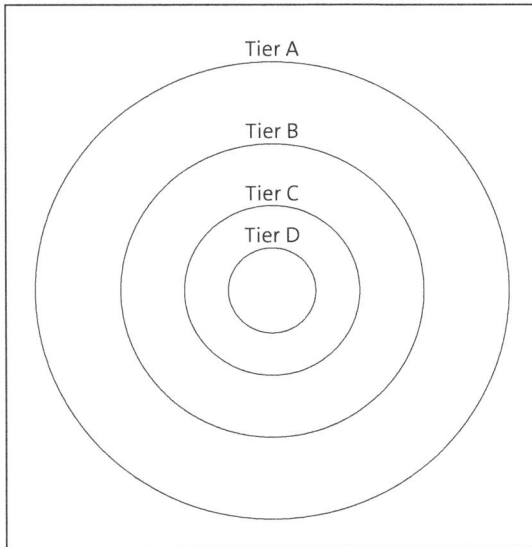

Figure 4.41 Membership Tiers

When it comes to listing features, the program must follow the guidelines in Table 4.7: Higher tiers must list not only their additional features but also all features of lesser tiers.

Tier	Must List	Must Also List
Tier D	Tier D features	—
Tier C	Tier C features	Tier D features
Tier B	Tier B features	Tier C + D features
Tier A	Tier A features	Tier B + C + D features

Table 4.7 Tier Feature Guideline

Listing 4.49 creates a small app that lists membership benefits. In this example, the `fallthrough` statements were cleverly planted into each `case` statement. That way, we ensure that once the initial highest-tier `case` is processed, its lesser tiers are also processed—thus listing all covered benefits cumulatively.

```
print("Membership tier (A-D): ", terminator: " ")
let userTier = readLine() ?? ""

switch userTier {
```

```
case "A":
    print("24/7 Premium Support")
    fallthrough
case "B":
    print("Exclusive Events Access")
    fallthrough
case "C":
    print("Monthly Discounts")
    fallthrough
case "D":
    print("Standard Content Access")
default:
    print("No Service")
}
```

Listing 4.49 Multiple fallthrough Statements in Single switch Condition

A sample output is shown in Figure 4.42. When "B" was entered for the tier, Swift processed the case blocks for tiers B, C, and D due to the fallthrough statements enforcing the succeeding blocks to be processed.

```
Membership tier (A–D):   B
Exclusive Events Access
Monthly Discounts
Standard Content Access
```

Figure 4.42 Output for Tier B

This realistic example also demonstrated a practical use case: In Swift, fallthrough statements are useful in switch cases when you want the execution to continue to the next case block, allowing multiple cases to share the same implementation without duplicating code. Although you will learn other ways to prevent code duplication, this is a quick local shortcut to do so.

4.3.4 Guard

The final control transfer statement you're going to discover here is guard, which can replace if conditions to improve readability sometimes. To understand the purpose and syntax of guard, let's write a code snippet twice: once without then once with the guard keyword.

Listing 4.50 contains the initial code, where an if statement has been used to control the flow. In this example, there is an array of money transfers; some are valid and some invalid. You iterate through those transfers, skipping invalid ones and executing valid ones. That's pretty straightforward, right? There's nothing out of the ordinary so far.

```
typealias MoneyTransfer = (recipient: String, amount: Double, isValid: Bool)

let moneyTransfers: [MoneyTransfer] = [
    ("Alice", 100.0, true),
    ("Bob", 200.0, false),
    ("Charlie", 300.0, true)
]

for moneyTransfer in moneyTransfers {
    print("-----")

    if !moneyTransfer.isValid {
        print("Skipping invalid transfer to \(moneyTransfer.recipient)")
        continue
    }

    // Pretending to send money
    print("Money sent to \(moneyTransfer.recipient)")

    // Pretending to send notification
    print("Notification sent to \(moneyTransfer.recipient)")
}
```

Listing 4.50 Money Transfer Code Without Guard

The output of this example is shown in Figure 4.43. Due to the code structure, the computer "guarded" against Bob's invalid money transfer and didn't execute it, while still processing other valid transfers.

```
-----
Money sent to Alice
Notification sent to Alice
-----
Skipping invalid transfer to Bob
-----
Money sent to Charlie
Notification sent to Charlie
```

Figure 4.43 Output of Money Transfer Demonstration

Now, how can you implement guard to improve this code snippet? It's easy: Check Table 4.8 for a comparison.

Code with if	Code with guard

`if !moneyTransfer.isValid {`
`    (...)`
`    continue`
`}`
	`guard moneyTransfer.isValid else {`
	`    (...)`
	`    continue`
	`}`

Table 4.8 Same Code Snippet Using if Versus guard Statement

In Swift, the `guard` statement is used to trigger an early exit from a code block (such as a loop) when certain conditions are not met. In the preceding example, we guarded the flow by ensuring that the money transfer is valid. Think of this as a gatekeeper: If the condition passes, then the gate opens and the flow continues, enabling the money transfer. Otherwise, the iteration is skipped, moving to the next transfer.

Although you can achieve the same functionality with an `if` condition too, `guard` is particularly useful to improve code readability. An `if` statement is a general-purpose condition, whereas `guard` has a specific purpose that becomes obvious at first sight.

For the record, Listing 4.51 contains the entire sample featuring the `guard` keyword. It produces the exact same output as you saw in Figure 4.43.

```
typealias MoneyTransfer = (recipient: String, amount: Double, isValid: Bool)

let moneyTransfers: [MoneyTransfer] = [
    ("Alice", 100.0, true),
    ("Bob", 200.0, false),
    ("Charlie", 300.0, true)
]

for moneyTransfer in moneyTransfers {
    print("-----")

    guard moneyTransfer.isValid else {
        print("Skipping invalid transfer to \(moneyTransfer.recipient)")
        continue
    }

    // Pretending to send money
    print("Money sent to \(moneyTransfer.recipient)")

    // Pretending to send notification
    print("Notification sent to \(moneyTransfer.recipient)")
}
```

Listing 4.51 Money Transfer Code with Guard

Should You Use an if or guard Statement?

In Swift, there isn't anything that can only be achieved with a guard statement that can't be accomplished with an if statement. Technically, you can rewrite any guard logic using if. However, guard offers advantages in code readability by highlighting the intent of the early exit, which if can't replicate as elegantly.

In Chapter 5, you'll learn about the usage of guard in functions too.

4.4 Summary

Congratulations, you've worked through one of the subjects of utmost importance! With knowledge of control flow statements, you can steer the code into any direction necessary.

As a wrap-up, we started with conditional statements like if and switch, which are used to execute alternative code snippets depending on given conditions.

That was followed by a section on loops, where we went through examples to explain the use of for-in, while, and repeat-while iterations to process collections.

Finally, you learned about control transfer statements like break, continue, fallthrough, and guard, which enable early exits from code blocks if necessary.

As you become more experienced in Swift, you can start to program more realistic examples too, like the ones in this chapter—small games, mini customer queues, and simple login mechanisms. This progress hopefully will make you more and more motivated; after all, you made the right choice to start with a solid Swift foundation instead of jumping to frameworks directly. Those realistic examples will be inspirational blocks for your solid apps in the future.

In the next chapter, you'll learn about functions, which are a central part of code modularization and reusability.

Chapter 5
Functions

Functions are the basic type of subroutines in Swift. They offer an encapsulation for reusable chunks of code, often enriched with input and output parameters.

We have arrived at another chapter of utmost importance, in which you will learn about *functions*—arguably the backbone of modular programming. In this chapter, you will learn how to define and call functions in Swift and about options for parameterization. But we won't stop there: Your journey will reach beyond vanilla functions. You will learn how to use higher-order functions that accept other functions as parameters, as well as closures that offer an express syntax. Generic functions, which allow flexibility in data types, are also in our scope.

For those of you new to programming, we'll start with a definition of functions, and then we'll proceed from there.

5.1   What Is a Function?

So far, you have developed sample Swift apps as single units, without any modularity whatsoever. Although you've manipulated the flow via conditions and loops, the program structures still looked like straight routes from the start until the end—as illustrated in Figure 5.1.

Functions, on the other hand, open new possibilities for programmers. In a nutshell, a *function* is a named block of code that performs a specific task. Basically, you define a function and pack a chunk of code within it. Once it's ready, you can invoke the function any time you need to perform its designated action.

As a conceptual example, Figure 5.2 showcases the flow of a function called `vibrate-Phone()`. The task of this function is to vibrate the iPhone of the user, of course. This function would contain some Swift code that accesses the haptics engine of iOS and triggers the required pattern to make the phone vibrate.

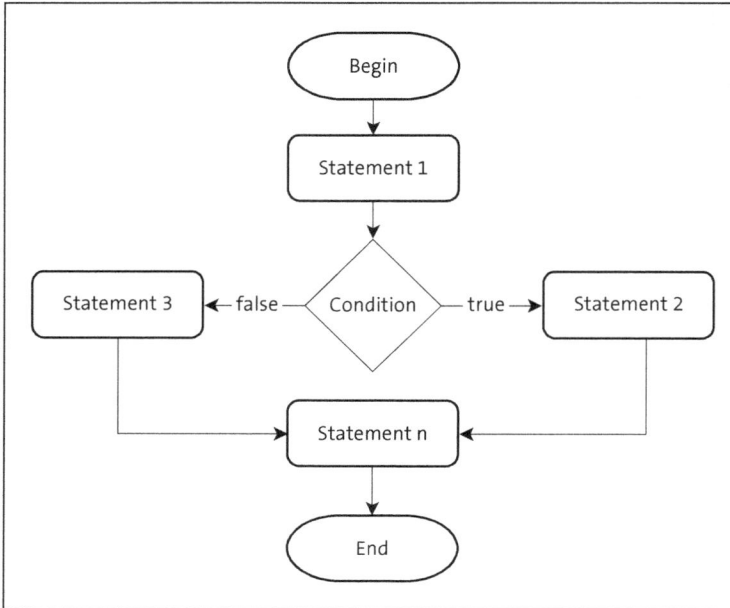

Figure 5.1 Nonmodular Program Flow

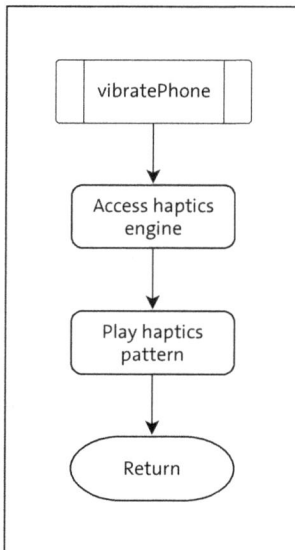

Figure 5.2 Sample Flow for vibratePhone Function

Once you have this function at hand, you can invoke vibratePhone() any time you need to vibrate the user's iPhone. That's very convenient as otherwise you would have to repeat the code inside vibratePhone() everywhere! Figure 5.3 shows the flow of the function next to the flow of a main program. Can you spot the places where vibratePhone() was invoked?

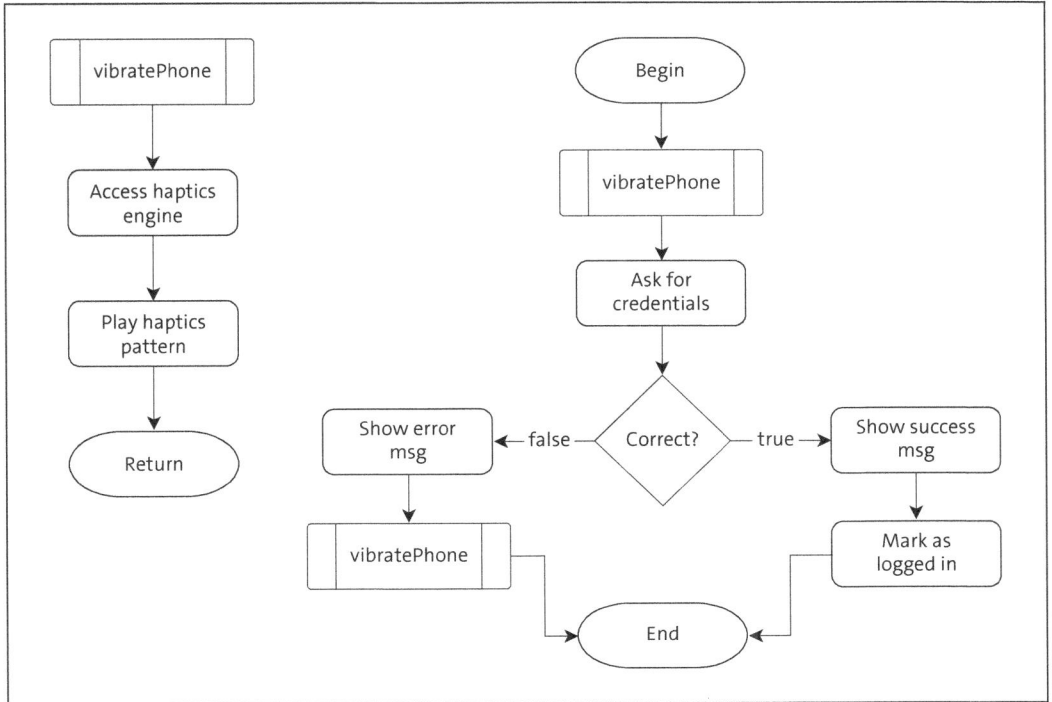

Figure 5.3 Invoking vibratePhone Function

As you can see, vibratePhone() has been invoked twice: once at the beginning of the program and once for a case in which incorrect credentials are supplied. Every time vibratePhone() is called, Swift will execute the code within the function and return to the main flow afterward. Neat, right?

In this chapter, we will go through many practical examples to solidify the logic of functions and explore their features. But before that, let's praise functions by going through some of their benefits in Table 5.1.

Benefit	Explanation	Example
Reusability	Functions prevent code duplication by centralizing specific tasks.	You can call vibratePhone() any time you want to invoke haptics; there's no need to duplicate its code.
Maintenance	Improvement of a function benefits the entire program.	If you fix an error in vibratePhone, every caller of the function is also fixed automatically.

Table 5.1 Some Benefits of Functions

Benefit	Explanation	Example
Encapsulation	Functions isolate specific tasks from the main flow, enabling a cleaner code structure.	Vibrating the phone is an isolated task, which doesn't really belong to the authorization flow.
Readability	Via descriptive names, functions make it easy to understand what the code does.	vibratePhone is an intuitive name that can be understood easily by any programmer.
Flexibility	Functions can import or export parameters, which empowers them to handle a flexible array of scenarios.	vibratePhone could accept a parameter for, for example, vibration type and behave accordingly.
Testability	Because functions are self-contained units, they can be invoked and tested independently.	You can invoke vibratePhone and see if the phone vibrates or not— independent from the authorization flow.

Table 5.1 Some Benefits of Functions (Cont.)

Now that you understand the logic of functions, it's time to get your hands dirty! Let's begin with simple Swift functions and move forward toward more advanced features, step by step.

5.2 Defining and Calling Functions

Let's get warmed up by defining functions in their simplest form and see how to call them. We will also discuss the subject of variable scopes for functions.

5.2.1 Defining a Function

In Swift syntax, a function is expressed with the func keyword, which is an obvious abbreviation. Listing 5.1 demonstrates a very simple function definition.

```
func sayHello() {
    print("Hello from the function!")
}
```

Listing 5.1 Simple Function

The func sayHello() expression is telling Swift that we are about to declare a new function, which is called sayHello. Following that, we put the function code into the {...} brackets. Easy, right? This is not unlike the way loops work, which you learned about in Chapter 4.

Although this simple function contains only one `print` statement, a function can contain any appropriate Swift code necessary. Listing 5.2 demonstrates a slightly longer function, which generates a random number and prints it out.

```
func printRandomNumber() {
    let randomNumber = Int.random(in: 1...100)
    print(randomNumber)
}
```

5

Listing 5.2 Function to Print Random Number

Naturally, a program can (and probably will) contain multiple functions too—as demonstrated in Listing 5.3.

```
func greet() {
    print("Hello!")
}

func bidFarewell() {
    print("Goodbye!")
}
```

Listing 5.3 Multiple Functions in Same Program

5.2.2 Calling a Function from the Program

The code of a function executes only when you call the function. A function that's never called is an obsolete component. Therefore, you will naturally want to call a function in the appropriate place in the program flow.

Listing 5.4 demonstrates a simple program in which the `sayHello` function is called at the beginning of the main flow.

```
// Function
func sayHello() {
    print("Hello from the function!")
}

// Main flow
sayHello()
print("Hello from the main flow!")
```

Listing 5.4 Calling Simple Function

As demonstrated, mentioning the name of the function in `sayHello()` is enough to call the function! Upon processing the `sayHello()` statement, the computer will execute the `print("Hello from the function!")` function code. Once the function is finished, the computer will follow with the next statement after the function call, which is `print("Hello from the main flow!")`.

The flow becomes evident in Figure 5.4, which displays the output of the program. The function's `print` statement is executed first, and the main flow's `print` statement is executed right after.

```
Hello from the function!
Hello from the main flow!
```

Figure 5.4 Output of Simple Function Example

To strengthen your grasp of the flow, Listing 5.5 showcases an enhanced version of the last example. Here, `sayHello` is called three times sequentially.

```
// Function
func sayHello() {
    print("Hello from the function!")
}

// Main flow
sayHello()
sayHello()
sayHello()
print("Hello from the main flow!")
```

Listing 5.5 Calling Function Multiple Times

The output of this example is shown in Figure 5.5. As you can see, the `sayHello` function was executed every time you called it—which is three times total.

```
Hello from the function!
Hello from the function!
Hello from the function!
Hello from the main flow!
```

Figure 5.5 Output of Multiple Function Calls

Now, let's see an example that calls different functions throughout the main flow. Listing 5.6 features a demonstration in which three different functions are called in the main flow.

```
// Functions
func greet() {
    print("Hello!")
}

func showLuckyNumber() {
```

```
    let luckyNumber = Int.random(in: 1...10)
    print("Your lucky number is \(luckyNumber) today.")
}

func bidFarewell() {
    print("Goodbye!")
}

// Main flow
greet()
showLuckyNumber()
bidFarewell()
```

Listing 5.6 Program Calling Different Functions Sequentially

The computer will execute the greet, showLuckyNumber, and bidFarewell functions sequentially throughout the main flow. Figure 5.6 shows a possible output of this program.

```
Hello!
Your lucky number is 7 today.
Goodbye!
```

Figure 5.6 Output of Sequential Function Calls

5.2.3   Calling a Function from Another Function

So far, you have learned how to call a function from the main program. However, it is possible to call a function from within another function too.

To see this feature in context, let's write a program to prepare an imaginary breakfast. The flow plan is demonstrated in Figure 5.7.

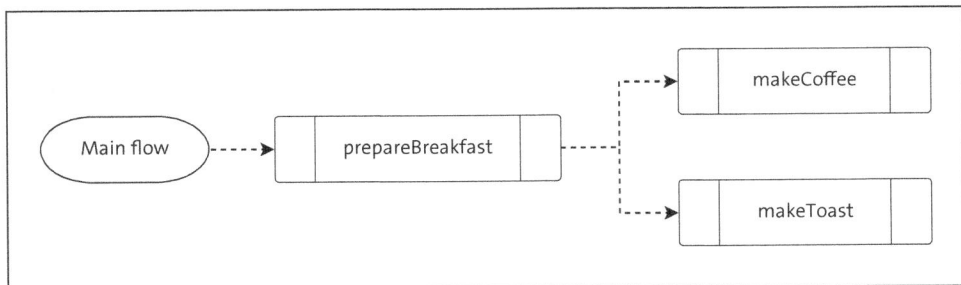

Figure 5.7 Breakfast Preparation Plan

In the main flow, you will merely call the prepareBreakfast function. But the code in prepareBreakfast itself will make two further function calls to makeCoffee and makeToast.

This makes sense, right? The code in a function is still Swift code. If the main flow can call functions, then a function can call functions too. The code implementation of this example is shown in Listing 5.7. You can follow through the code and see how it reflects the visual plan.

```swift
// Functions
func prepareBreakfast() {
    print("Starting breakfast preparation...")
    makeCoffee()
    makeToast()
    print("Breakfast is ready!")
}

func makeCoffee() {
    print("Grinding coffee beans")
    print("Brewing coffee")
    print("Coffee is done")
}

func makeToast() {
    print("Slicing bread")
    print("Toasting bread")
    print("Toast is done")
}

// Main flow
prepareBreakfast()
```

Listing 5.7 Breakfast Preparation Code

The output of the program is shown in Figure 5.8. The print statements in the program flow can be mapped with the output, but a more detailed breakdown of the program flow also is provide in Table 5.2.

```
Starting breakfast preparation...
Grinding coffee beans
Brewing coffee
Coffee is done
Slicing bread
Toasting bread
Toast is done
Breakfast is ready!
```

Figure 5.8 Output of Breakfast Preparation Program

Level	Source	Statement	Output
1	Main flow	prepareBreakfast()	–
2	prepareBreakfast	print()	**Starting breakfast preparation...**
2	prepareBreakfast	makeCoffee()	–
3	makeCoffee	print()	**Grinding coffee beans**
3	makeCoffee	print()	**Brewing coffee**
3	makeCoffee	print()	**Coffee is done**
2	prepareBreakfast	makeToast()	–
3	makeToast	print()	**Slicing bread**
3	makeToast	print()	**Toasting bread**
3	makeToast	print()	**Toast is done**
2	prepareBreakfast	print()	**Breakfast is ready!**

Table 5.2 Breakdown of Program Execution

5.2.4   Variable Scope in Functions

You have seen that functions can contain variables, but it's important to understand the distinction between global and localError! Bookmark not defined. variables. A *local variable* is a variable declared within a sub code block, such as a function, whereas a *global variable* is a variable declared in the main flow.

Listing 5.8 contrasts the difference between those variable types. localVar was declared inside of someFunction; therefore, it is a local variable of that function. On the other hand, globalVar was declared in the scope of the main flow; therefore, it is a global variable of the program.

```
// Global variables
let globalVar = "global"

// Functions
func someFunction() {
    let localVar = "local"
}
```

Listing 5.8 Contrasting Global and Local Variables

What difference does it make if a variable is global or local? The big difference lies in the visibility and access options for those variable types.

As their name suggests, global variables are visible to the main flow and all functions. You can access them and read their values anywhere you want. They are kept alive so long as the program is active.

On the other hand, local variables are visible only to their defining functions. As soon as a function ends, its local variables are thrown away. In other words, the lifespan of a local variable is limited to the execution time of its function.

This scope difference is demonstrated in Listing 5.9. The value of globalVar can be read in someFunction, as well as the main flow, whereas the value of localVar is only available to someFunction—not to the main flow (or other functions). That makes sense: Once someFunction finishes, localVar is scrapped. That's why the last line will generate an error if you try to execute it.

```
// Global variables
let globalVar = "global"

// Functions
func someFunction() {
    let localVar = "local"
    print(globalVar)    // "global"
    print(localVar)     // "local"
}

// Main flow
print(globalVar)        // "global"
print(localVar)         // ERROR!
```

Listing 5.9 Global and Local Variable Read Demonstration

In this example, we focused on reading global and local variables. How about changing them? This is simple for local variables. Because local variables are born, used, and scrapped in a function, the parent function is free to read or change them. They cannot be read nor changed outside of the function because they don't exist outside of the function. As their name suggests, they are local. Listing 5.10 contains a quick demonstration of this.

```
func someFunction() {
    var localVar = "local"
    localVar += " varible"
    print(localVar) // local variable
}
```

Listing 5.10 Changing Local Variable Inside Function

Global variables can be freely changed in the main flow, as we did many times before. In many code samples in previous chapters, we declared and changed global variables in the main flow—as demonstrated in Listing 5.11.

```
var globalVar = "global"
globalVar += " variable"
print(globalVar) // global variable
```

Listing 5.11 Changing Global Variable in Main Flow

Because global variables are global, they should be changeable by any function too, right? That's correct—yet we need to add a small attribute to the function to indicate that it may mutate global variables. Listing 5.12 features such an example; note the @MainActor attribute added to the function.

```
var globalVar = "global"

@MainActor
func someFunction() {
    globalVar += " variable"
    print(globalVar) // global variable
}
```

Listing 5.12 Changing Global Variable in Function

Functions with the @MainActor attribute are allowed to mutate global variables. Without this attribute, the function can read the value of globalVar, but not change it.

As a conclusion to this section, variable scopes are summarized in Table 5.3.

Variable Type	Scope	Readable	Changeable
Local	Function	Yes	Yes
Local	Main flow	No	No
Global	Function	Yes	Yes if it has @MainActor attribute
Global	Main flow	Yes	Yes

Table 5.3 Variable Scope Summary

So far, you have worked with raw functions to understand their basics. In the next section, you will learn how to work with function parameters.

5.3   Input Parameters

Functions empower programmers by acting like small code capsules. Instead of coding everything into the main flow, you can *encapsulate* chunks of code as functions to serve specific purposes. That helps with code organization and readability and prevents code duplication: Instead of repeating the same code over and over again, you can simply call the appropriate function that contains the common code snippet.

On top of these benefits, functions can also accept parameters from their callers and use them like local variables. This feature increases the flexibility of functions immensely, as you will see in this section.

5.3.1   Named Parameters

Let's start simple, by looking at functions with named parameters. This is the most straightforward way of adding input parameters to a function. Listing 5.13 features a function for name analysis, which accepts a single input parameter.

```
func analyseName(name: String) {
    let nameLength = name.count
    let vowelCount = name.lowercased().filter { "aeiou".contains($0) }.count

    print("\(name) has \(nameLength) characters and \(vowelCount) vowels")
}
```

Listing 5.13 Function to Analyze Name

Let's focus on the function definition: `func analyseName(name: String)` means that when you are calling this function, you must pass a `name` value to it. In the function body, the `name` variable has been used exactly like a local variable, and its length and vowel count are printed out.

Any `name` you pass to this function will be analyzed in the same way. You see how flexible this function has become? The usability of this code capsule is now much higher. Table 5.4 shows how the function will behave for different `name` values.

Passed Name Value	Calculated Length	Calculated Vowel Count
Alice	5	3
Bob	3	1
Charlie	7	3

Table 5.4 Function Behavior for Different Name Values

Having a function with input parameters is only half of the story. Now you need to call this function from the main flow, right? Listing 5.14 features an extended version of the example, where the function is called in the main flow.

```
// Function
func analyseName(name: String) {
    let nameLength = name.count
    let vowelCount = name.lowercased().filter { "aeiou".contains($0) }.count

    print("\(name) has \(nameLength) characters and \(vowelCount) vowels")
}

// Main flow
analyseName(name: "Alice")
```

Listing 5.14 Calling Function to Analyze Name

The magic happens on the analyseName(name: "Alice") line. This expression tells Swift to call the analyseName function and pass "Alice" as the value of name. In the function, the name variable will carry the "Alice" value; all operations will execute accordingly.

Figure 5.9 shows the expected output. As you can see, nameLength received the length of "Alice" and vowelCount received the number of vowels in "Alice"—both because name is "Alice" in this case. The entire function is executed as if a local variable was declared as in let name = "Alice".

```
Alice has 5 characters and 3 vowels
```

Figure 5.9 Program Output for Alice

Passing Parameters Is Obligatory

If a function has input parameters, the call is obliged to set their values when calling the function. In the preceding example, analyseName(name: "Alice") is a valid call, but analyseName() is not—because the parameter name has not been passed.

Some parameters can be made optional via default values, though; you will learn this technique soon.

To spice things up a little, Listing 5.15 features an enhanced version of the program, which calls the function multiple times with different names.

```
// Function
func analyseName(name: String) {
    let nameLength = name.count
    let vowelCount = name.lowercased().filter { "aeiou".contains($0) }.count
```

```
        print("\(name) has \(nameLength) characters and \(vowelCount) vowels")
}

// Main flow
let customers: [String] = ["Alice", "Bob", "Charlie"]

for customer in customers {
    analyseName(name: customer)
}
```

Listing 5.15 Calling Function to Analyze Multiple Names

Because you call the function in a for loop, it will be executed once for each customer in customers. On each iteration, name will become a different customer. The output of this version is shown in Figure 5.10.

```
Alice has 5 characters and 3 vowels
Bob has 3 characters and 1 vowels
Charlie has 7 characters and 3 vowels
```

Figure 5.10 Program Output for Multiple Names

Argument Immutability
Arguments passed to a function are immutable by default. This means that the function's code can read arguments but can't change them. It is possible to change this default behavior, though; you will learn how in the upcoming sections of this chapter.

Naturally, a function can accept multiple parameters too. Listing 5.16 demonstrates functions that accept two input parameters instead of one. Each function runs a simple calculation using the n1 and n2 parameters.

```
// Functions
func showSum(n1: Int, n2: Int) {
    let sum = n1 + n2
    print(sum)
}

func showDiff(n1: Int, n2: Int) {
    let diff = n1 - n2
    print(diff)
}
```

```
// Main flow
showSum(n1: 150, n2: 50)     // 200
showDiff(n1: 150, n2: 50)    // 100
```

Listing 5.16 Functions Accepting Multiple Parameters

Just for the fun of it, Listing 5.17 demonstrates an enhanced version of this example. Here you have a parent function called showAllCalculations. In this parent function, you call the showSum and showDiff child functions, executing both operations. Here's the fun part: The parent function is able to pass its own parameters as parameters of the child functions. Cool, right?

```
// Functions
func showSum(n1: Int, n2: Int) {
    let sum = n1 + n2
    print(sum)
}

func showDiff(n1: Int, n2: Int) {
    let diff = n1 - n2
    print(diff)
}

func showAllCalculations(num1: Int, num2: Int) {
    showSum(n1: num1, n2: num2)
    showDiff(n1: num1, n2: num2)
}

// Main flow
showAllCalculations(num1: 150, num2: 50) // 200, 100
```

Listing 5.17 Function Calling Other Functions with Parameters

5.3.2   Omitting Argument Labels

In the previous section, you implemented functions with named parameters. Each function parameter had a name, and you had to use those names while calling the function.

To keep this section self-contained, Listing 5.18 shows such a function, where the parameters are named num1 and num2. When calling this function, you must name both variables, as in showSum(num1: 100, num2: 40).

```
// Function
func showSum(num1: Int, num2: Int) {
    let sum = num1 + num2
```

```
    print(sum)
}
```

```
// Main flow
showSum(num1: 100, num2: 40) // 140
```

Listing 5.18 Function with Named Parameters

That's all good and well, but you also have the option to omit parameter names if you want. This syntax would simplify the function call, making it more readable. As a demonstration, Listing 5.19 features the same program with omitted function parameter names.

```
// Function
func showSum(_ num1: Int, _ num2: Int) {
    let sum = num1 + num2
    print(sum)
}
```

```
// Main flow
showSum(100, 40) // 140
```

Listing 5.19 Same Function with Omitted Parameter Names

Check the func showSum(_ num1: Int, _ num2: Int) function definition; you'll notice the _ token before the num1 and num2 parameter names. This means that both parameters, num1 and num2, use _ to omit their argument labels; therefore, you can call the function without naming them.

That's exactly what we did in the showSum(100, 40) expression. When the function is called that way, Swift will process the variables in their given order, mapping num1 = 100 and num2 = 40 automatically.

Obviously, showSum(100, 40) is more concise and readable than showSum(num1: 100, num2: 40).

When to Omit Parameter Names

If omitting parameter names improves readability, why not omit all parameter names? When should you use named parameters and when not?

This is a matter of ambiguity. In the preceding example, showSum(_ num1: Int, _ num2: Int) is a function to sum two numbers, so there is no place for ambiguity between num1 and num2. Giving those numbers in any order produces the same output. Therefore, parameter names can be omitted safely.

However, when a parameter's role is not obvious, it is safer to use labels for clarity. Imagine a function called drawRectangle(width: Int, height: Int). If you omitted the

---

> parameter names here, a caller like drawRectangle(50, 100) would be ambiguous for someone reading the code as it is not immediately clear if 50 is the width or height. The same ambiguity may cause errors while coding, too: You could easily switch the intended values of width and height.
>
> In this case, drawRectangle(width: 50, height: 100) is the safest bet to prevent ambiguity.

You also have the freedom to use a mixed approach and omit only some parameter names. In Listing 5.20, the overlayText function omits only the first parameter name, text.

```
// Function
func overlayText(_ text: String, length: Int, filler: Character) {
    var result = text

    while result.count < length {
        result += String(filler)
    }

    print(result)
}

// Main
overlayText("Hello", length: 10, filler: "-")    // Hello-----
overlayText("World", length: 7, filler: ".")     // World..
```

Listing 5.20 Omitting Some Parameter Names

When calling this function as overlayText("Hello", length: 10, filler: "-"), you set the first parameter, text, without its name, but provide parameter names for others (length and filler), effectively preventing any ambiguity or confusion. This technique is commonly used when major parameters are already obvious without needing a name, but support parameters are not.

This technique becomes especially useful when you provide default values for some function parameters, which is the subject of the next section.

5.3.3   Default Arguments

Input parameters of functions increase their flexibility dramatically. By passing parameters to a function, you can make it operate with given values and adapt to the specific requirement at hand. If a function expects input parameters, then the caller is obliged to pass their values; otherwise, the program won't compile and run.

Regarding parameter obligation, you have a certain degree of flexibility: It is possible to define default values for some (or all) parameters of a function, and if the caller doesn't pass any value for those, then their default values will be used. This technique simplifies function calls in many cases.

Listing 5.21 demonstrates a function called `printName` with some default arguments. This simple function will format a name with given options and print it out.

```
// Function
func printName(_ name: String, title: String = "", capitalize: Bool = true) {
    var formattedName = name

    if title != "" {
        formattedName = title + " " + formattedName
    }

    if capitalize {
        formattedName = formattedName.uppercased()
    }

    print(formattedName)
}
```

Listing 5.21 Function with Default Arguments

In this example, the `name` parameter is obligatory; the function expects to definitely receive a value for that. On the other hand, the `title` parameter has a default value of `""`. If the caller doesn't pass any value for `title`, then the function will assume that the `""` argument was passed—effectively making it an optional parameter. Likewise, `capitalize` is also an optional parameter; the function will assume it to be `true` if no argument is passed by the caller.

Listing 5.22 demonstrates different ways of calling this function.

```
// Function
func printName(_ name: String, title: String = "", capitalize: Bool = true) {
    var formattedName = name

    if title != "" {
        formattedName = title + " " + formattedName
    }

    if capitalize {
        formattedName = formattedName.uppercased()
    }
```

```
    print(formattedName)
}

// Main flow
printName("John Doe", title: "Dr.", capitalize: false)  // Dr. John Doe
printName("Jane Smith", title: "Agent")                  // AGENT JANE SMITH
printName("Alice Brown")                                 // ALICE BROWN
```

Listing 5.22 Different Ways of Calling Function with Optional Parameters

In the first call, for John Doe, you pass values for all parameters, as though they had no default values. The function happily accepts those values instead of the defaults and uses them to format the name.

In the second call, for Jane Smith, you only pass title but don't specify a value for capitalize. In that case, the function assumes capitalize = true due to its default value in the function signature.

In the third call, for Alice Brown, you pass none of the optional parameters. In that case, the function assumes title = "" and capitalize = true due to their default values.

Those function calls were summarized in Table 5.5 for an overview.

Name	Title	Capitalize	Output
"John Doe"	"Dr."	false	**Dr. John Doe**
"Jane Smith"	"Agent"	true (default)	**AGENT JANE SMITH**
"Alice Brown"	"" (default)	true (default)	**ALICE BROWN**

Table 5.5 Summary of Function Calls

Optionals

Another way of creating optional function parameters is to declare them as optionals. You will learn about optionals in their dedicated chapter, Chapter 6.

5.3.4   Guarding Functions

In Swift, the guard statement is used to trigger early exits from a code block when certain conditions are not met. You learned about the guard keyword in Chapter 4, where you used it to manipulate the program flow in loops. Likewise, a common technique in Swift is to use guard at the beginning of a function for an early exit, depending on arguments.

Listing 5.23 demonstrates such a function, where guard has been used to ensure that a new user is obliged to be at least 18 years old.

```
func createUser(_ userName: String, password: String, age: Int) {
    guard age >= 18 else {
        print("No minors accepted")
        return
    }

    print("User \(userName) created")
}
```

Listing 5.23 Demonstration of Guard for Early Function Exit

If age is greater than (or equal to) 18, the function creates the user. Otherwise, the function will exit early via the return keyword. To demonstrate this feature, Listing 5.24 contains two different calls to createUser. In the first case, age is 20, so the user can be created. In the second case, age is 15, so guard activates and prevents the user's creation.

```
// Function
func createUser(_ userName: String, password: String, age: Int) {
    guard age >= 18 else {
        print("No minors accepted")
        return
    }

    print("User \(userName) created")
}

// Main flow
createUser("john", password: "1234", age: 20)   // User john created
createUser("marc", password: "abcd", age: 15)   // No minors accepted
```

Listing 5.24 Calling createUser for Different Ages

In this example, we snuck the return keyword in, which was used to exit the function early. But that's not the main purpose of return; you will use it often to return a value from a function. This mechanism is the subject of the next section.

5.4   Returning Values

You know by now that functions can import parameters—both obligatory and optional ones. In this section, you will learn about a new feature: Functions can return values as well, which is often the result of the operation within the function.

5.4.1 Returning Single Values

To understand the syntax to return values, let's start with a very simple example. Listing 5.25 contains a function that calculates the sum of two numbers and returns the result. Heads up: Returning the result is the new part!

```
func sumNumbers(_ n1: Int, _ n2: Int) -> Int {
    let sum = n1 + n2
    return sum
}
```

Listing 5.25 Simple Function Returning Sum of Two Numbers

This code snippet has two parts worth noting. First, the signature ends with the -> Int expression, which indicates that this function will return an integer value to the caller. Second, the function body ends with the return sum expression, which returns the calculated value of sum to the caller.

Naturally, the type of the returned value (sum) must match the type in the signature (-> Int); otherwise, Swift will generate an error message. Listing 5.26 demonstrates how to call this function.

```
// Function
func sumNumbers(_ n1: Int, _ n2: Int) -> Int {
    let sum = n1 + n2
    return sum
}

// Main flow
let result = sumNumbers(10, 20)
print("Sum: \(result)")      // Sum: 30
```

Listing 5.26 Calling Function to Return a Value

The result = sumNumbers(10, 20) expression is worth noting here. It is telling Swift to call sumNumbers with the n1 = 10 and n2 = 20 values and "catch" the returned sum (30) in the result variable. That's easy and intuitive, right?

In previous chapters, you called standard Swift functions that way—many times. Now you've learned how to write a custom function in the same fashion too!

As a more comprehensive exercise, Listing 5.27 features two functions: getSmallestNumber returns the smallest number in an array, and getBiggestNumber returns the biggest number in an array—with simple algorithms, not the best ones.

```
// Functions
func getSmallestNumber(_ numbers: [Int]) -> Int {
    var result = numbers[0]
```

```
        for number in numbers { if number < result { result = number } }
        return result
}

func getBiggestNumber(_ numbers: [Int]) -> Int {
    var result = numbers[0]
    for number in numbers { if number > result { result = number } }
    return result
}

// Main flow
let myNumbers: [Int] = [1, 2, 3, 4, 5]
let small = getSmallestNumber(myNumbers)
let big = getBiggestNumber(myNumbers)

print("Smallest number: \(small))")     // Smallest number: 1
print("Biggest number: \(big))")        // Biggest number: 5
```

Listing 5.27 Functions to Find Smallest and Biggest Numbers in Array

In the main flow, you call those functions and store the smallest number in small and biggest number in big. Straightforward, right?

Here's a cool trick: A function returning a value can be used as an in-line expression, like a variable. In Listing 5.28, you call getVowelCount(name) right within the print statement. The resulting value of this function call will be used.

```
// Function
func getVowelCount(_ name: String) -> Int {
    return name.lowercased().filter { "aeiou".contains($0) }.count
}

// Main flow
let name = "Abraham"
print("\(name) has \(getVowelCount(name)) vowels")  // Abraham has 3 vowels
```

Listing 5.28 Calling In-line Function

Here's an even cooler trick: A function that returns a Boolean value is often used in-line following a condition. As a demonstration, Listing 5.29 features a familiar example from previous chapters: a password strength validation app. Let's overlook the realism level and focus on the syntax instead.

```
// Function
func isPasswordStrong(_ password: String) -> Bool {
    if password.count < 8 { return false }
```

```
    if !password.contains(where: { ("a"..."z").contains($0) }) { return false }
    if !password.contains(where: { ("A"..."Z").self.contains($0) }) { return
false }
    if !password.contains(where: { ("0"..."9").contains($0) }) { return false }
    return true
}

// Main flow
print("Enter password candidate: ", terminator: " ")
let candidatePassword = readLine() ?? ""

if isPasswordStrong(candidatePassword) {
    print("Password is strong")
} else {
    print("Password is weak")
}
```

Listing 5.29 App to Test Password Strength

In this example, isPasswordStrong returns a Boolean value. If the given password is strong, then the function will return true; otherwise, it will return false.

In the main flow, you call this function directly in a condition as if isPasswordStrong. Swift will use the return value of the function to decide whether the true block or the false block should be executed.

Figure 5.11 demonstrates the output of the program for a weak password. Here, isPasswordStrong("Abc123") returned false; therefore, the if condition flowed toward the negative result and printed **Password is weak**.

```
Enter password candidate:   Abc123
Password is weak
```

Figure 5.11 Program Output for Weak Password

In-line Functions and Clean Coding

Use in-line function calls only when they are simple and self-explanatory. If the function call is complex, consider breaking it out into separate variables:

- Good example: if isUserActive(user) {…}
- Bad example: if calcUserScore(getUserValues(getUserByName("bob"))) > 50 {…}

Remember: Human readability is always a deciding factor. Your code quality today affects the program's maintainability tomorrow.

5.4.2 Returning Multiple Values

So far, we've looked at sample functions that return a single value. However, functions can also be set up to return multiple values. In this section, you will learn how.

Returning a Tuple

You already know that tuples are basic data structures that group multiple variables. If you write a function that returns a tuple, you can in effect return multiple values grouped as a tuple. Listing 5.30 demonstrates such an example, where getUserInfo returns the User tuple, which groups the user's name and age.

```
// Type
typealias User = (name: String, age: Int)

// Function
func getUserInfo() -> User {
    let result = User(name: "Bob", age: 40)
    return result
}

// Main flow
let currentUser = getUserInfo()
print("\(currentUser.name), \(currentUser.age)")     // Bob, 40
```

Listing 5.30 Function Returning a Tuple

To make the demonstration a bit more realistic, Listing 5.31 features an enhanced version of the same code snippet. This time, getUserInfo doesn't return a hard-coded tuple. Instead, it asks for the user information via the terminal. But the basic principle doesn't change: The function is still returning a tuple that groups multiple variables.

```
// Type
typealias User = (name: String, age: Int)

// Function
func getUserInfo() -> User {
    var result = User(name: "", age: 0)

    print("Enter your name:", terminator: " ")
    result.name = readLine() ?? ""

    print("Enter your age:", terminator: " ")
    result.age = Int(readLine() ?? "0") ?? 0
```

```
    return result
}
```

```
// Main flow
let currentUser = getUserInfo()
print("\(currentUser.name), \(currentUser.age)")
```

Listing 5.31 Function Querying User Info and Returning a Tuple

A visual demonstration of this small program is provided in Figure 5.12.

```
Enter your name: Kerem
Enter your age: 45
Kerem, 45
```

Figure 5.12 Terminal Output of Example

Returning a Collection

Arrays, sets, and dictionaries are basic collections of values. A function that returns a collection is effectively returning multiple values, right? Let's go over a couple of examples to see this in action.

Listing 5.32 demonstrates a function that returns an array of random numbers. Note that the core syntax of the function doesn't change much. All we did differently here was to end the function signature with -> [Int], indicating that the function will return an array of integers.

```
// Function
func generateRandomNumbers(numCount: Int) -> [Int] {
    var result: [Int] = []

    while result.count < numCount {
        result.append(Int.random(in: 0..<100))
    }

    return result
}
```

```
// Main flow
let randomNumbers = generateRandomNumbers(numCount: 10)
print(randomNumbers)    // Random output
```

Listing 5.32 Sample Function Returning an Array

Figure 5.13 shows a sample output of this program, featuring an array of random numbers.

```
[43, 46, 23, 81, 84, 75, 23, 62, 97, 70]
```

Figure 5.13 Sample Array Returned by Function

Using a similar syntax, you can write functions that return sets too. Listing 5.33 demonstrates a function that returns a set of unique product IDs. To ensure the uniqueness of each ID, we made use of the built-in UUID class. The logic of the function is nearly the same; the only natural difference is the final part, -> Set<String>, which indicates that the function will return a set of unique strings.

```swift
import Foundation

// Function
func generateProductGUIDs(quantity: Int, category: String) -> Set<String> {
    var guids: Set<String> = []
    let prefix = category.uppercased()

    while guids.count < quantity {
        let guid = UUID().uuidString
        let categorizedGUID = "\(prefix)-\(guid)"
        guids.insert(categorizedGUID)
    }

    return guids
}

// Main flow
let productGUIDs = generateProductGUIDs(quantity: 3, category: "CLOTH")
for guid in productGUIDs { print(guid) }
```

Listing 5.33 Sample Function Returning a Set

Figure 5.14 shows a sample output of this program, featuring a set of unique IDs.

```
CLOTH-458C8E93-C7F0-4028-A3CE-5B9B369740AE
CLOTH-06ABB979-1F98-4078-86BA-86EC52D15C5C
CLOTH-90D4471D-9311-4AFB-8CF0-E7A983C9B937
```

Figure 5.14 Sample Set Returned by Function

Finally, you can make a function return a dictionary too, as demonstrated in Listing 5.34. This time, the function signature ends with -> [String: String], indicating that it will return a dictionary with key/value pairs of type String.

```swift
// Function
func getFavoriteColors() -> [String: String] {
    let colors = ["Alice"    : "Blue",
```

```
                "Bob"      : "Green",
                "Charlie"  : "Red"]

    return colors
}

// Main flow
let favoriteColors = getFavoriteColors()
print(favoriteColors["Alice"] ?? "")  // Blue
```

Listing 5.34 Sample Function Returning a Dictionary

Now that you've warmed up to the subject, let's do a mind exercise with a more complex example. In Listing 5.35, there is a function called getInventoryByCategory, which returns a dictionary of product categories and their inventory counts—indicated by -> [String: Int]. In this function, you simulate fetching inventory data using hard-coded values and return stocks greater than 0. In the main flow, the entire inventory is printed, followed by the inventory count of "Clothing" individually.

```
// Function
func getInventoryByCategory(){
    // Simulate fetching inventory data
    let inventoryData = [
        "Clothing": 128,
        "Books": 73,
        "Furniture": 22,
        "Toys": 89,
        "Electronics": 0
    ]

    // Build result
    var result: [String: Int] = [:]

    for (category, count) in inventoryData {
        guard count > 0 else {continue}
        result[category] = count
    }

    // Return
    return result
}

// Main flow
let inventory = getInventoryByCategory()
print(inventory)
```

```
let clothingCount = inventory["Clothing"] ?? 0
print("Clothing items in stock: \(clothingCount)")
```

Listing 5.35 Inventory Demonstration, Featuring a Function Returning a Dictionary

The output of this program is shown in Figure 5.15. Although the example is a bit more complex than former ones, the core logic of the function returning a collection didn't change much. Once again, you specify the data type returned by the function and return the final value with the return keyword.

```
["Clothing": 128, "Books": 73, "Toys": 89, "Furniture": 22]
Clothing items in stock: 128
```

Figure 5.15 Output of Inventory Demonstration

Other Complex Types

Functions can return other complex types too, like structs and objects. You'll see examples of such functions in Chapter 8 for structs and Chapter 9 for classes.

Now that you know about the basics of importing and exporting function parameters, we can move forward to more advanced features. Let's start with variadic parameters.

5.5 Variadic Parameters

In Swift, *variadic parameters* allow a function to accept a variable number of input parameters of the same type. This is a useful feature for cases in which you don't know how many values will be passed to the function.

As a simple demonstration, Listing 5.36 features the sumNumbers function, which accepts the numbers variadic parameter. If you look closely at the parameter definition, numbers: Int..., you will see that it ends with three dots after the data type. Those dots mark the parameter as variadic, indicating that multiple integers can be sent.

```
// Function
func sumNumbers(_ numbers: Int...) -> Int {
    return numbers.reduce(0, +)
}

// Main flow
let total1 = sumNumbers(4)              // 4
let total2 = sumNumbers(1, 2, 3, 4)     // 10
let total3 = sumNumbers(10, 20)         // 30
```

Listing 5.36 Function with Variadic Parameter

In the main flow, you can call sumNumbers with any number of integers now! That way, you have the flexibility to calculate the total of any number of values. Once the function receives those numbers, they are internally converted to an array; therefore, in the function body, you can access numbers as if it was an array in the first place.

Variadic or Array?

In the preceding example, you could set the function as sumNumbers(_ numbers: [Int]) too, indicating that the function expects an array with a variable number of integers. If that's the case, why use variadic parameters instead?

Technically, both work! But in the main flow, variadic parameters allow you to pass values directly without explicitly creating an array first, making the function call more intuitive and concise. It is a small but meaningful quality-of-life improvement.

A function can have singular and variadic parameters together as well. Listing 5.37 demonstrates a function to send mobile messages, which receives a single message but a variable number of recipients.

```
// Function
func sendSMS(_ message: String, recipients: String...) {
    // Send the message to all recipients
}

// Main flow
sendSMS("hello", recipients: "+1234567890", "+0987654321")
```

Listing 5.37 Function with Singular and Variadic Parameters

As a complex exercise, Listing 5.38 features a function to send an email message to multiple recipients. Both to and cc are variadic parameters, accepting variable numbers of email addresses.

```
// Function
func sendEmail(to: String..., cc: String..., subject: String, body: String) {
    // E-mail simulation
    print(to)
    print(cc)
    print(subject)
    print(body)
}

// Main flow
sendEmail(to: "alice@example.com", "bob@example.com",
        cc: "john@example.com",
```

```
        subject: "Test",
        body: "This is just a test")
```

Listing 5.38 Function with Multiple Variadic Parameters

The output of this example is shown in Figure 5.16. Note that we have simply printed out the values here instead of sending an actual email.

```
["alice@example.com", "bob@example.com"]
["john@example.com"]
Test
This is just a test
```

Figure 5.16 Output of Email Simulation

Here's a nice feature: You don't have to pass values to a variadic parameter; doing so is optional by default. In that case, Swift will assume that you have passed an empty array instead. Check the sample in Listing 5.39, where the cc function parameter has not been provided in the main flow.

```
// Function
func sendEmail(to: String..., cc: String..., subject: String, body: String) {
    // E-mail simulation
    print(to)
    print(cc)
    print(subject)
    print(body)
}

// Main flow
sendEmail(to: "alice@example.com", "bob@example.com",
          subject: "Test",
          body: "This is just a test")
```

Listing 5.39 Skipping "cc" Variadic Parameter

As a result, the program will produce the output shown in Figure 5.17. Check the output of print(cc); it's merely an empty array.

```
["alice@example.com", "bob@example.com"]
[]
Test
This is just a test
```

Figure 5.17 Program Output Without "cc" Variadic Parameter

Cool, right? Thanks to that feature, you don't need to spend any effort to make variadic parameters optional. Our next topic is arguably even cooler than that, though: It will enable you to declare functions with alternative signatures.

5.6   Function Overloading

So far, we've offered examples in which a function was declared only once, with a single signature. This is the common and usual way of declaring functions. However, Swift also features the option of *function overloading*, declaring the same function multiple times with different parameter signatures.

Listing 5.40 demonstrates such an example, where the `login` function is overloaded with three different signatures.

```
func login(userName: String, password: String) { }
func login(email: String, password: String) { }
func login(forgotPasswordCode: String) { }
```

Listing 5.40 Overloaded Function Example

In this example,

- you can use the first version if you have a user name and password at hand;
- you can use the second version if you have an email address and password at hand; or
- you can use the third version if the user provides an account recovery code.

Neat, right? You could also use differing function names, like `loginWithUsername`, `login-WithEmail`, and `loginWithCode`, but taking advantage of function overloading makes the code arguably a bit tidier. After all, all the functions have a common goal: logging in.

Listing 5.41 demonstrates an extended example, in which the overloaded function returns a value. In this case, there are overloads for `areZodiacSignsCompatible`. The first variant would return the compatibility of the given zodiac signs, and the second variant would find the zodiac signs from the given dates and return their compatibility.

```
import Foundation

func areZodiacSignsCompatible(_ sign1: String, _ sign2: String) -> Bool {
    return true // Placeholder
}

func areZodiacSignsCompatible(_ date1: Date, _ date2: Date) -> Bool {
    return true // Placeholder
}
```

Listing 5.41 Overloaded Function Returning a Value

If you have overloaded functions with return values, then you must ensure that all variants return the same data type. In the preceding example, all variants of `areZodiacSigns-Compatible` must return a Boolean value—following the first variant as a template. Otherwise, Swift will generate an error and refuse to compile the app.

Another limitation centers on omitted argument labels. For example, consider Listing 5.42. Here, doesUserExit has two variants that accept a single string parameter, but the argument labels are different, username and emailAddress. In the main flow, Swift can spot the intended variant accurately due to the distinction of argument labels.

```
// Function
func doesUserExist(userName: String) -> Bool {
    return true // Placeholder
}

func doesUserExist(emailAddress: String) -> Bool {
    return true // Placeholder
}

// Main flow
doesUserExist(userName: "DarkNight")
doesUserExist(emailAddress: "darknight@example.com")
```

Listing 5.42 Valid Overloaded Function with Similar Data Types

However, Listing 5.43 is an invalid example that Swift will refuse to compile. Both variants of doesUserExist have omitted argument labels here. That makes it impossible to spot the intended variant in the main flow because both calls are passing a single string. Remember, Swift craves clarity!

```
// Function
func doesUserExist(_ userName: String) -> Bool {
    return true // Placeholder
}

func doesUserExist(_ emailAddress: String) -> Bool {
    return true // Placeholder
}

// Main flow
doesUserExist("DarkNight")
doesUserExist("darknight@example.com")
```

Listing 5.43 Invalid Overloaded Function with Omitted Argument Labels

In such a case, you can return to specific parameter names, as in Listing 5.42, or populate distinct functions, as in Listing 5.44.

```
// Function
func doesUserWithNameExist(_ userName: String) -> Bool {
    return true // Placeholder
```

```
}

func doesUserWithEmailExist(_ emailAddress: String) -> Bool {
    return true // Placeholder
}

// Main flow
doesUserWithNameExist("DarkNight")
doesUserWithEmailExist("darknight@example.com")
```

Listing 5.44 Populated Functions Instead of Overloading

So, which approach is better—function overloading or populating functions with alternative names? That's a controversial topic that could be discussed at length, but for our purposes, Table 5.6 summarizes some basic factors to nudge you in the right direction for your case.

Factor	Winner	Reason
Conciseness	Overloading	Callers have to deal with a smaller number of functions.
Unified Intent	Overloading	Overloading signals that all variants lead to the same action.
Scalability	Overloading	Overloading enables new functions without bloating the namespace.
Explicitness	Population	Distinct function names are more descriptive, documenting themselves.
Debugging	Population	Distinct function names make debugging a little easier.
Portability	Population	Not all languages support overloading.

Table 5.6 Overloading Versus Population

It's nice to have both options in the toolbox and have the freedom to pick the right tool for the right job, right? Another significant tool is the ability to change parameter values in functions, which you'll learn about in the next section.

5.7   Inout Parameters

In Swift functions, the inout keyword allows a function to modify a parameter's value. In the examples so far, we have either imported read-only parameters to a function or returned a brand-new value built in the function. inout features a middle ground, where you can read *and* change the value of a parameter.

Listing 5.45 demonstrates a simple function that accepts score as an inout parameter. Without the inout keyword, Swift would generate an error message indicating that score can't be mutated within the function. Adding the inout parameter tells Swift that you intentionally made score mutable, allowing the function code to change its value.

```
func addPoints(_ points: Int, score: inout Int) {
    score += points
}
```

Listing 5.45 Simple Function with inout Parameter

Now, let's see how this function is called. In Listing 5.46, the function has been called with the addPoints(20, score: &playerScore) statement, increasing the value of player-Score from 50 to 70.

```
// Function
func addPoints(_ points: Int, score: inout Int) {
    score += points
}

// Main flow
var playerScore = 50
print("Initial score: \(playerScore)") // Initial score: 50
addPoints(20, score: &playerScore)
print("Final score: \(playerScore)")   // Final score: 70
```

Listing 5.46 Calling Function with inout Parameters

You have undoubtably spotted the & token in addPoints(20, score: &playerScore); the variable has been mentioned as &playerScore instead of playerScore. That is not a decoration, but the required way to pass a variable to an inout function parameter.

The difference between the playerScore and &playerScore expressions is illustrated in Figure 5.18, with the former pointing to the value and the latter pointing to its memory address.

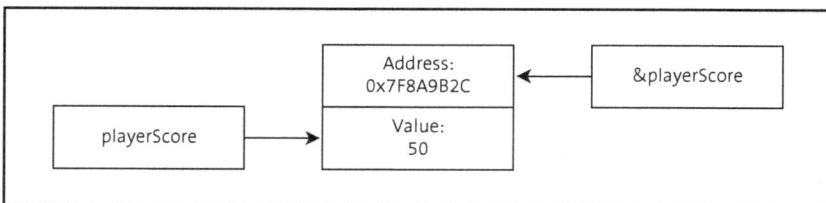

Figure 5.18 Difference Between playerScore and &playerScore Expressions

In Swift, function arguments are normally passed by *value*, meaning that the function gets a read-only copy of the passed argument. With inout, though, you are telling Swift to accept the *reference* of the variable instead—that is, its memory address. That way, the function can change the data at that memory location.

In simpler terms, &playerScore points to the memory address of playerScore, which is passed to the function so that it can change the value in that address. That's how we say, "Hey, I'm OK with this variable being modified—and here's its address."

If that's a bit too technical for you, don't fret! For the time being, you can just memorize the rule that inout parameters need to be passed with an & in front of them. After all, the syntax within the function doesn't change at all—except for adding the benefit that the inout variable is mutable.

As an additional example, Listing 5.47 features the strengthenPassword function, which accepts the password inout parameter. This function will add random characters to password until it reaches the length of 10, mutating it all along the way.

```
// Function
func strengthenPassword(_ password: inout String) {
    guard password.count < 10 else { return }

    while password.count < 10 {
        password += String(Int.random(in: 0...9))
    }
}

// Main flow
var pwd = "abcde"
strengthenPassword(&pwd)
print(pwd)  // abcde41067 (or similar)
```

Listing 5.47 Another Sample Function with inout Parameter

Finally, Listing 5.48 demonstrates a sample function that modifies an array in its body. The function can mutate the array because it was passed as an inout parameter.

```
// Function
func addInactiveUsers(_ users: inout [String]) {
    users.append("Bob")
    users.append("Mary")
}

// Main flow
var users = ["Emma", "Adam"]
addInactiveUsers(&users)
print(users) // Emma, Adam, Bob, Mary
```

Listing 5.48 Array as inout Parameter

Let's continue your journey with nested functions, which will allow you to declare functions inside functions—*Inception* style!

5.8   Nested Functions

Yeah, you heard right: Swift allows you to nest child functions into parent functions, not unlike a dream within a dream as in the *Inception* movie. In case you missed the movie, Listing 5.49 features a simple example in which the parent isNameValid function contains nested child functions isLongEnough, hasNoSpaces, and hasNoNumbers.

```
// Function
func isNameValid(_ name: String) -> Bool {
    func isLongEnough(_ name: String) -> Bool {
        return name.count >= 3
    }

    func hasNoSpaces(_ name: String) -> Bool {
        return !name.contains(" ")
    }

    func hasNoNumbers(_ name: String) -> Bool {
        return !name.contains { $0.isNumber }
    }

    return isLongEnough(name) && hasNoSpaces(name) && hasNoNumbers(name)
}

// Main flow
isNameValid("John")     // true
isNameValid("Bob12")    // false
isNameValid("Al ice")   // false
```

Listing 5.49 Nested Function Example

Following the child function declarations, the return statement is the actual body of isNameValid, which calls the child functions to decide if name is valid or not.

Note that the child isLongEnough, hasNoSpaces, and hasNoNumbers functions are only usable within the context of the parent isNameValid function. They are encapsulated and hidden from the main flow and other functions, as sketched in Figure 5.19.

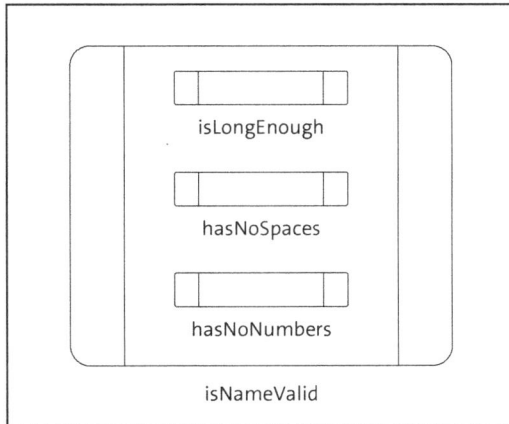

Figure 5.19 Encapsulation of Nested Functions

Although that may sound like a limitation, it's the exact purpose of nested functions! Someone reading your code can undoubtedly be sure that child functions are only used within their parents. That improves the readability of the code and keeps the name-space clean, preventing unintended use outside their context. But if you need to call the child functions in other contexts too, then you should declare them as standalone functions.

Alternatives for Nested Functions

If your parent function becomes bloated due to nested functions, it would be wise to reevaluate your code structure. In many cases, it could be preferable to declare a struct or class instead of a complex parent function as they offer many benefits like polymorphism, encapsulation, inheritance, abstraction, state persistence, extensibility, and testability.

You will learn about structs in Chapter 8 and classes in Chapter 9 and get to know such features well.

Having that said, nested functions are still a nice feature for quick, localized tasks like one-off validations or calculations hidden from the outside world.

So nested functions are protected and invisible from the outside world. But what about the encapsulation within the parent function? The simple answer: You can assume that the parent function has a main flow of its own and the child functions are normal functions, so to speak. Therefore, you can assume the following:

- Child functions can access variables of the parent function but not vice versa.
- Child functions can't access each other's local variables.
- Sibling functions can call each other.

To materialize the internal encapsulation logic, Listing 5.50 demonstrates a comprehensive function called makeRandomGreeting, which has two variables and three child functions.

```
// Function
func makeRandomGreeting(_ name: String) -> String {
    let intro1 = "Hello"
    let intro2 = "Hi"

    func getBaseGreeting() -> String {
        let intro = Int.random(in: 0...1) == 0 ? intro1 : intro2 // Parent
variables
        return "\(intro), \(name)!" // Uses parent parameter
    }

    func getQuestGreeting() -> String {
        let baseGreeting = getBaseGreeting( ) // Calls sibling function
        return "\(baseGreeting) How are you?"
    }

    func getExclGreeting() -> String {
        let baseGreeting = getBaseGreeting( ) // Calls sibling function
        return "\(baseGreeting) Great to see you!"
    }

    return Int.random(in: 0...1) == 0 ? getQuestGreeting( ) : getExclGreeting(
)
}

// Main flow
print(makeRandomGreeting("Alice"))
print(makeRandomGreeting("Bob"))
print(makeRandomGreeting("Charlie"))
```

Listing 5.50 Demonstration of Internal Encapsulation of Nested Functions

The purpose of this program is to generate random personalized greetings. A sample output is shown in Figure 5.20.

```
Hi, Alice! How are you?
Hello, Bob! How are you?
Hi, Charlie! Great to see you!
```

Figure 5.20 Some Random Greetings Generated by makeRandomGreeting Function

If you look closely at makeRandomGreeting, you'll see that child functions were able to access the parent function's variables and parameters as well as call each other. Table 5.7 summarizes the access map within the parent function.

Child Function	Access	Comment
getBaseGreeting	intro1	Parent function variable
getBaseGreeting	intro2	Parent function variable
getBaseGreeting	name	Parent function parameter
getQuestGreeting	getBaseGreeting	Sibling function
getExclGreeting	getBaseGreeting	Sibling function

Table 5.7 Access Map Within makeRandomGreeting

Nested functions offer a useful feature to enhance a regular function, enriching it with child functions. Another enhancing feature is the function type, which is the topic of the next section.

5.9   Function Types

Remember type aliases? In Chapter 2, you learned how to use type aliases for tuples. That way, you were able to standardize tuples like cookie cutters, ensuring that all matching tuples conform to the same template.

As further support for standardization, functions also can be forced to conform to a standard template using type aliases. This powerful feature adds immense flexibility to functional programming, empowering functions with polymorphism-like behavior without introducing objects.

But let's not get ahead of ourselves: In this section, we'll walk you step by step through this concept and demonstrate cases in which it would be useful.

5.9.1   Basic

For starters, let's get warmed up with a comfortable example without function types. Listing 5.51 features two simple functions that conduct math operations.

```
func add(_ a: Int, _ b: Int) -> Int {
    return a + b
}

func multiply(_ a: Int, _ b: Int) -> Int {
    return a * b
}
```

Listing 5.51 Two Simple Functions Conducting Math Operations

Although the functions are simple, they are also very similar. Both functions have the same signature: Importing two integers and returning an integer. This similarity opens the door for function standardization via type aliases (like in tuples). In Listing 5.52, the example was extended with `typealias MathFunction`, where the common signature of both functions has been expressed as a cookie-cutter template.

```
// Define a function type
typealias MathFunction = (Int, Int) -> Int

// Matching functions
func add(_ a: Int, _ b: Int) -> Int {
    return a + b
}

func multiply(_ a: Int, _ b: Int) -> Int {
    return a * b
}
```

Listing 5.52 Defining Type Alias for Common Function Signatures

Now that you have a standard type alias for both functions, you gain the ability to store either function in a variable. Thus, you can dynamically pick the function to execute, increasing the flexibility of the program.

That surely was a mouthful, so let's break it down and see it in action. Listing 5.53 extends the example with the main flow, which will be the focus now.

```
// Define a function type
typealias MathFunction = (Int, Int) -> Int

// Matching functions
func add(_ a: Int, _ b: Int) -> Int {
    return a + b
}
```

```
func multiply(_ a: Int, _ b: Int) -> Int {
    return a * b
}

// Main flow
var operation: MathFunction

operation = add
operation(10, 20)    // 30

operation = multiply
operation(10, 20)    // 200
```

Listing 5.53 Assigning Function to a Variable

At the top of the main flow, you make a variable declaration, var operation: MathFunction. That's the debut of the added flexibility. The operation variable can symbolize any function that conforms to the cookie-cutter template of MathFunction. So you can make operation symbolize the add or multiply function—because either conforms to Math-Function.

In the middle of the main flow, you make the operation = add assignment. From that point on, any mention of operation will be assumed as a mention of add. In the following line, operation(10, 20) is assumed as add(10, 20), calculating the result of 30—that is, operation became an alias for add.

At the end of the main flow, you make the operation = multiply assignment. From that point on, any mention of operation will be assumed as a mention of multiply. In the following line, operation(10, 20) is assumed as multiply(10, 20), calculating the result of 200—that is, operation became an alias for multiply.

Cool, right? With function signature standardization, you can juggle similar functions dynamically by assigning them to variables.

As a second example, Listing 5.54 features a mock-up game engine snippet in which orcs, goblins, or spiders are added to the map. Obviously, we don't have a real game here; the code simply prints the name of the enemy at this time.

```
// Function type
typealias EnemyFactory = (Int) -> Void

// Functions to add enemies to game
func addOrc(_ health: Int) {
    print("Orc added to map")
}
```

```
func addGoblin(_ health: Int) {
    print("Goblin added to map")
}

func addSpider(_ health: Int) {
    print("Spider added to map")
}

// Main flow: Add random enemies
var factory: EnemyFactory

for _ in 0..<5 {
    let rnd = Int.random(in: 0..<3)

    switch rnd {
    case 0: factory = addOrc
    case 1: factory = addGoblin
    default: factory = addSpider
    }

    factory(100)
}
```

Listing 5.54 Game Example Using Function Types

Let's break this down together. At the top, you have a type alias called EnemyFactory. This is a template for any function that will spawn a new enemy on the map. The -> Void suffix means that those functions won't return any value.

In the middle part, there are three functions conforming to this type alias: addOrc, addGoblin, and addSpider, each spawning a familiar Middle-earth creature.

In the bottom part, you have the main flow—where the magic happens. First you declare the factory variable of the EnemyFactory type. This means that factory can be used to symbolize any function conforming to EnemyFactory.

The code continues with a for loop. Within that loop, factory gets a random assignment of addOrc, addGoblin, or addSpider. Finally, factory(100) is used to call the assigned random function, spawning an enemy. Figure 5.21 shows a possible output of this sample program.

```
Spider added to map
Orc added to map
Goblin added to map
Goblin added to map
Orc added to map
```

Figure 5.21 Output of Randomly Spawned Enemies

So far, so good. You've learned that you can use function type aliases to make variables symbolize functions, but the real benefits will shine when you toss those functions around like you do variables—as parameters, for instance.

5.9.2   Function Types as Parameter Types

If you can make a variable contain a function, you should be able to pass a function like a parameter, right? It's true! And you're about to learn how! Start by inspecting Listing 5.55.

```
// Type alias
typealias Greeter = (String) -> String

// Greeters
func friendlyGreeter(_ name: String) -> String {
    return "Howdy, \(name)!"
}

func chattyGreeter(_ name: String) -> String {
    return "Hello, \(name)! How are you?"
}

func formalGreeter(_ name: String) -> String {
    return "Greetings, \(name)."
}

// Main function
func greet(name: String, greeter: Greeter) {
    print(greeter(name))
}
```

Listing 5.55 Passing Function as Parameter

In this example, you have a function type called Greeter, which can be used as a cookie-cutter template for other functions. The following functions, friendlyGreeter, chattyGreeter, and formalGreeter, conform to this template. So far, so good; there's nothing new here.

The main greet function is worth your utmost attention. It accepts two parameters: name, which is a string, and greeter, which is a Greeter function. This means that you can pass any of your Greeter functions as the greeter parameter! Therefore, greeter(name) will invoke the passed function.

To show this feature in action, Listing 5.56 extends the program with the main flow.

```
// Type alias
typealias Greeter = (String) -> String

// Greeters
func friendlyGreeter(_ name: String) -> String {
    return "Howdy, \(name)!"
}

func chattyGreeter(_ name: String) -> String {
    return "Hello, \(name)! How are you?"
}

func formalGreeter(_ name: String) -> String {
    return "Greetings, \(name)."
}

// Main function
func greet(name: String, greeter: Greeter) {
    print(greeter(name))
}

// Main flow
greet(name: "Alice", greeter: chattyGreeter)
greet(name: "Bob", greeter: formalGreeter)
greet(name: "Charlie", greeter: friendlyGreeter)
```

Listing 5.56 Adding Main Flow to Program

In the first main flow line, you pass the greeter: chattyGreeter parameter. In a sense, the greeter parameter becomes an alias for the chattyGreeter function. In the greet function, any time greeter is mentioned, the computer will call chattyGreeter instead. Therefore, print(greeter(name)) will execute as print(chattyGreeter(name)).

Likewise, formalGreeter and friendlyGreeter were passed as parameters in the following line. The overall output is shown in Figure 5.22, where people are greeted using the mentioned function.

```
Hello, Alice! How are you?
Greetings, Bob.
Howdy, Charlie!
```

Figure 5.22 Output with Alternative Greeting Styles

> **Dependency Injection**
>
> In object-oriented programming, there is a technique called *dependency injection*, in which a protocol/interface is passed as an argument. Passing a function as a parameter is a similar mechanism, increasing the flexibility in a similar way.
>
> If you aren't experienced in object-oriented programming, don't fret: There are upcoming chapters dedicated to that subject.

5

5.9.3   Function Types as Return Types

If you can use functions as import parameter types, you should naturally be able to use them as return types too, right? Right! To demonstrate that feature in a familiar environment, Listing 5.57 features an inverted version of the greeting program. This time, the getRandomGreeter() -> Greeter function randomly returns one of the Greeter functions.

```
// Type alias
typealias Greeter = (String) -> String

// Greeters
func friendlyGreeter(_ name: String) -> String {
    return "Howdy, \(name)!"
}

func chattyGreeter(_ name: String) -> String {
    return "Hello, \(name)! How are you?"
}

func formalGreeter(_ name: String) -> String {
    return "Greetings, \(name)."
}

// Main function
func getRandomGreeter() -> Greeter {
    let rnd = Int.random(in: 0..<3)

    if rnd == 0 {
        return friendlyGreeter
    } else if rnd == 1 {
        return chattyGreeter
    } else {
        return formalGreeter
    }
```

```
}

// Main flow
for name in ["Alice", "Bob", "Charlie"] {
    let greeter = getRandomGreeter()
    print(greeter(name))
}
```

Listing 5.57 Returning Function as Parameter

In the main flow, the `let greeter = getRandomGreeter()` expression receives and stores a random `Greeter` function in the `greeter` variable. Following that, `print(greeter(name))` will invoke the returned function. Cool, right? A generated sample output is shown in Figure 5.23.

```
Howdy, Alice!
Hello, Bob! How are you?
Hello, Charlie! How are you?
```

Figure 5.23 Output with Alternative Greeting Styles via Returned Function

In summary, this section taught you how to standardize functions using type aliases. This gives you the flexibility of making variables act as aliases for functions—and you can toss those variables/functions around like any other variables. They can act, for example, as function input or return parameters. You can also put them into collections, for instance—which you can try as an exercise.

The next section is about closures, which can be considered cousins of type-alias-standardized functions as they offer a similar mechanism.

5.10   Closures

Closures are a powerful feature in Swift; in a nutshell, they are self-contained blocks of code that can be passed around and executed later. In the sense that closures can act like executable variables, they are similar to function types—but though their functionality overlaps, their typical use cases differ.

Function types are ideal when you need a standardized signature for multiple functions that can substitute for each other in different cases. Each compliant function must conform to the function type like a protocol/interface.

Closures, on the other hand, are ideal when you need an immediate implementation without the necessity of a standardized signature. Although closures can conform to a function type, that's not necessarily their selling point.

This differentiation will solidify as we go through the closure examples in the following sections.

> **Lamba Functions**
>
> Those of you experienced in other programming languages can consider closures as Swift's version of lambda functions. Both are anonymous functions that can be passed around or stored. Closures offer the additional feature of capturing variables from their surrounding scope, which you'll discover in the upcoming examples.

5.10.1   Closure Expressions

A *closure expression* is the basic building block—an inline function with a specific syntax. To reveal the syntax of a closure, Listing 5.58 showcases a function and a closure, with both doing the exact same thing.

```
// Function
func applyDiscountFunc(price: Double) -> Double {
    let discount = price * 0.2
    return price - discount
}

// Closure
let applyDiscountClosure = { (price: Double) -> Double in
    let discount = price * 0.2
    return price - discount
}
```

Listing 5.58 Identical Pair: Function and Closure

As you can see, the bodies of the function and the closure are identical. Although the syntax of a closure's signature differs, it is intuitive as it's similar enough to a regular function. Now that you understand how to declare a closure, we can move forward to its specialties.

In most cases, Swift can infer the return type of a closure by looking at the return statements. Therefore, the -> Double expression in the closure signature can be omitted, shortening it as in Listing 5.59.

```
let applyDiscount = { (price: Double) in
    let discount = price * 0.2
    return price - discount
}
```

Listing 5.59 Closure with Omitted Return Type

Now that you have seized the closure in the `applyDiscount` variable, let's look at a simple way to execute it. Listing 5.60 features a demonstration in which the closure is executed like a regular named function.

```
// Closure definition
let applyDiscount = { (price: Double) in
    let discount = price * 0.2
    return price - discount
}


// Main flow
let originalPrice: Double = 100
let discountedPrice = applyDiscount(originalPrice)
print("Discounted price: \(discountedPrice)")    // 80
```

Listing 5.60 Executing Closure Like Named Function

So far, you've seen closure features that enable them to be used like functions. Now, check the example in Listing 5.61, in which a closure really shines as an anonymous, nameless code snippet.

```
// Function that takes a price and a closure to format it
func formatPrice(_ amount: Double, formatter: (Double) -> String) {
    let formatted = formatter(amount)
    print("Formatted price: \(formatted)")
}


// Main flow
formatPrice(29.99, formatter: { price in return "\(price) USD" } )  // 29.99 USD
formatPrice(25.88, formatter: { price in return "$\(price)" } )     // $25.88
```

Listing 5.61 Passing Closure as Function Parameter

In this example, check the function signature first. Here, `formatPrice` expects an amount as well as a closure (code snippet) called `formatter`, which supports the `(Double) -> String` signature. This means that the closure to be passed as `formatter` must import a double and return a string.

And that's exactly what happens in the main flow.

In the first call, you pass the `{ price in return "\(price) USD" }` closure. This closure imports a `price` value and returns it with the `" USD"` suffix. The call produces the output **29.99 USD**.

In the second call, you pass the `{ price in return "\$(price)" }` closure. This closure imports a `price` value and returns it with the `"$"` prefix. The call produces the output **$25.88**.

Higher-Order Function

In Swift terminology, higher-order functions are simply functions that take a function (closure) as an argument or return one. In the preceding example, formatPrice is a higher-order function because it imports another function in formatter.

Naturally, a closure can fulfill the expectation of a function type too, so long as it matches the template of the type alias. Listing 5.62 demonstrates the same example with a small twist: The signature of the formatter function parameter is standardized with a type alias now. Note that you can pass a closure instead of a named function.

```swift
// Type alias
typealias Formatter = (Double) -> String

// Function that takes a price and a closure to format it
func formatPrice(_ amount: Double, formatter: Formatter) {
    let formatted = formatter(amount)
    print("Formatted price: \(formatted)")
}

// Main flow
formatPrice(29.99, formatter: { price in return "\(price) USD" } )  // 29.99
USD
formatPrice(25.88, formatter: { price in return "$\(price)" } )    // $25.88
```

Listing 5.62 Closure Fulfilling Expectation of Function Type

Closures also can read and mutate variables in their environments, such as global variables. For functions, you had to declare the function with @MainActor to enable such behavior. Closures have this feature naturally with no extra effort needed.

Listing 5.63 demonstrates such an example, in which the riggedDice closure can access and mutate the riggedValue global variable. Every time the dice are rolled, riggedDice returns subsequent values. That's not good for a real game, but it can be a nice engine for unit tests.

```swift
// Variables
var riggedValue = 1

// Function
func rollDices(roller: () -> Int) -> [Int] {
    let dice1 = roller()
    let dice2 = roller()
    return [dice1, dice2]
}
```

```
// Closure
let riggedDice: () -> Int = {
    riggedValue += 1
    if riggedValue > 6 { riggedValue = 1 }
    return riggedValue
}

// Main flow
rollDices(roller: riggedDice)    // 2, 3
rollDices(roller: riggedDice)    // 4, 5
rollDices(roller: riggedDice)    // 6, 1
rollDices(roller: riggedDice)    // 2, 3
```

Listing 5.63 Closure Used as Rigged Dice

For a little fun and mental exercise, Listing 5.64 features the same example but with the niceDice closure passed as the dice roller. In this closure, you really generate a random number between 1 and 6, emulating an actual dice roll.

```
// Function
func rollDices(roller: () -> Int) -> [Int] {
    let dice1 = roller()
    let dice2 = roller()
    return [dice1, dice2]
}

// Closure
let niceDice: () -> Int = { return Int.random(in: 1...6) }

// Main flow
rollDices(roller: niceDice)
rollDices(roller: niceDice)
rollDices(roller: niceDice)
rollDices(roller: niceDice)
```

Listing 5.64 Closure Used as Nice Dice

The contrast between the last two examples has highlighted the flexibility of closures. You can have multiple different closures to roll dice and pass the appropriate one depending on the difficulty level. If this was a backgammon game, then

- at an easy level, you could use a closure producing some artificial dice rolls that benefit the player;
- at a medium level, you could use a closure producing totally random dice rolls; or
- at a hard level, you could use a closure producing some artificial dice rolls that benefit the computer.

Dependency Injection

As noted when we discussed function types, in object-oriented programming, there is a technique called *dependency injection*, in which a protocol/interface is passed as an argument. Passing a closure as a parameter, like passing a function as a parameter, is a similar mechanism, increasing the flexibility.

In an object-oriented environment, you could have a protocol/interface called Dice-Roller and class implementations called RiggedDiceRoller and NiceDiceRoller, which would be dependency-injected into the function. Closures feature a similar mechanism when objects are not present.

If you aren't experienced with object-oriented programming, it will be the core subject of upcoming chapters; such terminology will be clarified at that point.

Now that you've learned the basics of closures, we can move forward with some advanced features.

5.10.2   Trailing Closures

A *trailing closure* in Swift is a special syntax in which a closure passed as the last argument to a function can be written outside the function's parentheses. This is completely optional syntactic sugar, really. It makes the code cleaner and improves readability, especially for short, expressive closures.

Listing 5.65 demonstrates such an example, in which an imaginary SMS is being sent. The higher-order sendSMS function expects cleanser, which ought to be a closure to cleanse the phone number, ensuring its compliance for messaging.

```
import Foundation

// Function
func sendSMS(_ text: String, phone: String, cleanser: (String) -> String) {
    let cleansedPhone = cleanser(phone)
    print("Sending \(text) to \(cleansedPhone)")
}

// Main flow
sendSMS("Hi there", phone: "+13334455") { ph in
    return ph.replacingOccurrences(of: "+", with: "00")
}
```

Listing 5.65 Trailing Closure Example

Check the main flow here. Instead of passing the closure as the named cleanser parameter, you can place it at the end of the sendSMS call. That's definitely more readable than what's shown in Listing 5.66, where the closure has been passed as the named `cleanser` parameter.

```
import Foundation

// Function
func sendSMS(_ text: String, phone: String, cleanser: (String) -> String) {
    let cleansedPhone = cleanser(phone)
    print("Sending \(text) to \(cleansedPhone)")
}

// Main flow
sendSMS("Hi there", phone: "+13334455", cleanser: { ph in return
ph.replacingOccurrences(of: "+", with: "00")} )
```

Listing 5.66 Same Example Without Trailing Closure

Although their syntaxes differ, both code samples produce the same output, shown in Figure 5.24.

```
Sending Hi there to 0013334455
```

Figure 5.24 Output of SMS Closure Example

> **Trailing Closures in Collections**
>
> Fun fact: Back in Chapter 3, you used trailing closures for collections without knowing what they were! For example, the `bigNumbers = numbers.filter { $0 > 3 }` expression is actually a call to the `numbers.filter` function, where the `{ $0 > 3 }` trailing closure has been sent as an argument. From that point of view, the syntax makes more sense, right? As an exercise, you can go through that chapter and spot other similar trailing closures.

5.10.3 Escaping Closures

By default, when you pass a closure to a function, Swift assumes that it will be executed within the body of the function, before the function exists. That's an intuitive expectation, right? This is technically called a *nonescaping closure*.

However, sometimes you need a closure to *escape* the function's body, meaning that it's executed *after* the function returns. The typical case is a function performing an asynchronous task like a network call: You may want the closure to execute after the network call is completed. For such scenarios, Swift supports *escaping closures*.

Let's build an escaping closure example step by step for a solid understanding. We'll start with a regular higher-order function accepting a nonescaping closure for starters—as supplied in Listing 5.67.

```
func sampleFunc(closure: (String) -> Void) {
    print("Function started")
    closure("Closure executed")
    print("Function finished")
}
```

Listing 5.67 Sample Function Accepting Nonescaping Closure

There's nothing fancy so far—just a function with three lines! The top and bottom lines print outputs directly, while the middle line prints an output via printClosure. To solidify the example, Listing 5.68 adds a main flow to it.

```
// Function
func sampleFunc(closure: (String) -> Void) {
    print("Function started")
    closure("Closure executed")
    print("Function finished")
}

// Main flow
print("Main flow started")
sampleFunc() { (log: String) in print(log) }
print("Main flow ended")
```

Listing 5.68 Nonescaping Closure Supplied via Main Flow

The order of events will be printed out as shown in Figure 5.25. As an exercise, follow through the code and compare the flow to the output. This understanding is about to be challenged, so you should first understand how the nonescaping closure behaved in a straightforward way.

```
Main flow started
Function started
Closure executed
Function finished
Main flow ended
```

Figure 5.25 Output of Nonescaping Closure Example

So far, so good: You've built the framework for the real example. Now: What if the function contained an async operation? Check the extended code in Listing 5.69.

```
import Foundation

// Function
func sampleFunc(closure: (String) -> Void) {
    print("Function started")
    closure("Closure executed")
```

275

```
    DispatchQueue.main.asyncAfter(deadline: .now() + 2) {
        print("Async finished")
    }

    print("Function finished")
}

// Main flow
print("Main flow started")
sampleFunc() { (log: String) in print(log) }
print("Main flow ended")
```

Listing 5.69 Function with async Operation

This version adds an `async` code snippet to the higher-order function's body. Because you haven't learned about concurrency yet, you can overlook the details; the only important thing to understand is that `print("Async finished")` will be executed *asynchronously* (in a parallel thread) with a delay of two seconds.

> **Concurrency in Swift**
> Concurrency is a topic of its own, covering asynchronous operations, multithreading, and so on. This topic will be addressed in Chapter 15. For now, we will merely dip our toes into the water for the minimum functionality needed here.

The output of this version is displayed in Figure 5.26. Although **Async finished** was printed within `sampleFunc`, it was executed after the main flow finished—because we asked Swift to execute it with a delay of two seconds. Once again, you are encouraged to follow the code and match the flow with the output for a solid understanding.

```
Main flow started
Function started
Closure executed
Function finished
Main flow ended
Async finished
```

Figure 5.26 Output of Nonescaping Closure with async Code

Now, let's raise the stakes even higher: What if you needed the closure to execute within the `async` code block? Because the `async` block is clearly executed after the function has finished, it outlives the lifespan of the function. Thus the closure should also outlive the lifespan of the function.

Listing 5.70 is an incorrect attempt at `async` closure execution. Because the closure is nonescaping by default, it won't outlive the function and the `async` code block will fail—because the closure will no longer be "alive."

```
import Foundation

func sampleFunc(closure: (String) -> Void) {
    print("Function started")

    DispatchQueue.main.asyncAfter(deadline: .now() + 2) {
        closure("Closure executed")
        print("Async finished")
    }

    print("Function finished")
}
```

Listing 5.70 Incorrect Attempt at async Closure Execution

To make the closure outlive the function, you need to declare it as an escaping closure—
which means that the closure can "escape" the function, outliving its lifespan. Luckily,
this is very easy to do: You can achieve it simply by adding the @escaping attribute to
the closure definition, as shown in Listing 5.71.

```
import Foundation

func sampleFunc(closure: @escaping (String) -> Void) {
    print("Function started")

    DispatchQueue.main.asyncAfter(deadline: .now() + 2) {
        closure("Closure executed")
        print("Async finished")
    }

    print("Function finished")
}
```

Listing 5.71 Correct Attempt at async Closure Execution

With that function signature, the entire sample code looks like Listing 5.72. We snuck in
a secondary attribute called @Sendable to solidify the code, but you can ignore that for
now; it's part of the concurrency topic, to be addressed in Chapter 15.

```
import Foundation

// Function
func sampleFunc(closure: @escaping @Sendable (String) -> Void) {
    print("Function started")

    DispatchQueue.main.asyncAfter(deadline: .now() + 2) {
```

```
        closure("Closure executed")
        print("Async finished")
    }

    print("Function finished")
}

// Main flow
print("Main flow started")
sampleFunc() { (log: String) in print(log) }
print("Main flow ended")
```

Listing 5.72 Entire Escaping Closure Example

It's time to let the drums roll and execute the final example! As an exercise, you can follow through the code flow and guess the output it will produce. When you're done, compare your results with Figure 5.27.

```
Main flow started
Function started
Function finished
Main flow ended
Closure executed
Async finished
```

Figure 5.27 Output of Final Escaping Closure Example

See what we achieved here? We were able to sneak the closure execution into the async code block by making the closure outlive the function! The **Closure executed** and **Async finished** outputs came after **Main flow ended**.

This technique is particularly useful when you need a chunk of code to stick around after a function finishes. Some sample use cases are listed in Table 5.8.

Topic	Case	Escaping Closure Solution
Networking	HTTP requests typically complete with an unforeseeable delay.	You can sneak an escaping closure in to be executed after the request is completed.
Timers	Some actions need to be executed after the timer finishes.	Delay actions can be passed as an escaping closure to be executed after the timer.
GUI events	You might have actions to be executed after a user input, like a button click.	Such callback codes can be passed as an escaping closure to be executed on the event.

Table 5.8 Some Escaping Closure Use Cases

5.10.4   Autoclosures

Finally, it's time to learn about autoclosures in Swift—which are much simpler than escaping closures, we promise! An *autoclosure* is a special kind of closure that's automatically created from an expression passed to a function. You won't need to explicitly write a full closure within the {...} block.

What makes an autoclosure special is that it lets you delay evaluation. The code inside won't be run until the closure is called. Delaying evaluation is useful in some cases—such as inevitably slow code with performance costs.

Let's see it all in an example! Check the code snippet in Listing 5.73.

```
// Functions
func isOverBudget(_ spending: Double, budget: @autoclosure () -> Double) ->
Bool {
    guard spending > 10 else { return false }
    return spending > budget()
}

func getBudget() -> Double {
    let expenses = [50.0, 10.0, 40.0] // Pretend those came from a database
    return expenses.reduce(0, +)
}

// Main flow
isOverBudget(15, budget: getBudget()) // false
```

Listing 5.73 Autoclosure Example

Let's inspect isOverBudget first—which is a regular higher-order function. The only syntactical difference is that the budget closure is marked as @autoclosure; keep that in mind. Within the function, if the spending is less than 10, the function exits without even checking budget(). Otherwise, it checks if the spending is over budget()—and that's where the closure is addressed.

In the main flow, you pass the value 15 as the spending amount and the getBudget() function as the budget closure. This has two advantages:

- You didn't have to build a closure for budget. Swift has accepted the getBudget() expression, automatically converting it to a closure.
- getBudget() is passed as an argument without being executed. It will only be executed if the isOverBudget function calls the closure. Therefore, the expensive database operation is not invoked unless absolutely needed.

In that sense, autoclosure features a mechanism for lazy evaluation. Cool, right?

This concludes our content on closures, which add immense flexibility to functions by passing code snippets as arguments. Now we will continue with generics, which will extend this flexibility even further by using placeholder types instead of specific ones.

5.11   Generic Functions

Generic functions is a powerful Swift feature that allows you to write flexible and reusable code. Instead of repeating similar but slightly different code blocks, you can generalize them with help of generics. In this section, we'll go over examples demonstrating generics and highlighting their use cases.

5.11.1   Generic Functions

To understand generic functions, let's start with a case study in which two mostly similar but slightly different functions are present. Listing 5.74 demonstrates swapInts and swapStrings for that purpose.

```
// Functions
func swapInts(_ a: inout Int, _ b: inout Int) {
    let tmp = a
    a = b
    b = tmp
}

func swapStrings(_ a: inout String, _ b: inout String) {
    let tmp = a
    a = b
    b = tmp
}

// Main flow
var name1 = "John"
var name2 = "Jane"
swapStrings(&name1, &name2)
print("name1: \(name1)")     // name1: Jane
print("name2: \(name2)")     // name2: John
```

Listing 5.74 Two Similar Functions for Different Types

If you look closely, both functions perform a similar task: swapping the values of given variables. The difference is that swapInts swaps integers, while swapStrings swaps strings (obviously), as demonstrated in the main flow.

Because it has a strict type system for code security and reliability, Swift enforces declaring types of function parameters in most cases—which might lead to a point where you have to duplicate very similar code blocks due to type differences, as in swapInts and swapStrings.

To mitigate such cases, Swift offers generic functions, in which you can merge such similar functions into singular functions with abstract data types. As a follow-up to the previous example, Listing 5.75 demonstrates the generic swapValues function, which can fulfill the task of both swapInts and swapStrings.

```
func swapValues<T>(_ a: inout T, _ b: inout T) {
    let tmp = a
    a = b
    b = tmp
}
```

Listing 5.75 Merged Functions as Generic Function

Note that the body of swapValues is exactly the same as swapInts and swapStrings, but its signature is a bit unusual. Let's break it down.

<T> is a generic type parameter. It's a placeholder for a type that will be determined when the function is called. You can imagine it like a blank spot to fill later with a concrete type, such as Int or String. Following that, parameters a and b are also declared as generic type T. Here you're telling Swift that you don't know the type of a and b yet, but whatever the type is, it must be the same for both.

Before this gets too complex, let's check the main flow of Listing 5.76 to see how the generic function is called.

```
// Function
func swapValues<T>(_ a: inout T, _ b: inout T) {
    let tmp = a
    a = b
    b = tmp
}

// Main flow
var name1 = "John"
var name2 = "Jane"
swapValues(&name1, &name2)   // T becomes String
print(name1, name2)          // Jane, John

var grade1 = 100
var grade2 = 200
```

```
swapValues(&grade1, &grade2)   // T becomes Int
print(grade1, grade2)          // 200, 100
```

Listing 5.76 Generic Function Call

In the first part, you call swapValues with two strings. Swift cleverly figures out that the generic type <T> should be evaluated as String for this call and executes swapValues accordingly—just like it would execute swapStrings. In the end, two strings are swapped.

In the second part, you call swapValues with two integers. Swift cleverly figures out that the generic type <T> should be evaluated as Int for this call and executes swapValues accordingly—just like it would execute swapInts. In the end, two integers are swapped.

Cool, right? We were able to merge two functions into one by abstracting their parameter types.

In the last example, the generic function featured only one type, swapValues<T>. However, Swift is flexible enough to allow generic functions that feature multiple types too—as demonstrated in Listing 5.77.

```
// Function
func buildTuple<T, U>(_ first: T, _ second: U) -> (T, U) {
    return (first, second)
}

// Main flow
let numbers = buildTuple(10, 15)    // (10, 15)
let letters = buildTuple("a", "b")  // ("a", "b")
```

Listing 5.77 Generic Function Featuring Multiple Types

In this example, buildTuple features two types instead of one, expressed as <T, U>. It will accept the first parameter of the T type (whatever it might be) and the second parameter of the U type (whatever it might be). The function is simple; it does nothing more than build a tuple—but it can build a tuple out of any pair of data types, as demonstrated in the main flow.

The basic form of a generic function offers a certain degree of flexibility. However, Swift will behave conservatively about what you can do in the function's body. In the preceding example, you can't perform a calculation like first + second because Swift doesn't know whether these are strings, numbers, dates, or something else.

To work around that limitation and gain a little more flexibility, you can use type constraints in generic functions—which is our next topic.

5.11.2   Type Constraints

In the preceding generic function examples, the functions conducted basic operations that are available for any data type—like swapping their values or grouping them into a tuple. In many cases, though, you will want to be a little more specific with data types to enable core operations while still being vague enough to keep type flexibility.

For such cases, you can declare <T> using an umbrella type of constraint (such as Numeric) that covers all corresponding subtypes underneath it (such as Int, Double, and Float). That would enable arithmetic operations available for any Numeric type because Swift can ensure that a non-numeric argument won't be passed to the function.

We will go over some common type constraints and examples in the following sections to solidify this approach.

Numeric

Let's start with numeric type constraints. Listing 5.78 demonstrates a basic function for that.

```
func calcSum<T: Numeric>(_ a: T, _ b: T) -> T {
    return a + b
}
```

Listing 5.78 Basic Numeric Type Constraint Function

As you can see, the function signature didn't deviate much from the former syntax. All we did was to throw in a type constraint, <T: Numeric>. This tells Swift that the function can be called for any numeric type. In return, Swift enables all common numeric operations to be used in the function's body, such as math calculations. Figure 5.28 illustrates this idea for a better understanding.

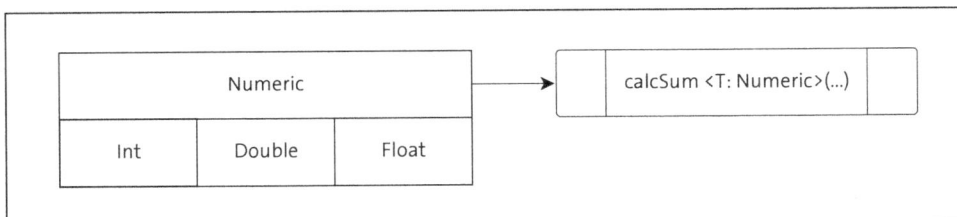

Figure 5.28 Numeric Type Constraint Visualized

If you defined the function merely as calcSum<T> (without the Numeric type constraint), then math operations in the function's body wouldn't be allowed. Swift wouldn't be able to ensure whether the caller passed a number or an incorrect argument unsuitable for math operations (like a String).

Speaking of callers, Listing 5.79 shows the extended code with a couple of callers.

```
// Function
func calcSum<T: Numeric>(_ a: T, _ b: T) -> T {
    return a + b
}

// Main flow
let a: Int = 10
let b: Int = 20
calcSum(10, 20)      // 30

let c: Double = 3.5
let d: Double = 4.5
calcSum(c, d)        // 8
```

Listing 5.79 Numeric Type Constraint Caller Example

As expected, calcSum works with Int and Double now because both types fall under the umbrella of Numeric. Cool, right? The function signature is still flexible, and you have more empowerment in the function's body.

Now that you understand the core logic of type constraints, let's move on to other constraint types.

Collection

In Swift, the Collection protocol is a common abstraction for types representing a sequence of elements. You learned about collections in Chapter 3. Arrays, dictionaries, sets, and ranges fall under the Collection umbrella.

You can use Collection as a type constraint too, as demonstrated in Listing 5.80. In this example, summarizeCollection is a function to detect and return properties of a Collection as a string.

```
// Function
func summarizeCollection<T: Collection>(_ coll: T) -> String {
    // Basic properties
    let count = coll.count
    let isEmpty = coll.isEmpty

    // Build a string representation of up to 3 elements
    var desc = "Elements: "
    desc += isEmpty ? "none" : coll.prefix(3).map { "\($0)" }.joined(separator:
", ")
    if count > 3 { desc += ", ... (\(count - 3) more)" }
```

```
    // Compile summary
    return """
    Count   : \(count)
    Is Empty: \(isEmpty)
    Elements: \(desc)
    """
}

// Main flow
let numbersArray = [10, 20, 30, 40, 50]
let arraySummary = summarizeCollection(numbersArray)
print("Array Summary:\n\(arraySummary)\n")

let namesSet: Set<String> = ["Alice", "Bob", "Charlie"]
let setSummary = summarizeCollection(namesSet)
print("Set Summary:\n\(setSummary)")
```

Listing 5.80 Collection Type Constraint Demonstration

In this sample code, summarizeCollection has a parameter called coll, which can be any Collection. This enables you to access common Collection components, such as coll.count, coll.isEmpty, and coll.Prefix. That makes sense, right? Both arrays and sets have those components available, so therefore you can access them.

In the main flow in the sample, you take advantage of that flexibility by calling the function with an array as well as a set. Each qualifies as a Collection. The output of the example is shown in Figure 5.29, which proves that the function worked perfectly in both cases!

```
Array Summary:
Count   : 5
Is Empty: false
Elements: Elements: 10, 20, 30, ... (2 more)

Set Summary:
Count   : 3
Is Empty: false
Elements: Elements: Bob, Alice, Charlie
```

Figure 5.29 Output of Collection Type Constraint Demonstration

Protocols as Type Constraints

In the preceding examples, both Numeric and Collection were *built-in protocols*—also known as *interfaces* in some other languages. Just like you used those built-in protocols as type constraints, you can use your custom protocols as type constraints too. This topic will be covered in Chapter 10.

Equatable

This protocol compares data types, which can be tested for equality via == or inequality via !=. And that's a really broad definition: Many built-in data types are intuitively Equtable. This simple feature is demonstrated in Listing 5.81.

```
// Function
func areEqual<T: Equatable>(_ a: T, _ b: T) -> Bool {
    return a == b
}

func areDifferent<T: Equatable>(_ a: T, _ b: T) -> Bool {
    return a != b
}

// Main flow
areEqual(1, 1)          // true
areDifferent("x", "x")  // false
```

Listing 5.81 Equatable Type Constraint Demonstration

As expected, you can call the areEqual and areDifferent functions with different data types supporting the Equtable protocol.

Are Equatables Useful?

You may be wondering why an Equatable type constraint is used at all. In the preceding example, you could have compared 1 == 1 or "x" != "x" directly. Such a doubt is understandable in this simple example, but the value of this approach will be clear shortly when you learn about generic where clauses in Section 5.11.3.

The same applies to the next topic, Comparable.

Comparable

The Comparable protocol is an extended version of Equatable. On top of the == and != operators, Comparable supports further comparison operators, like <, >, <=, and >=. Because you've already learned about those operators, Listing 5.82 should be an intuitive example.

```
// Function
func isBigger<T: Comparable>(_ a: T, _ b: T) -> Bool {
    return a > b
}

func isSmaller<T: Comparable>(_ a: T, _ b: T) -> Bool {
    return a < b
```

```
}

// Main flow
isBigger(2, 1)        // true
isSmaller("x", "x")  // false
```

Listing 5.82 Comparable Type Constraint Demonstration

As expected, isBigger and isSmaller can be used with any data type supporting the Comparable protocol. In this example, we have called them with integers and strings.

Hashable

Another sample protocol for type constraints is Hashable. In Chapter 3, you learned about the hash-based Set and Dictionary collection types. Any data type that can be put into such collections must conform to the Hashable protocol. That makes sense, right? If you can't hash a value, you can't put it into a Set in the first place.

Many core data types, such as Int and String, are already Hashable. That leads to the conclusion that you can use the Hashable type constraint to write flexible functions that would work for any Set, for instance.

Case in point: Listing 5.83 demonstrates a code example in which the appendIfMissing function can be called for any Set—not just a Set of a specific data type. This function will check if the set argument already contains the given element. If not, element will be inserted into set.

```
// Function
func appendIfMissing<T: Hashable>(_ set: inout Set<T>, _ element: T) {
    guard !set.contains(element) else { return }
    set.insert(element)
}

// Main flow
var intSet: Set<Int> = []
appendIfMissing(&intSet, 1)
appendIfMissing(&intSet, 1)
appendIfMissing(&intSet, 2)
print(intSet)  // 1, 2 (order varies)

var strSet: Set<String> = []
appendIfMissing(&strSet, "a")
appendIfMissing(&strSet, "a")
appendIfMissing(&strSet, "b")
print(strSet)  // a, b (order varies)
```

Listing 5.83 Hashable Type Constraint Demonstration

In the main flow, you were able to call the same function for a set of integers as well as a set of strings, demonstrating the flexibility of this option.

5.11.3 Generic Where Clauses

In the previous section, you used various type constraints to improve the flexibility of generic functions. For instance, you can use the <T: Collection> type constraint to imply that the function can be executed for any Collection type, such as an array or set.

However, in this instance, imagine that you don't have much control over the elements of the collection. The caller might have passed a string array, a dictionary array, a tuple array, or maybe even an array of a crazy custom objects: Who knows? Due to that uncertainty, Swift won't let you perform any serious operation with collection elements in the function body.

But what if there was a way to declare a type constraint on the collection's elements? If that was possible, you would have more flexibility to play around with the elements too, right?

As you probably have guessed already, Swift does in fact give you this flexibility—via generic where clauses. This is easier demonstrated than explained, so let's jump right into Listing 5.84. This example features the isSorted function, which is supposed to determine if the given collection is sorted or not.

```
// Function
func isSorted<T: Collection>(_ coll: T) -> Bool where T.Element: Comparable {
    for i in 0..<(coll.count - 1) {
        let current = coll[coll.index(coll.startIndex, offsetBy: i)]
        let next = coll[coll.index(coll.startIndex, offsetBy: i + 1)]
        if current > next {
            return false
        }
    }
    return true
}


// Main flow
isSorted(["a", "b", "c"])   // true
isSorted([3, 2, 1])         // false
```

Listing 5.84 Generic where Clause Demonstration

The function's signature prefix is familiar: isSorted<T: Collection>(_ coll: T) -> Bool declares that the function expects coll as a parameter, which can be any type of Collection (array, set, etc.).

However, there is a newcomer in the signature suffix. The `where T.Element: Comparable` expression declares that each element of the collection must be comparable. This enforces `coll` to contain comparable variables (strings, integers, etc.) instead of weird incomparable objects. Now that the scope of `coll` is narrowed down, Swift will let you use comparison operators for `coll` elements like `==`, `!=`, `<`, `>`, `<=`, and `>=`.

This flexibility is already exploited in the main flow, where you are able to call `isSorted` for a string array as well as an integer array. Because both `String` and `Integer` are `Comparable` types, Swift happily accepts them as arguments and lets the function do its job.

5.12   Summary

This chapter was all about subroutines. You learned how to declare and call functions, which are executable and reusable code blocks. Functions can import and return parameters too, increasing their flexibility. It is also possible to overload the same function with multiple signatures and to create nested functions.

On the advanced side, you learned about standardizing functions as cookie-cutter templates using function types, making them interchangeable. Closures enable you to declare quick in-line functions, while generics allow you to flex Swift's strict type system for function parameter types.

If you're a beginner, you might not need those advanced features immediately. Many apps can be developed without them—but they surely ease the lives of programmers. If you found them a bit confusing, you can return to this chapter when you have more experience and revisit the topics for a better understanding.

The next chapter will address another core feature of Swift: optionals. That will be our last prerequisite stop before moving forward to advanced structures like enums, structs, and classes.

Chapter 6
Optionals

Optionals are a signature feature of the Swift language, through which the possibility of a variable having an empty value is managed via simple operators.

Swift's optionals provide a type-safe way to handle the presence or absence of a value. They prevent unsafe access to null values, which might otherwise crash an app easily. Optionals are simple and easy to learn, but they are one of the cornerstones of Swift's safety features.

In this chapter, you will learn about optionals, their syntaxes, and their features via sensible examples—as usual.

6.1 What Are Optionals?

In previous chapters, we went over many topics dealing with values. You know by now that variables and collections can hold values, while functions can import, calculate, and return values. Your intuition might trick you into a false expectation that each variable will definitely hold a value, but that might not always be the case.

Optionals enable you to mark variables as optional, giving you (and the compiler) a standardized way of dealing with cases in which they might be empty. In this section, we'll go over optionals-related examples of different Swift components for a good understanding.

6.1.1 Tuple Optionals

In Listing 6.1, you declared a simple tuple called Person, which holds someone's name and the brand of their car.

```
typealias Person = (name: String, carBrand: String)
```

Listing 6.1 A Simple Tuple

Here is the problem: Both name and carBrand are obligatory variables in this tuple. You can't declare a Person without providing values for both. But realistically speaking, can you assume that every person has a car? Do you exclude people without a car from the app? Of course not: It's better to be inclusive.

This leads to the result that carBrand should be an optional variable in this tuple. For people with cars, you can assign a value to carBrand. For people without cars, you should be able to leave it empty. That makes sense, right?

This is exactly what optionals are for. By applying the ? token, you can mark a variable as optional in its context. In Listing 6.2, the carBrand variable is an optional because its type has been defined as String?, instead of the vanilla String. The question mark suffix flags carBrand as an optional variable.

```
typealias Person = (name: String, carBrand: String?)
```

Listing 6.2 Simple Tuple with Optional

Now let's declare some variables of the Person type. In Listing 6.3, p1 has values for both variables (name and carBrand). Therefore, declaration of p1 is done in the usual way. However, p2 has only a name value—no carBrand. In that case, you can use the nil keyword to keep carBrand without a value.

```
typealias Person = (name: String, carBrand: String?)

let p1: Person = ("John", "Toyota")
let p2: Person = ("Mary", nil)
```

Listing 6.3 Usage of nil for Optionals

Easy, right? The ? token marks a variable as optional, and nil can be used to set no value for an optional variable.

Nil Versus Empty Value

Note that nil doesn't mean an *empty* value; instead, it means *no* value. There is a difference. If you make the assignment carBrand = "", it is no longer empty; it holds the value of a String with no characters. If you make the assignment carBrand = nil, though, it becomes truly empty: uninitialized, with no value.

The same idea exists in other languages too. SQL, Java, and JavaScript call it null, while Python calls it None.

Now you know how to declare optional variables and (optionally) assign values to them. But how can you read them safely? If you attempt to access an optional value directly as in Listing 6.4, that would be an unsafe operation. Because carBrand might be nil (and is nil for p2), it's risky to access it while assuming that it will contain a value.

```
typealias Person = (name: String, carBrand: String?)
let p2: Person = ("Mary", nil)
print(p2.carBrand)
```

Listing 6.4 Unsafe Access to Optional Variable

The easiest way to get to the safe side is to use the nil-coalescing operator ??, which you used in previous chapters. This operator enables you to provide a fallback value in case the primary variable is nil. A demonstration is shown in Listing 6.5, where you provide the fallback value "No car" while printing carBrand.

```
typealias Person = (name: String, carBrand: String?)

let p1: Person = ("John", "Toyota")
let p2: Person = ("Mary", nil)

print(p1.carBrand ?? "No car")  // Toyota
print(p2.carBrand ?? "No car")  // No car
```

Listing 6.5 Usage of Nil-Coalescing Operator

For p1, the output will be **Toyota** because p1.carBrand has that value. For p2, the output will be **No car** because p2.carBrand is nil. Easy, right? Here's a quick token summary: ? is used to declare an optional value and ?? is used to declare a default fallback value.

Defensive Programming

When accessing an optional value, some measure of defensive programming is strongly recommended. An app crashing due to a nil value is not a nice sight. Swift will guide you to such defensive measures via warnings and compilation errors anyway. Providing a default fallback value with ?? is the least you can do. Other options will be highlighted in upcoming sections.

6.1.2 Simple Variable Optionals

Although we made a head-start with tuples, simple standalone variables naturally can also be optionals. The syntax and general rules won't change at all. As a demonstration, Listing 6.6 features a small app that splits the user's name into parts.

```
var firstName: String?
var midName: String?
var lastName: String?

print("Enter your name:", terminator: " ")
let fullName = readLine() ?? ""
```

```
let splitNames = fullName.split(separator: " ")

switch splitNames.count {
case 1:
    firstName = String(splitNames[0])
case 2:
    firstName = String(splitNames[0])
    lastName = String(splitNames[1])
case 3:
    firstName = String(splitNames[0])
    midName = String(splitNames[1])
    lastName = String(splitNames[2])
default:
    break
}

print("Your first name: \(firstName ?? "")")
print("Your middle name: \(midName ?? "")")
print("Your last name: \(lastName ?? "")")
```

Listing 6.6 Splitting Name Input into Optional Variables

Here is the tricky point: Although you might expect values for `firstName`, `midName`, and `lastName`, you can't guarantee what kind of input will be provided by the user. Therefore, you must declare those standalone variables as optionals. Table 6.1 demonstrates how different user inputs will be reflected with optional variables.

Input	firstName	midName	lastName
(empty)	nil	nil	nil
John	John	nil	nil
John Doe	John	nil	Doe
John David Doe	John	David	Doe

Table 6.1 Sample User Inputs and Assignment to Optionals

In the last part, you print each optional with the `?? ""` suffix. If the optional is `nil`, then the fallback value of an empty string will be printed—which is much better than an application crash, right? Figure 6.1 features a sample output for the case in which the user typed "John Doe". Because there is no middle name, the fallback value of "" has been printed instead.

```
Enter your name: John Doe
Your first name: John
Your middle name:
Your last name: Doe
```

Figure 6.1 Sample Output, Missing Middle Name

6.1.3 Collection Optionals

Collections support optionals too! Now that you have the hang of it, we can directly jump to the example in Listing 6.7.

```
let myArray: [String?] = ["Hello", nil, "World"]

for entry in myArray {
    print(entry ?? "xxx")
}
```

Listing 6.7 Array with Optionals

In this example, the array type has been declared as [String?], which means that each entry in the array is an optional String. The value assignment also matches the declaration: There is a nil value in the middle.

As you loop through the array to print values, each entry might be nil. Therefore, you can provide the fallback default value of "xxx" in the print statement. Whenever the computer encounters a nil value, it will print "xxx" instead—as shown in Figure 6.2.

```
Hello
xxx
World
```

Figure 6.2 Array with Optionals Output

6.1.4 Function Optionals

Functions are one of the most significant Swift elements that support optionals. Because functions can both import and return values, they support optionals on both fronts.

Listing 6.8 demonstrates the sumUser function, which builds a short string out of user data. Because you might not know the gender and location of each user, those parameters have been marked as optional.

```
func sumUser(name: String, gender: String?, location: String?) -> String {
    var sum = name
    sum += ", " + (gender ?? "N/A")
    sum += ", " + (location ?? "N/A")
    return sum
```

```
}

sumUser(name: "John", gender: "male", location: "Home") // John, male, Home
sumUser(name: "Jane", gender: nil, location: "Work")    // Jane, N/A, Work
sumUser(name: "Bob", gender: "male", location: nil)     // Bob, male, N/A
```

Listing 6.8 Sample Function Importing Optional Parameters

The function's body has the usual approach: A default fallback value of "N/A" has been provided for cases in which the optional arguments might be nil. The output of each sumUser call is built accordingly.

Functions can return optional values too. In Listing 6.9, a function spots and returns the number disrupting a supposedly sorted number sequence. But if all numbers are ordered correctly, then the function would have nothing to return. Therefore, the return type is declared as Int?, indicating that the function might or might not return a value.

```
// Function
func getUnsortedValue(_ numbers: Int...) -> Int? {
    for i in 1..<numbers.count {
        if numbers[i] < numbers[i-1] { return numbers[i-1] }
    }

    return nil
}

// Main flow
getUnsortedValue(1, 2, 13, 4)   // 13
getUnsortedValue(1, 2, 3, 4)    // nil
```

Listing 6.9 Sample Function Returning Optional Value

In the main flow, getUnsortedValue(1, 2, 13, 4) returns the value 13 because it disrupts the number order. On the other hand, getUnsortedValue(1, 2, 3, 4) returns nil because all numbers are ordered correctly; the function has nothing to return.

6.2   Optional Binding

In Swift, *optional bindings* enable you to safely unwrap an optional value using a condition. In this common technique, if the optional contains a value, then the code in the condition is executed; otherwise, it is not. Optional bindings are typically used in conjunction with if and guard statements.

Listing 6.10 features a simple demonstration using an `if` condition. Note that `midName` is an optional.

```
var name: String
var midName: String?
var surname: String

// No mid name
name = "John"
surname = "Doe"

if let mNam = midName {
    print("\(name) \(mNam) \(surname)")
} else {
    print("\(name) \(surname)")
}
```

Listing 6.10 Optional Binding with if Statement

The `if` statement is where the magic happens. `if let mNam = midName` means

- if `midName` is not `nil`, then assign its value to `mNam` and execute the following {...} code;
- otherwise, execute the `else` {...} code.

Neat, right? The output of this code sample will be **John Doe** because the optional `midName` is `nil`; no value was assigned to it. The `else` {...} code will be executed.

Whereas the output of Listing 6.11 will be **John Michael Doe** because the optional `midName` has a value now. The immediate {...} code will be executed, including `midName` in the output.

```
var name: String
var midName: String?
var surname: String

// Mid name
name = "John"
midName = "Michael"
surname = "Doe"

if let mNam = midName {
    print("\(name) \(mNam) \(surname)")
} else {
    print("\(name) \(surname)")
}
```

Listing 6.11 Optional Binding with midName Value

297

A common use case for optional binding is functions with optional parameters. In the function's body, you can use optional binding to determine if an argument is nil or not and act accordingly. Listing 6.12 demonstrates the buildEmailDict function, which builds a dictionary out of arguments. Note that cc and subj are optional parameters.

```
typealias EMailDict = Dictionary<String, String>

// Function
func buildEmailDict(_ to: String, _ body: String, cc: String? = nil, subj:
String? = nil) -> EMailDict {
    var result = Dictionary<String, String>()

    result["to"] = to
    result["body"] = body

    if let resultCC = cc {
        result["cc"] = resultCC
    }

    if let resultSubject = subj {
        result["subject"] = resultSubject
    }

    return result
}

// Main flow
print(buildEmailDict("hi@sample.com", "Hello"))
print(buildEmailDict("hi@sample.com", "Hello", cc: "info@sample.com"))
```

Listing 6.12 Optional Binding with Optional Function Parameters

The if let statements in the function's body enable you to handle optional arguments easily. If cc has a value, it is added to the dictionary. The same applies to subj.

The output of the main flow is shown in Figure 6.3. In the first call, neither cc nor subj is included in the dictionary because both were nil. In the second call, cc was included but not subj because subj was nil again.

```
["body": "Hello", "to": "hi@sample.com"]
["body": "Hello", "to": "hi@sample.com", "cc": "info@sample.com"]
```

Figure 6.3 Produced Email Dictionaries

Using optional bindings with guard conditions is also a common practice. Listing 6.13 demonstrates a function importing an array with optional strings.

```
// Function
func printValidNames(_ names: [String?]) {
    for name in names {
        guard let validName = name else {
            print("Skipping nil name")
            continue
        }
        print("Processing: \(validName)")
    }
}


// Main flow
let names = ["Alice", nil, "Bob", nil]
printValidNames(names)
```

Listing 6.13 Optional Bindings with Guard

In the function's body, `guard let validName = name` works the way you would expect. If `name` is `nil`, then the immediate code in the {...} block will be executed, skipping that empty `name`. Otherwise, the function body will continue, printing out the `validName`.

In the main flow, we have deliberately passed an array with some `nil` values. The output will appear as shown in Figure 6.4, proving that the optional binding of the `guard` statement has worked as expected.

```
Processing: Alice
Skipping nil name
Processing: Bob
Skipping nil name
```

Figure 6.4 Output of Optional Bindings with Guard

6.3   Implicitly Unwrapped Optionals

Although it's dangerous, it is an occasional necessity to implicitly unwrap an optional. Using this technique, you can tell Swift that it can assume an optional to contain a value and behave accordingly, risking a runtime crash if accessed when `nil`.

Let's start discovering this feature with the regular case in Listing 6.14. Here you have the optional `name` variable. Somewhere along the way, you assign the value "Alice" to it—but because the variable is optional, it might have been `nil` too.

```
var name: String?
// ...
name = "Alice"
// ...
print(name ?? "")
```

Listing 6.14 Accessing Optional with Fallback Value

When you access name in the last line, you have to provide a fallback value with ?? in case name is nil. That's nothing new; it's simply readdressing what you learned before. The new toy is demonstrated in Listing 6.15 as an alternative approach, where you implicitly unwrap the optional name.

```
var name: String?
// ...
name = "Alice"
// ...
print(name!)
```

Listing 6.15 Implicitly Unwrapping an Optional

The name! expression means, "Hey, although name is an optional, I guarantee that it will have a value at this point. If not, I accept responsibility for an app crash." It's hard to believe that a single token like ! would have such a deep meaning, right? But it does.

Unsafe Code

Implicitly unwrapped optionals should be used sparingly as they bypass Swift's safety checks. Their utility lies in unavoidable scenarios, where a variable must be declared as an optional but is guaranteed to be non-nil afterwards. For app safety, it is advised to avoid implicit unwrap operations as much as possible.

6.4   Optional Chaining

Optional chaining in Swift is an extended version of optional binding, which lets you access the properties of an optional in case it is not nil.

Inspect Listing 6.16 for a demonstration, where getEmptyRooms() returns an optional array of strings. This is actually an array of empty rooms in a hotel. If all rooms are booked, then the function will return nil.

```
// Function
func getEmptyRooms() -> [String]? {
    return nil // Simulation
}
```

```
// Main flow
if let emptyRoom = getEmptyRooms()?[0] {
    print("Empty room: \(emptyRoom)")
} else {
    print("No empty rooms, sorry")
}
```

Listing 6.16 Optional Chaining with nil Value

In the main flow, an extended optional binding syntax is present: `if let emptyRoom = getEmptyRooms()?[0]`. If `getEmptyRooms()` returns a proper array, then this expression will put the first empty room into the `emptyRoom` variable and execute the following {...} code block. Otherwise, the `else {...}` block will be executed.

In its current state, the code will produce the output shown in Figure 6.5 because `getEmptyRooms()` returns `nil`. As expected, the `else {...}` block has been executed.

```
No empty rooms, sorry
```

Figure 6.5 Output for No Empty Rooms

Now let's change the function as shown in Listing 6.17. This time, it returns an array of empty rooms instead of `nil`.

```
// Function
func getEmptyRooms() -> [String]? {
    return ["13", "21", "34"] // Simulation
}

// Main flow
if let emptyRoom = getEmptyRooms()?[0] {
    print("Empty room: \(emptyRoom)")
} else {
    print("No empty rooms, sorry")
}
```

Listing 6.17 Optional Chaining with Non-Nil Value

In this case, `if let emptyRoom = getEmptyRooms()?[0]` behaves differently. Because `getEmptyRooms()` returns a proper array, its first element in `[0]` will be assigned to the `emptyRoom` variable, and the following {...} block will be executed, therefore printing out the room number as shown in Figure 6.6.

```
Empty room: 13
```

Figure 6.6 Output for Empty Room

Optional chaining allows programmers to write more concise code blocks with improved readability.

6.5   Summary

In this chapter, you learned about optionals—a staple of Swift in terms of type safety. It is possible to mark a variable as optional, allowing it to carry nil values. When it comes to accessing an optional, Swift empowers programmers with easy syntax options while preserving safety—such as the nil-coalescing operator, optional binding, and optional chaining.

It is also possible to implicitly unwrap an optional, enforcing the unsafe assumption that it contains a value—but this technique should be avoided whenever possible.

So far, you've used variables as data holders. Simple variables, collections, and optionals: All of them hold data but don't have any behavior. You have mostly used functions to implement behaviors in the program. In the next chapter, we will venture further into advanced data types, which can encapsulate data and behavior under the same umbrella.

Chapter 7
Enumerations

Enumerations define a group of constant values in a type-safe way. They can include cases, associated values, and functions. Enums are powerful for modeling fixed sets of options, like directions.

Starting in this chapter, we will go through a series of Swift types that can hold data and functions simultaneously. Enumerations will be our first stop, which will be followed by structs and classes in the following two chapters.

We will start by offering a basic understanding of enumerations and follow that by examining their declaration syntax. You will see how to add computed properties and functions to enumerations and how to iterate over their values—similar to collections. We won't finish the chapter before addressing recursion in enumerations as well as application of generic types to improve flexibility.

Sounds good? Great! Let's start with a basic definition.

7.1 What Are Enumerations?

In Swift, *enumerations* are types that allow you to define a group of constant values centrally. They improve the readability of the code and reduce the risk of coding errors due to literal duplication. On top of that, enumerations can also contain functions and even conform to protocols, offering flexibility for a wide range of use cases.

An introductory use case would be for directions in a map-related app. At some point, you would have to distinguish north, south, east, and west in your source code. Using the knowledge you have so far, you might declare those constants as distinct global strings using `let` statements—as in, `let dirNorth = "North"`. However, enumerations offer a standardized way of declaring such constant groups centrally, which should be preferred over multiple individual declarations in most cases. Check Figure 7.1 for an illustration of this concept.

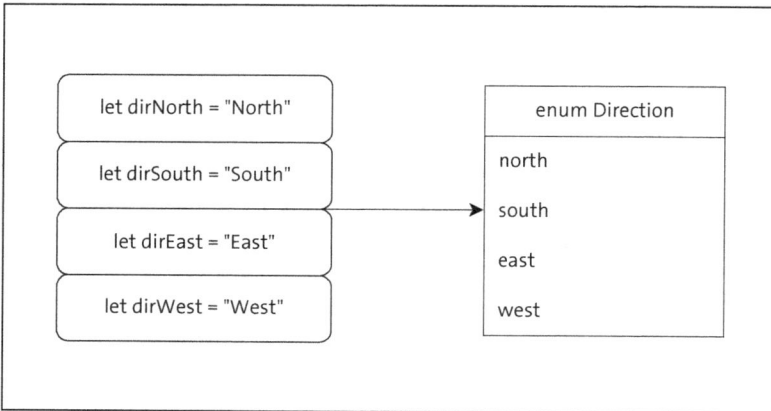

Figure 7.1 Grouping Individual Constants as Enumeration

For a better understanding, Table 7.1 showcases other examples in which using an enumeration would be appropriate. Note that in each case, we have a group of constant values.

Enumeration	Values
Direction	north, south, east, west
HttpMethod	get, post, put, delete
ContactPreference	email, sms, none
UserRole	admin, editor, viewer
GameStatus	playing, paused, stopped
PaymentStatus	pending, completed, failed

Table 7.1 Some Enumeration Examples

Some benefits of preferring enumerations over individual constants are highlighted in Table 7.2.

Benefit	Description	Example
Type safety	Enumerations restrict values to a predefined set, eliminating semantic errors.	You can't make up a fifth value which is not present in Direction.
Grouped semantics	Enumerations logically group related values, self-documenting your code.	By checking UserRole, you can see all possible roles.

Table 7.2 Some Benefits of Enumerations over Individual Constants

Benefit	Description	Example
Case integrity	Swift ensures that `switch` statements cover all cases in an enumeration.	If you missed `GameStatus.paused` in a `switch` statement, Swift will let you know.
Behavior	Enumerations can contain functions and computed properties.	You may implement the `Direction.translate()` function right in the enumeration.
Organization	Enumerations encapsulate cases in a namespace, reducing conflicts.	`ContactPreference.email` won't be ambiguous with a variable called `email`.
Error reduction	Typos or invalid values are prevented by the compiler.	`Direction.nroth` typo will be caught by the compiler, but the literal typo `"Nroth"` won't.

Table 7.2 Some Benefits of Enumerations over Individual Constants (Cont.)

Now that we are through with the theoretical part, let's jump to Xcode and learn about the enumeration syntax through code examples.

7.2 Declaring Enumerations

The declaration of enumerations comes in multiple flavors. In this section, you will discover the options in that regard and we will evaluate the ideal use cases of each one.

7.2.1 Case Values

Case enumeration values represent the most basic form. These are individual cases as type-safe identifiers. Listing 7.1 demonstrates the syntax to declare such an enumeration via the `enum` keyword.

```
enum Direction {
    case north
    case south
    case east
    case west
}
```

Listing 7.1 Enumeration with Case Values

This code snippet should be mostly intuitive: `enum Direction` declares the enumeration named `Direction`, and each possible value is declared using the `case` keyword. As a shorter syntax option, Listing 7.2 demonstrates the same declaration, but with cases separated by commas.

```
enum Direction {
    case north, south, east, west
}
```

Listing 7.2 Short Form of Case Values

Once the enumeration is declared, accessing its values is pretty straightforward. Listing 7.3 shows how to assign enumeration cases to variables.

```
enum Direction {
    case north, south, east, west
}

let carDirection = Direction.east
let windDirection = Direction.south
```

Listing 7.3 Assigning Enumeration Cases to Variables

Accessing enumerations in a switch statement is a fun endeavor as Swift will enforce you to cover all cases of the enumeration. In Listing 7.4, you may not exclude any Direction value in the switch statement. You may try to do so as an exercise and see the resulting compiler error.

```
enum Direction {
    case north, south, east, west
}

let carDirection = Direction.east

switch carDirection {
case .north:
    print("Car is heading north.")
case .south:
    print("Car is heading south.")
case .east:
    print("Car is heading east.")
case .west:
    print("Car is heading west.")
}
```

Listing 7.4 Accessing Enumeration in switch Statement

If you don't have special logic for all enumeration cases, you can take advantage of the `default` keyword. As with any `switch` condition, you can declare your fallback case that way. The compiler will use the `default` code for all missing enumeration cases, as demonstrated in Listing 7.5.

```
enum Direction {
    case north, south, east, west
}

let carDirection = Direction.east

switch carDirection {
case .north:
    print("Car is heading north.")
case .south:
    print("Car is heading south.")
default:
    print("Car is going elsewhere.")
}
```

Listing 7.5 Enumeration Fallback in switch Statements

Naturally, it's possible to check enumeration cases in individual `if` conditions too. Listing 7.6 demonstrates such a code snippet.

```
enum Direction {
    case north, south, east, west
}

let carDirection = Direction.east

if carDirection == .north {
    print("You are driving north.")
} else {
    print("You are not driving north.")
}
```

Listing 7.6 Enumeration in if Condition

Case Limitation

Note that in the preceding examples, `carDirection` can't be checked against a case that is not defined in `Direction`. This limitation is actually a benefit, ensuring code consistency and prevention of literal typos.

The case values form can be used for simple cases, where grouping constants under a common umbrella is enough for the task at hand. Now let's walk through how this approach can be enhanced for different requirements.

7.2.2 Raw Values

Raw enumeration values are built upon case values, with an enhancement. By using raw values, you can declare a specific type for each case: integer, string, double, and so on. Listing 7.7 shows the Direction enumeration with raw String values.

```
enum Direction: String {
    case north = "N"
    case south = "S"
    case east = "E"
    case west = "W"
}
```

Listing 7.7 Directions with Raw Values

That enhancement allows you to assign a String literal for each enumeration case. Despite the enhancement, you can still access and use the enumeration as before. Case in point: The body of Listing 7.8 is exactly the same as the former implementation in Listing 7.6, and Direction can be accessed the same way if you want.

```
enum Direction: String {
    case north = "N"
    case south = "S"
    case east = "E"
    case west = "W"
}

let carDirection = Direction.east

if carDirection == .north {
    print("You are driving north.")
} else {
    print("You are not driving north.")
}
```

Listing 7.8 Directions with Raw Values in if Condition

What you gain on top of that is the ability to access the raw values of the enumeration cases—for this example, "N", "S", "E", and "W". Listing 7.9 demonstrates the syntax to access those raw values; all you need to do is to invoke the rawValue property of the enumeration case.

```
enum Direction: String {
    case north = "N"
    case south = "S"
    case east = "E"
    case west = "W"
}

let carDirection = Direction.east
print(carDirection.rawValue)     // E

let windDirection = Direction.south
print(windDirection.rawValue)    // S
```

Listing 7.9 Accessing Raw Values

Here's a neat trick: For raw `Int` values, Swift can autoassign incremental values starting from 0 if you don't specify them. Listing 7.10 demonstrates the manual declaration of raw integer values.

```
enum Planet: Int {
    case mercury = 0
    case venus = 1
    case earth = 2
    case mars = 3
}

print(Planet.mercury.rawValue)  // 0
print(Planet.mars.rawValue)     // 3
```

Listing 7.10 Manual Declaration of Raw Integer Values

Instead of using the long form, you can let Swift autoassign values using the shortened syntax, as shown in Listing 7.11. In both cases, the program will generate the exact same output.

```
enum Planet: Int {
    case mercury, venus, earth, mars
}

print(Planet.mercury.rawValue)  // 0
print(Planet.mars.rawValue)     // 3
```

Listing 7.11 Automatic Declaration of Raw Integer Values

Raw values are particularly useful when you need to set human-readable values for cases or interface with external systems. Listing 7.12 demonstrates an example in which you declare a user-friendly raw value for each supported operating system. Those values can be shown in the user interface directly.

```
enum OperatingSystem: String {
    case macOS = "Apple macOS"
    case linux = "Linux"
    case windows = "Microsoft Windows"
}
```

Listing 7.12 User-Friendly Raw Values

Listing 7.13 demonstrates an example in which the technical identifiers for device types were declared as raw values. Those values might be considered foreign keys in a remote database and used to build SQL statements.

```
enum DeviceType: String {
    case mobilePhone = "MP"
    case tablet = "TB"
    case desktop = "DP"
    case laptop = "LP"
    case unknown = "UN"
}
```

Listing 7.13 Raw Values with Database Correspondence

Although raw types enhance the functionality of enumerations, they are limited, with a single value and a fixed type for each case. For scenarios in which further values are needed per case, associated values could be the answer.

7.2.3   Associated Values

In Swift enumerations, *associated values* allow cases to store additional data. Unlike raw values, in which you can assign a single value with a fixed type, associated values offer a higher level of flexibility. Listing 7.14 demonstrates a Food enumeration featuring associated values.

```
enum Pizza {
    case cheesy(extra: Bool)
    case veggy(spicy: Bool, thin: Bool)
    case meaty(pepperonis: Int)
}
```

Listing 7.14 Associated Value Example

As is evident in the code sample, each enumeration case has different associated values:

- cheesy can be defined with an extra option (for extra cheese).
- veggy can be defined with spicy and thin options.
- meaty can be defined with pepperonis, indicating the number of pepperoni slices on top.

Flexibility of Associated Values

Note that each enumeration case may have different numbers/types of associated values, improving the flexibility beyond the singularity of raw values.

To assign enumeration cases to a variable, you need to set values for their associated values as well. In Listing 7.15, myPizza and herPizza store different pizza types with totally different options.

```
enum Pizza {
    case cheesy(extra: Bool)
    case veggy(spicy: Bool, thin: Bool)
    case meaty(sausages: Int)
}

let myPizza = Pizza.cheesy(extra: true)
let herPizza = Pizza.veggy(spicy: false, thin: true)
```

Listing 7.15 Variable Assignment with Associated Values

See? You can make enumerations almost configurable with associated values. You declared not only what kind of pizza each customer wants but also the options for their pizza of choice.

Accessing such enumerations in a switch statement requires a slightly different syntax, which is demonstrated in Listing 7.16.

```
enum Pizza {
    case cheesy(extra: Bool)
    case veggy(spicy: Bool, thin: Bool)
    case meaty(sausages: Int)
}

let customersPizza = Pizza.cheesy(extra: true) // Input simulation

// ...

switch customersPizza {
case .cheesy(let extra):
```

```
    print("Cheesy pizza ordered")
    print("Extra cheese: \(extra)")
case .veggy(let spicy, let thin):
    print("Veggy pizza ordered")
    print("Spicy: \(spicy)")
    print("Thin: \(thin)")
case .meaty(let sausages):
    print("Meaty pizza ordered")
    print("Sausages: \(sausages)")
}
```

Listing 7.16 Accessing Associated Values in switch Statement

Note that for each case in the switch statement, you have to declare associated values too. An interpretation of those case expressions is given in Table 7.3.

Expression	Effect
case .cheesy(let extra)	If customersPizza is cheesy, its associated extra value will be assigned to the extra variable for the following code block.
case .veggy(let spicy, let thin)	If customersPizza is veggy, its associated spicy and thin values will be assigned to the spicy and thin variables for the following code block.
.meaty(let sausages)	If customersPizza is meaty, its associated sausages value will be assigned to the sausages variable for the following code block.

Table 7.3 Interpretation of switch case Blocks with Associated Values

In this case, customersPizza is cheesy. Therefore, the first case block will be executed, and it will produce the output shown in Figure 7.2. Note that you can use the associated extra value in the code block and print it out.

```
Cheesy pizza ordered
Extra cheese: true
```

Figure 7.2 Output of switch Statement

Naturally, enumerations with associated values can be subject to if conditions too. As shown in Listing 7.17, the syntax is very similar to that of switch conditions.

```
enum Pizza {
    case cheesy(extra: Bool)
    case veggy(spicy: Bool, thin: Bool)
    case meaty(sausages: Int)
```

```
}

let customersPizza = Pizza.cheesy(extra: true) // Input simulation

// ...

if case let .cheesy(extra) = customersPizza {
    print("Cheesy pizza with extra cheese: \(extra)")
} else {
    print("Not a cheesy pizza")
}
```

Listing 7.17 Accessing Associated Values in an if Statement

Scope of Associated Values

It is wise to use associated values mindfully. They improve the flexibility of enumerations, which makes them great for finite options with small value attachments—but their scope shouldn't be expanded to replace structs or classes. Like each kitchen utensil has a different purpose, each type has a different purpose too. It's important to pick the correct tool for the correct purpose.

You'll learn about structs in Chapter 8 and classes in Chapter 9, after which the differences and use cases will be clearer.

Now that we're through with declaration and consumption of enumerations, we can move forward to their more advanced features.

7.3   Computed Properties

Computed properties in Swift enumerations allow you to derive and return values dynamically. Those computed properties are typically calculated based on the selected case.

Listing 7.18 demonstrates a simple example via the Coffee enumeration. There are three different cases here—espresso, americano, and coldBrew—which could correspond to coffee options in a small shop.

```
enum Coffee {
    case espresso
    case americano
    case coldBrew

    var slogan: String {
        switch self {
```

```
        case .espresso:
            return "Espresso, small but mighty"
        case .americano:
            return "Americano, the classic choice"
        case .coldBrew:
            return "Cold brew, what a breeze"
        }
    }
}
```

Listing 7.18 Enumeration with Computed Property

In this example, there is a *computed property* called `slogan`. Although it is declared like a variable, the following {...} code block must return a value as if it was a function. In a way, computed properties look like variables, but under the hood, they are functions.

Note that we have returned a different string for each coffee type. In the property body, `self` symbolizes the current case determined by the caller—so it's `espresso`, `americano`, or `coldBrew`.

Now, how do you consume this computed property? See the main flow in Listing 7.19 for the answer!

```
// Enum
enum Coffee {
    case espresso
    case americano
    case coldBrew

    var slogan: String {
        switch self {
        case .espresso:
            return "Espresso, small but mighty"
        case .americano:
            return "Americano, the classic choice"
        case .coldBrew:
            return "Cold brew, what a breeze"
        }
    }
}

// Main flow
let myCoffee = Coffee.espresso
print(myCoffee.slogan) // Espresso, small but mighty
```

```
let herCoffee = Coffee.americano
print(herCoffee.slogan) // Americano, the classic choice
```

Listing 7.19 Computed Property Consumption

The `myCoffee.slogan` or `herCoffee.slogan` syntax demonstrates how to access computed properties. Because `slogan` is not declared as a function, you don't need to put parentheses at the end of the expression; `myCoffee.slogan` is enough to access the property.

So far, so good: You were able to set up simple literals as computed properties. But what if you have a complex calculation, based on associated values? If you had to calculate the price of each coffee, it would depend on the number of shots, cup size, decaf option, and so on—right? Such associated values need to be taken into consideration.

Let's revisit the coffee example with associated values then. Listing 7.20 demonstrates the enhanced version of the `Coffee` enumeration, in which each case offers different options. We left `slogan` out to save some headspace as we won't focus on it right now.

```
// Enum
enum Coffee {
    case espresso(shots: Int)
    case americano(size: Int, milk: Bool)
    case coldBrew(decaf: Bool)
}

// Main flow
let myCoffee = Coffee.espresso(shots: 2)
let herCoffee = Coffee.americano(size: 1, milk: false)
```

Listing 7.20 Coffee Enumeration with Associated Values

Now it's time to calculate the coffee prices based on those options. Listing 7.21 demonstrates a made-up price-calculation algorithm for each case, making use of the associated values.

```
// Enum
enum Coffee {
    case espresso(shots: Int)
    case americano(size: Int, milk: Bool)
    case coldBrew(decaf: Bool)

    var price: Double {
        switch self {
        case .espresso(let shots):
            return 1 + (Double(shots) * 0.5)
```

```
        case .americano(let size, let milk):
            var aPrice = 2 + (Double(size) * 0.5)
            if milk { aPrice += 0.5 }
            return aPrice
        case .coldBrew(let decaf):
            return decaf ? 4 : 3
        }
    }
}

// Main flow
let myCoffee = Coffee.espresso(shots: 2)
print(myCoffee.price)    // 1 + (2 * 0.5) = 2

let herCoffee = Coffee.americano(size: 1, milk: false)
print(herCoffee.price)   // 2 + (1 * 0.5) = 2.5
```

Listing 7.21 Price Calculation, Based on Associated Values

The price calculation scheme can be seen in the {…} code block after var price: Double. For each coffee type, there is a different calculation based on its options. The calculations are described in Table 7.4.

Coffee Type	Calculation
espresso	Base price is 1, plus 0.5 added for each shot.
americano	Base price is 2, plus 0.5 added for each size. Milk costs another 0.5 on top.
coldBrew	Decaf costs 4, and regular costs 3.

Table 7.4 Coffee Price Calculation Explained

In the main flow in Listing 7.21, you simply provide options for different coffee orders, and the price is calculated by the enumeration. Cool, right?

Computed properties are useful in the sense that they are centrally encapsulated, right into the enumeration. This prevents scattered logic across the codebase, making the app easier to maintain and understand. Refactoring becomes easier too: If the coffee price algorithm changes, it is enough to implement the change in the Coffee enumeration, without touching anything else.

Now, let's go one step further and explore enumeration functions, which offer similar benefits with a different scope.

7.4   Functions in Enumerations

Just like you can plant functions into the main flow, you can plant functions right into an enumeration too. Like computed properties, enumeration functions can access the current case via the self keyword.

Listing 7.22 features such an example through an enumeration called Mood. This enumeration contains the reflectToText function, which transforms the given text based on the mood at hand.

```
// Enum
enum Mood {
    case happy
    case neutral
    case angry

    func reflectToText(_ text: String) -> String {
        switch self {
        case .happy:
            return text + ", hooray!"
        case .neutral:
            return text
        case .angry:
            return text.uppercased() + "!!!"
        }
    }
}


// Main flow
let myMood = Mood.happy
print(myMood.reflectToText("I'm learning Swift"))  // I'm learning Swift, hooray!

let hisMood = Mood.angry
print(hisMood.reflectToText("I am hungry"))        // I AM HUNGRY!!!
```

Listing 7.22 Enumeration Containing a Function

In the main flow, you can see how the enumeration function is invoked. Syntax-wise, the function is invoked as variable.function():

- When myMood.reflectToText is invoked, the function will execute with the value self = happy, because myMood is set as Mood.happy.
- When hisMood.reflectToText is invoked, it will execute with the value self = angry, because hisMood is set as Mood.angry.

Easy, right? You can see that reflectToText is a proper function and can do everything that any other function can. The only significant extra is that self can be used to access the current case of the variable.

Naturally, an enumeration can contain multiple functions if needed. Plus, a function may access computed properties too. Listing 7.23 features an enhanced version of the mood example, where the computed emoticon property has been added. This returns an old-school smiley representing the mood at hand.

```
// Enum
enum Mood {
    case happy
    case neutral
    case angry

    var emoticon: String {
        switch self {
        case .happy:
            return ":)"
        case .neutral:
            return ":|"
        case .angry:
            return ">:("
        }
    }

    func reflectToText(_ text: String) -> String {
        switch self {
        case .happy:
            return text + ", hooray!"
        case .neutral:
            return text
        case .angry:
            return text.uppercased() + "!!!"
        }
    }

    func addEmoticonToText(_ text: String) -> String {
        return text + " " + emoticon
    }
}

// Main flow
```

```
let myMood = Mood.happy
print(myMood.addEmoticonToText("I'm learning Swift"))   // I'm learning Swift
:)
```

Listing 7.23 Enumeration with Multiple Functions and Computed Property Access

In this example, the second function, addEmoticonToText, accesses a computed property to get the mood's emoticon value and append it to the given text. In the main flow, it is plain to see how it worked.

Other function types and features can also be used in enumerations if needed, such as variadic parameters and generic functions. Those topics were addressed in Chapter 5; as an exercise, you can go through the features you'd like to address and implement them in a sample enumeration.

7.5   Iterating over Enumerations

In some cases, you might need to iterate through all cases of an enumeration. For instance, you might want to fill all options in a combo box of a user interface. For such cases, Swift supplies the CaseIterable protocol to enable iterations. Let's discover the options for enumeration iteration in Swift.

7.5.1   Iterating over Case Values

It's very easy to iterate over case values in an enumeration. All you need to do is to add the :CaseIterable suffix, as shown in Listing 7.24.

```
enum Priority: CaseIterable {
    case low
    case medium
    case high
}
```

Listing 7.24 Iterable Enumeration

And voilà! The enumeration magically can be iterated now! But how? The answer can be seen in Listing 7.25.

```
// Enumeration
enum Priority: CaseIterable {
    case low
    case medium
    case high
}
```

```
// Main flow
for priority in Priority.allCases {
    print("Priority: \(priority)")
}
```

Listing 7.25 Iteration Through Enumeration Cases

A simple for loop over Priority.allCases is all you need for iteration. In enumerations, allCases represents a collection of cases, which can be iterated thanks to the CaseIterable protocol. The output is shown in Figure 7.3.

```
Priority: low
Priority: medium
Priority: high
```

Figure 7.3 Output of Enumeration Iteration

7.5.2   Iterating over Raw Values

The preceding example featured a simple enumeration with case values. Iteration can be done with other enumeration types too, such as raw values. Listing 7.26 demonstrates an enhanced example, in which Priority now has two suffixes: String and CaseIterable.

```
// Enumeration
enum Priority: String, CaseIterable {
    case low = "Trivial"
    case medium = "Normal"
    case high = "Urgent"
}
```

```
// Main flow
for priority in Priority.allCases {
    print("Priority: \(priority.rawValue)")
}
```

Listing 7.26 Iteration Through Enumeration with Raw Values

The String suffix marks Priority as an enumeration with raw string values, while CaseIterable marks it as iterable. Iterating through enumeration cases has the same syntax—as seen in the main flow. The output of the raw values is shown in Figure 7.4.

```
Priority: Trivial
Priority: Normal
Priority: Urgent
```

Figure 7.4 Raw Value Output of Enumeration Iteration

7.5.3   Iterating over Associated Values

Finally, let's go through an example that iterates over associated values. This requires a bit more work than the previous two options, but it's not overly complex.

Listing 7.27 demonstrates such an example, in which UserStatus symbolizes the status of the user, denoting them as either member or guest. If the user is a member, then they could be an admin too: This is symbolized by the isAdmin variable.

```
// Enum
enum UserStatus: CaseIterable {
    case member(isAdmin: Bool)
    case guest

    static var allCases: [UserStatus] {
        [
            .member(isAdmin: true),
            .member(isAdmin: false),
            .guest
        ]
    }
}

// Main flow
for status in UserStatus.allCases {
    switch status {
    case .member(isAdmin: let isAdmin):
        print("Member status, is admin: \(isAdmin)")
    case .guest:
        print("Guest status")
    }
}
```

Listing 7.27 Iteration Through Enumeration with Associated Values

To add the iteration feature, the CaseIterable protocol should be applied as usual. But we need an additional step here: A computed property named allCases must return all possible values supported by UserStatus. This implementation enables you to iterate through a finite set of cases instead of an infinity of possibilities.

Object-Oriented Territory

Application of protocols and keywords like static belong to the object-oriented realm. You will learn about these topics thoroughly in the upcoming chapters. For now, you can accept them as boilerplate code. It will all make more sense in due time.

In the main flow, a `for` loop is used to access all cases as usual. Within the loop, you will see a `case` statement that should be familiar from Section 7.2.3.

Note that for the `member` case, you can access its associated `isAdmin` value too—which is expected. Due to the values provided in `allCases`, the `for` loop will enter `case .member` twice: once as `isAdmin = true` and once as `isAdmin = false`.

For the `guest` case, the logic is simple: It will be entered once because it has no associated values and it was returned from `allCases` only once. The output in Figure 7.5 fulfills those expectations, right?

```
Member status, is admin: true
Member status, is admin: false
Guest status
```

Figure 7.5 Associated Value Iteration Output

Now, let's move to a topic that might be a bit challenging for those of you without recursion experience. But don't worry: It's not rocket science, after all.

7.6 Recursive Enumerations

So far, you've learned that enumeration cases can hold attachments that are called *associated values*. In the preceding examples, those associated values were basic data types. Naturally, an associated value's type can be another enumeration too.

Listing 7.28 demonstrates such an example. `Severity` is a simple enumeration, indicating a risk level. The following `MessageType` enumeration contains a case named `error`, which has an associated value of type `Severity`. In the main flow, the access to the associated type is also demonstrated.

```
// Enumerations
enum Severity {
    case low
    case medium
    case high
}

enum MessageType {
    case info
    case warning
    case error(severity: Severity)
}

// Main flow
```

```
let msgType1 = MessageType.info
let msgType2 = MessageType.error(severity: .medium)
```

Listing 7.28 Associated Value with Enumeration Type

The usage logic is sketched in Figure 7.6 for clarity.

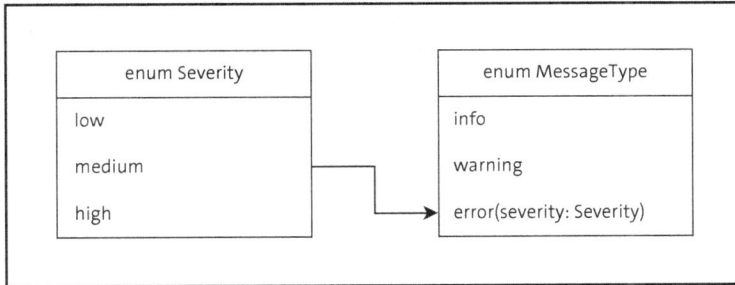

Figure 7.6 Sketch of Enumeration as Associated Value Data Type

To make the example more realistic, Listing 7.29 demonstrates an enhanced code snippet in which the logMessage function has been thrown into the mix. Through this function, you can log messages with declared types, and you can declare severity levels for error messages.

```
// Enumerations
enum Severity {
    case low
    case medium
    case high
}

enum MessageType {
    case info
    case warning
    case error(severity: Severity)
}

// Functions
func logMessage(_ msg: String, type: MessageType) {
    // Code to log the message
}

// Main flow
logMessage("User logged in", type: .info)
logMessage("Disk out of space", type: .error(severity: .high))
```

Listing 7.29 Enhanced Version, Featuring Simple Message Log

This example was straightforward and easy to follow, right? All we did was use Severity as a data type for MessageType.error's associated value. But things get interesting when you introduce recursion to this concept. Through recursive enumerations, an associated value can be declared in the type of the enumeration itself.

Recursion

Recursion is a technique in which a programming element refers to itself. The most typical example is a function that calls itself to iterate through a tree structure.

That sounds more complex than the practical syntax seems, as you will see for yourself in Listing 7.30.

```
indirect enum FamilyMember {
    case father(spouse: FamilyMember? = nil)
    case mother(spouse: FamilyMember? = nil)
    case daughter
    case son
}
```

Listing 7.30 Recursive Enumeration

The declaration of such enumerations must start with the indirect suffix to enable recursion. Within the enumeration, the father case has an associated value called spouse, which is also a FamilyMember! That's the recursion in action. The mother case is the same.

You logically know that father's spouse is going to be FamilyMember.mother and mother's spouse is going to be FamilyMember.father in general—but as a simple case study, the example demonstrates how recursion applies to enumerations.

Listing 7.31 demonstrates two methods to assign values to recursive associated values. In the Early assignment part, wife's spouse: husband is assigned during the initialization. In the Late assignment part, husband's spouse: wife is assigned after the initialization.

```
// Enum
indirect enum FamilyMember {
    case father(spouse: FamilyMember? = nil)
    case mother(spouse: FamilyMember? = nil)
    case daughter
    case son
}

// Early assignment
var husband = FamilyMember.father()
var wife = FamilyMember.mother(spouse: husband)
```

```
// Late assignment
if case .father(_) = husband {
    husband = .father(spouse: wife)
}
```

Listing 7.31 Assignment of Recursive Enumerations

Logically, late assignment is unavoidable here. You can't make wife an associated value without declaring it first. It has to be assigned to husband.spouse only after you have declared both wife and husband.

Before we conclude this chapter on enumerations, let's take a look at generic enumerations and how they can increase your flexibility for associated values.

7.7   Generic Enumerations

In Chapter 5, you learned about generic functions. You are encouraged to revisit that discussion if you need to as we will reuse that knowledge and syntax intuition here.

Back already? Good! Using a similar syntax as generic functions, enumerations can contain associated values of a generic type—typically symbolized as <T>. Listing 7.32 contains a simple enumeration called TreasureChest, which corresponds to types of chests in a video game.

```
enum TreasureChest<T> {
    case empty
    case treasure(T)
}
```

Listing 7.32 Enumeration with Generic Associated Values

Just like generic functions, this generic enumeration has the suffix <T>, indicating that it is flexible enough to accept different data types. TreasureChest has two cases:

- empty means that the chest is empty, and the player wasted their time opening it.
- treasure(T) means that the chest contains some kind of treasure of type T.

According to the imaginary game rules, if T is an integer (like 50), then the chest contains gold of that quantity. If T is a string (like "Excalibur"), then the chest contains that item. As a demonstration of that, Listing 7.33 shows an enhanced version of the code sample, which contains declarations of various chest types.

```
// Enumeration
enum TreasureChest<T> {
    case empty
    case treasure(T)
}
```

```
// Different chests
let emptyChest: TreasureChest<Int> = .empty
let goldChest: TreasureChest<Int> = .treasure(1337)
let swordChest: TreasureChest<String> = .treasure("Excalibur")
```

Listing 7.33 Declaration of Different Chest Types

In this version, goldChest was declared as a TreasureChest with <T> as Int, which contains 1337 gold. Meanwhile, swordChest was declared as a TreasureChest with <T> as String, which contains the famous "Excalibur" sword.

Lucky players finding that sword are sure to hack and slash their way through the game—but we need to address one more topic beforehand: pattern matching. The example code is enhanced once more in Listing 7.34, this time by adding pattern-matching code samples.

```
// Enumeration
enum TreasureChest<T> {
    case empty
    case treasure(T)
}


// Different chests
let emptyChest: TreasureChest<Int> = .empty
let goldChest: TreasureChest<Int> = .treasure(1337)
let swordChest: TreasureChest<String> = .treasure("Excalibur")

// Pattern matching
switch goldChest {
case .empty:
    print("No gold found")
case .treasure(let chestContent):
    print("Found \(chestContent) gold coins!")
}

switch swordChest {
case .empty:
    print("No item found")
case .treasure(let chestContent):
    print("Found the item \(chestContent)!")
}
```

Listing 7.34 Pattern Matching in Generic Enumerations

The syntax is nearly the same as for regular enumerations, right?

Generic enumerations in Swift are a powerful feature, but they are somewhat niche compared to other Swift constructs like generic structs, classes, or protocols. But you are in luck: Those three topics are the focus of the upcoming three chapters, in that exact order!

7.8   Summary

In this chapter, you learned about enumerations and their benefits for programmers. In a nutshell, enumerations are groups of fixed constant values. Instead of repeating literal values all over the codebase, it makes more sense to group them as enumerations because it makes code maintenance much easier.

Enumerations caxome in three flavors: case values, raw values, and associated values. Naturally, it is possible to iterate through cases—but beyond that, associated values also support recursion and generic types. Enumerations also offer advanced features like computed properties and functions, and you can conduct operations over enumeration cases.

Starting from this chapter, we will walk you through a series of Swift types that can hold data and functions simultaneously. Next up, we'll introduce structs, classes, and protocols in the following three chapters.

Be aware that these constructs have overlapping features. Initially, it might be unclear how to determine which one to pick for a given case. As you learn about what each construct can or can't do and gain more experience, such decisions will become easier. It's good to focus on the differences of these constructs to understand them better.

7

Chapter 8
Structs

Structs are special types in Swift; they are similar to classes but offer a different mechanism. Therefore, this topic is placed before classes to offer a gradual warm-up. It is a challenge to split features between this topic and classes as the data structure mechanisms greatly overlap. Our strategy is to keep this topic at a relatively basic level and move to more advanced topics in the class and protocol chapters. Features that also work for structs will be highlighted and demonstrated there as well.

In the previous chapter, you learned about enumerations, which are ideal to consolidate constants centrally, under a common umbrella. They also offer empowering features like computed properties and functions. In the next chapter, you will learn about classes, which are the core pillars of object-oriented Swift. The best spot to address structs is right between those topics—so here we are!

Following an introduction to structs, you will learn how to declare and access them. You'll learn the syntax to add various kinds of properties and functions to structs and apply generics for type flexibility. Finally, we will address advanced struct features like lazy properties, property observers, and subscripts.

Let's begin with the basic introduction and proceed from there.

8.1   What Is a Struct?

In Swift, a *struct* is a data container template to store multiple properties of an entity. That entity could be, for example, an employee, an invoice, a mobile phone, a book in a library, and so on. A struct to represent books could have the properties shown in Table 8.1.

Property	Type	Sample Value
name	String	"My Autobiography"
author	String	"John Doe"
genre	enum Genre	Genre.nonFiction

Table 8.1 Sample Properties of Book Struct

Property	Type	Sample Value
pageCount	Int	150
chapters	Array[String]	["Intro", "Childhood", "Adulthood", "Epilogue"]

Table 8.1 Sample Properties of Book Struct (Cont.)

Once you have a struct called Book, you can use it as a template to instantiate various books. For instance, you can declare three different variables of type Book, which store properties of three different books—as sketched in Figure 8.1.

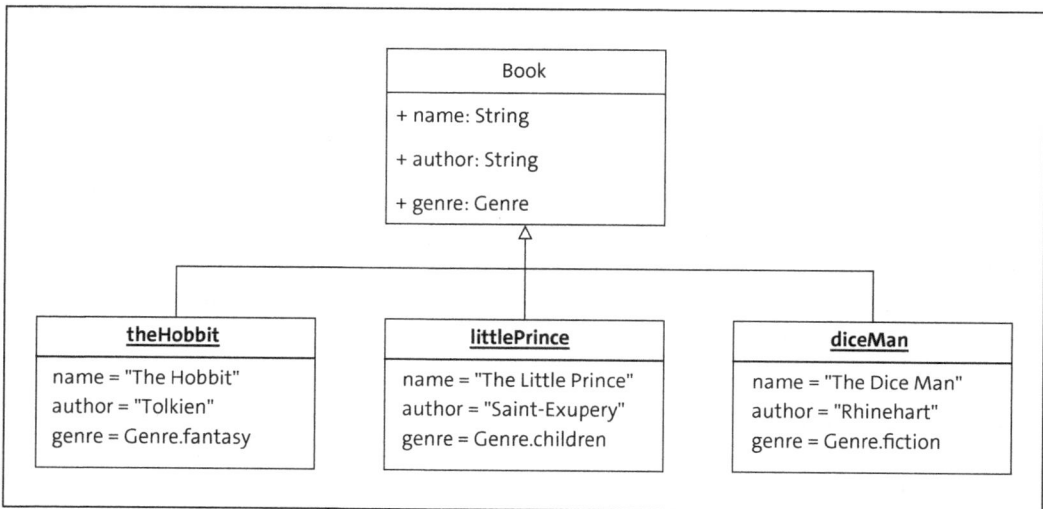

Figure 8.1 Three Variables of Type Book

As demonstrated visually, the Book struct is a template data container, which can store properties of any book. This struct has been instantiated as theHobbit, littlePrince, and diceMan; each variable holds the properties of different books. Nifty, right? That way, you get an organized custom data structure to hold multiple values of an entity.

To help explain the place of structs in the Swift language, Table 8.2 contains short comparisons to data types you learned about before.

Data Type	Typical Purpose	Struct Comparison
Variable	Holds a single property	Holds multiple properties
Array, Set	Contains multiple entities of the same type	Represents a single entity
Enumeration	Groups constant universal values	Groups properties of a single entity

Table 8.2 Core Comparison of Former Data Types with Structs

As we go through struct code samples in this chapter, this distinction will become more concrete. Without further ado, let's start by declaring some structs!

Tuple or Struct?

Both tuples and structs can be used to group related data, so what is the difference? Although tuples and structs have overlapping features, they serve different purposes.

Tuples are lightweight, anonymous types that loosely group multiple values, without the logical expectation of them describing a common entity. They are typically used for temporarily grouping a small number of values without defining a formal type. Tuples are limited in the sense that they can't have functions or conform to protocols.

Structs are proper, named types that tightly group multiple values, with the logical expectation of them describing a common entity. They are typically used as blueprints for entity property sets. Structs are reusable types that can have functions and conform to protocols.

In a case in which you're unsure which one to pick, we recommend preferring structs over tuples: They are reusable with better structure and offer better extensibility if the necessity arises in the future.

8.2 Declaring Structs

This is the initial hands-on coding section for structs. We'll start by declaring structs and create instances for them, demonstrating how to set and read their properties. Once we're through with those fundamental skills, we can move forward to the more advanced features they offer.

8.2.1 Basic Struct Declaration

Because we started with an example centered on books, let's continue with that for the first hands-on code samples too. Listing 8.1 contains a basic declaration example of a struct.

```
struct Book {
    let name: String
    let author: String
    let pageCount: Int
}
```

Listing 8.1 Simple Struct for Books

At this point, with your experience in Swift so far, this code snippet should be very intuitive, right? The struct Book expression means that you are declaring a struct called Book, and the following {...} code block contains a list of book properties.

Now let's instantiate this struct to store properties of various books! In Listing 8.2, two distinct Book instances for two distinct books are created: one for *The Hobbit* and one for *The Little Prince*.

```
struct Book {
    let name: String
    let author: String
    let pageCount: Int
}

let theHobbit = Book(name: "The Hobbit",
                     author: "Tolkien",
                     pageCount: 320)

let prince = Book(name: "The Little Prince",
                  author: "Saint-Exupery",
                  pageCount: 100)

print("The Hobbit has \(theHobbit.pageCount) pages") // The Hobbit has 320 pages
print("Little Prince has \(prince.pageCount) pages") // Little Prince has 100 pages
```

Listing 8.2 Two Book struct Instances

After the declaration part, theHobbit and prince become two separate Book instances, holding property values of their own. Those entities are completely isolated from each other; their properties aren't mixed up. For example, theHobbit.pageCount is 320 and prince.pageCount is 100 in this instance, which is printed out in the end.

Structs and Code Organization

Even in this simple example, the organizational benefits of structs become tangible. theHobbit.name or prince.author are self-explanatory expressions. You can't get more obvious than that! However, code tidiness is not the only benefit of structs; you'll see more features in the following pages.

8.2.2   Mutable Properties

Note that each "property" is actually a variable. If a property needs to be immutable, then it should be declared with the let keyword as in the example; otherwise, the var keyword can be used—as for any variable.

In a library or bookstore application, the availability of books may change, right? If someone borrows or buys the last copy of a book, then it becomes unavailable. Therefore, a property like isAvailable would need to be mutable. Listing 8.3 demonstrates this in action.

```
struct Book {
    let name: String
    let author: String
    let pageCount: Int
    var isAvailable: Bool
}

var theHobbit = Book(name: "The Hobbit",
                     author: "Tolkien",
                     pageCount: 320,
                     isAvailable: true)

theHobbit.isAvailable = false // Last copy is taken
```

Listing 8.3 Struct with Mutable Properties

Naturally, the theHobbit variable needed to be mutable as well; otherwise, you can't change any of its properties.

8.2.3 Complex Properties

Although structs are perfectly fine for storing simple values, they can store complex values too. Listing 8.4 demonstrates an enhanced version of the Book struct, storing an enumeration (genre) and an array (chapters) as properties.

```
enum Genre {
    case fantasy
    case fiction
    case nonFiction
}

struct Book {
    let name: String
    let author: String
    let pageCount: Int
    let genre: Genre
    let chapters: [String]
    var isAvailable: Bool
}

var myBook = Book(
    name: "My Autobiography",
    author: "John Doe",
    pageCount: 150,
```

```
    genre: .nonFiction,
    chapters: ["Intro", "Childhood", "Adulthood", "Epilogue"],
    isAvailable: true)
```

Listing 8.4 Struct with Complex Properties

Data Sources

In a typical real-world app, properties of myBook would be filled from a data source such as a database, configuration file, or RESTful API. Hard-coded property values could be seen in some unit tests though.

Because our main focus is Swift syntax, we used hard-coded values here to keep the code as clean and understandable as possible.

In this last example, the typical use case difference between enumerations and structs became even more evident. The Genre enumeration contains a fixed list of universal constants, representing different categories of books. Genre would not be instantiated, and those categories won't change very often. On the other hand, the Book struct is a blueprint for all books: For each book in the catalog, a new instance of Book can be created, such as myBook, storing the properties of that particular book.

8.2.4   Nested Structs

Structs can even contain other structs as properties! Listing 8.5 has a struct called Author, which is contained as a property in the Book struct.

```
struct Author {
    let name: String
    let surname: String
}

struct Book {
    let name: String
    let author: Author
    let pageCount: Int
}

let tolkien = Author(name: "J.R.R", surname: "Tolkien")

let lordOfTheRings = Book(
    name: "Lord of the Rings",
    author: tolkien,
    pageCount: 1000)
```

```
let hobbit = Book(
    name: "Hobbit",
    author: tolkien,
    pageCount: 300)

print(lordOfTheRings.author.name)    // J.R.R
print(hobbit.author.surname)         // Tolkien
print(tolkien.surname)               // Tolkien
```

Listing 8.5 Structs as Struct Properties

Once you declare `tolkien` as `Author`, you could assign it to `lordOfTheRings.author` and `hobbit.author`, preventing the repetition of the literals `"J.R.R."` and `"Tolkien"` for each book.

In the final part of the example, the book author's properties are accessed using a chain syntax: `hobbit.author.surname`. But accessing `tolkien.surname` is *almost* equally valid at that point.

Why almost? Well, there is a critical detail to be aware of that deserves your utmost attention. The `hobbit.author = tolkien` assignment creates a copy of `tolkien`. Once you make that assignment, `hobbit.author` and `tolkien` become two completely isolated variables. If you change the value of `tolkien.surname`, it won't change the value of `hobbit.author.surname` because those are duplicated and isolated values—as demonstrated in Figure 8.2.

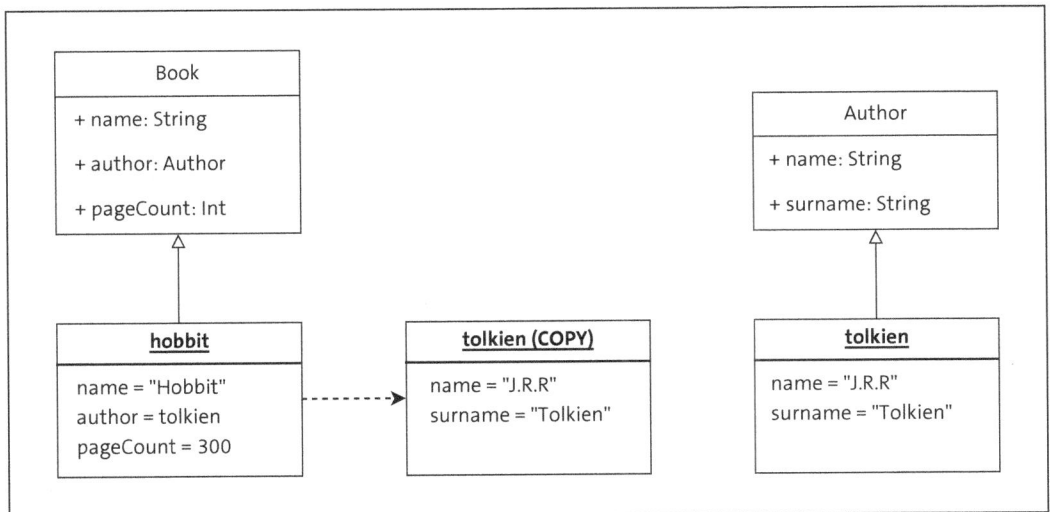

Figure 8.2 Duplication of tolkien Variable

To see this concept in action, check out Listing 8.6. Initially, you declare `tolkien` as `Author` and set `hobbit.author = tolkien`. In the first printout, both `hobbit.author.surname` and `tolkien.surname` produce the same output because both copies hold the same value.

```
struct Author {
    var name: String
    var surname: String
}

struct Book {
    let name: String
    let author: Author
    let pageCount: Int
}

// Initial values
var tolkien = Author(name: "J.R.R", surname: "Tolkien")

let hobbit = Book(
    name: "Hobbit",
    author: tolkien,
    pageCount: 300)

print(hobbit.author.surname)    // Tolkien
print(tolkien.surname)          // Tolkien

// Value change
tolkien.surname = "Tolkien, Jr."
print(hobbit.author.surname)    // Tolkien
print(tolkien.surname)          // Tolkien, Jr.
```

Listing 8.6 Demonstration of tolkien Duplication

However, things get interesting in the second part. When you change `tolkien.surname` to `"Tolkien, Jr."`, only the value of that variable is changed; the independent `hobbit.author.surname` copy remains untouched.

In summary, nested structs hold copied values and become independent entities seperate from their original source.

Composition

The value duplication mechanism of structs is neither an advantage nor a disadvantage; it is merely a feature to be aware of. If you want `hobbit.author` to act as a mirror of `tolkien` without copying values, you would be aiming for *composition*. That's a feature of classes, not structs—which will be addressed in Chapter 9.

Because structs aren't classes, struct instances shouldn't be considered proper objects; they are merely value holders.

8.2.5   Default Property Values

Just like functions can have default parameter values, structs can be declared with default property values too. If an instantiation doesn't set any values for those properties, then the default values are assigned automatically.

Listing 8.7 shows a demonstration of this feature, where Book.genre and Book.isAvailable have default values.

```
enum Genre {
    case unknown
    case fantasy
    case fiction
    case nonFiction
}

struct Book {
    let name: String
    let author: String
    let pageCount: Int
    var genre: Genre = .unknown
    var isAvailable: Bool = true
}

var myBook = Book(
    name: "My Autobiography",
    author: "John Doe",
    pageCount: 150)

print(myBook.genre)        // unknown
print(myBook.isAvailable)  // true
```

Listing 8.7 Default Struct Property Values

In this example, myBook is declared without providing values for genre and isAvailable; therefore, the default values, genre = .unknown and isAvailable = true, were assigned automatically. Easy, right?

8.2.6   Optional Properties

You learned about Swift optionals in Chapter 6, which you may revisit if you need to refresh your memory. Naturally, structs support optional properties, and any feature of optionals can be applied to struct properties too. Listing 8.8 features the Author struct, which contains an optional property called birthYear.

```
struct Author {
    let name: String
    let surname: String
    var birthYear: Int?
}
```

Listing 8.8 Struct with Optional Properties

The optionality of this property is expressed with the ? token at the end of its type: Int?. If you don't know the birth year of an author, you can leave this value as nil without assigning a made-up value.

To solidify this concept, Listing 8.9 features two distinct Author instances: We provided a value for tolkien.birthYear and no value for eldric.birthYear.

```
struct Author {
    let name: String
    let surname: String
    var birthYear: Int? = nil
}

let tolkien = Author(
    name: "J.R.R",
    surname: "Tolkien",
    birthYear: 1892)

let eldric = Author(
    name: "Eldric",
    surname: "Varnholt")

print(tolkien.birthYear ?? "Unknown")   // 1892
print(eldric.birthYear ?? "Unknown")    // Unknown
```

Listing 8.9 Struct Instances with / without Optional Property Values

This set eldric.birthYear as nil, meaning that it doesn't hold any value. That's not too different from any other optional, right?

Congratulations—you're done with struct basics and the various available property types! Now that you have a solid framework at hand, we will move on to advanced features of structs.

8.3   Static Properties

So far, you have seen instance properties of structs. Each instance holds its distinct isolated properties, which are never mixed. In Figure 8.3, name and health are instance properties of the Monster struct, while goblin.health and dragon.health are two distinct, isolated variables that belong to either goblin or dragon. They aren't common in any way.

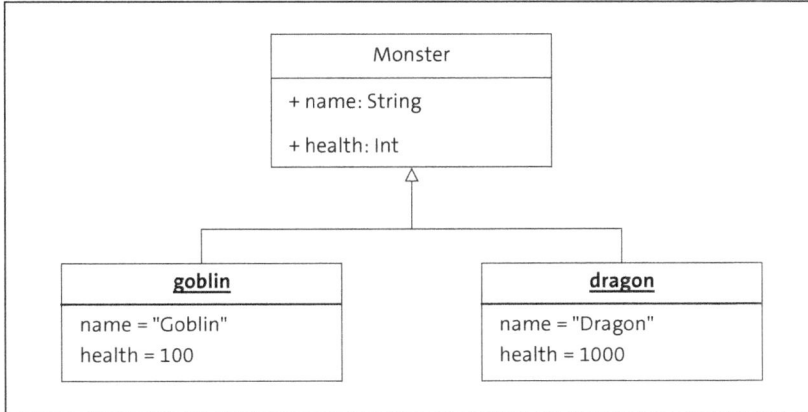

Figure 8.3 Instance Properties of Monster

This is nothing new, so far. All struct properties you have seen to this point were instance properties anyway. But hold on: We're just setting the stage for static properties. Listing 8.10 is also part of this setup, providing a short reminder of how to initialize and change instance properties.

```
struct Monster {
    let name: String
    var health: Int
}

var goblin = Monster(name: "Goblin", health: 100)
var dragon = Monster(name: "Dragon", health: 1000)

goblin.health -= 20
```

Listing 8.10 Code Sample for Instance Properties

In this sample, you only have two monsters at hand—but in a real game, there could be dozens of them.

Now, what if there was a common property that was viable for all monsters simultaneously? isPaused could be such a property: If the player clicks **Pause**, you want all monsters to stop—not just individual ones like goblin and dragon (plus dozens of others).

For such cases, Swift offers the feature of static properties. Unlike *instance properties*, which are scoped per instance (like goblin and dragon), *static properties* are scoped per struct (like Monster)—independent from individual instances. Check the sketch in Figure 8.4: Can you spot the newcomer?

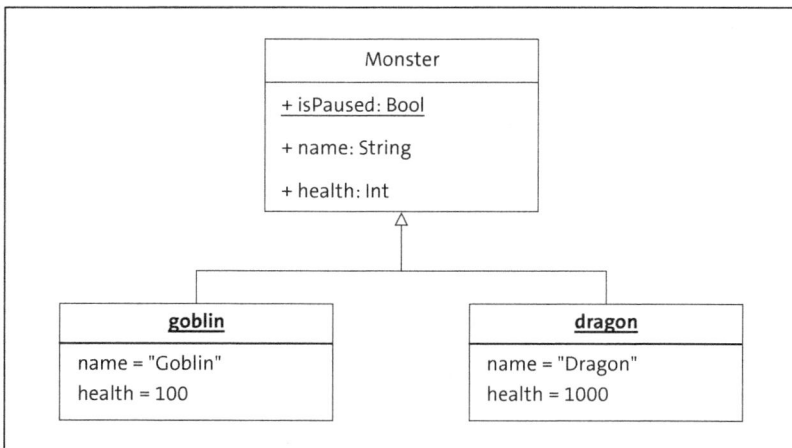

Figure 8.4 Static Property of Monster

In this example, we welcome the static Monster.isPaused variable, instead of instance variables goblin.isPaused and dragon.isPaused. You can set up the rest of the game in such a way that if Monster.isPaused is true, the animation engine would cease moving any monster. That's much easier, faster, simpler, and more straightforward than checking each and every Monster instance (like goblin and dragon), right?

Listing 8.11 shows the enhanced code sample containing the static Monster.isPaused property. You can ignore the @MainActor expression for now; it is not relevant to the topic at hand.

```
struct Monster {
    @MainActor
    static var isPaused: Bool = false

    let name: String
    var health: Int
}

var goblin = Monster(name: "Goblin", health: 100)
var dragon = Monster(name: "Dragon", health: 1000)

goblin.health -= 20
Monster.isPaused = true
```

Listing 8.11 Code Sample for Static Properties

In contrast to instance properties, static properties need to have the `static` prefix when declared. When accessing those values, they need to be expressed using the struct name, like `Monster.isPaused`, instead of the instance's name. That's easy and intuitive, right?

> **Static Properties Versus Global Variables**
>
> Alternatively, you could declare a global variable called `areAllMonstersPaused` outside of the scope of `Monster`—but that would reduce the code clarity. When you consolidate all monster-related global values under the `Monster` struct, the codebase becomes easier to follow and understand, reducing the risk of coding errors in the long run.

The conceptual difference between instance and static versions applies to other struct features too, such as computed properties and functions. As we go through those topics, we will assume that the difference is clear to you. At any point of confusion, you can return to this topic and refresh your memory.

8.4   Computed Properties

In the previous chapter, you learned about computed properties for enumerations, remember? Structs feature computed properties in nearly the same way. Based on existing properties of a struct, you can compute new values when requested by the caller. Likewise, you can accept a computed value from the caller to interpret and break down into real properties. In this section, you will see examples of both.

8.4.1   Getters

This is going to be a familiar topic, similar to enumerations—so expect an easy ride! In a struct, a *getter* is a computed property that computes and returns a value based on existing properties. In Listing 8.12, `isDead` is a getter of the `Monster` struct.

```
struct Monster {
    @MainActor
    static var isPaused: Bool = false
    let name: String
    var health: Int

    var isDead: Bool {
        return health <= 0
    }
}
```

Listing 8.12 Monster Struct with isDead Getter

Note that isDead is not a real property holding a value. Instead, it only *appears* to be a variable to the callers—but every time isDead is called, its value is calculated on the fly. If the monster's health is depleted, isDead will be true; otherwise, it will be false.

Check the caller demonstration in Listing 8.13. When goblin.health is 80, goblin.isDead returns false. When goblin.health is 0, goblin.isDead returns true due to the computation in the struct.

```
struct Monster {
    @MainActor
    static var isPaused: Bool = false
    let name: String
    var health: Int

    var isDead: Bool {
        return health <= 0
    }
}


var goblin = Monster(name: "Goblin", health: 100)

goblin.health -= 20
print(goblin.health, " - ", goblin.isDead)  // 80 - false
goblin.health -= 80
print(goblin.health, " - ", goblin.isDead)  // 0 - true
```

Listing 8.13 Accessing a Getter of a Struct

Naturally, getters can return complex data types too—like a collection or another struct. But the core syntax will be the same as is shown in Listing 8.13.

In the last example, we demonstrated an instance getter, but you can implement static getters too, which are compute properties over static properties. Listing 8.14 shows the static Monster.stateDescription getter.

```
struct Monster {
    @MainActor
    static var isPaused: Bool = false
    let name: String
    var health: Int

    @MainActor
    static var stateDescription: String {
        Monster.isPaused ? "Paused" : "Running"
    }
}
```

```
Monster.isPaused = true
print(Monster.stateDescription) // Paused

Monster.isPaused = false
print(Monster.stateDescription) // Running
```

Listing 8.14 Struct Static Getter Demonstration

Like a static property, a static getter needs to be declared with the static prefix. In the body of stateDescription, a string has been computed and returned based on Monster. isPaused.

Also like a static property, the caller needs to access the static getter using the name of the struct: Monster.stateDescription. That's clear, right?

Now that we're through with getters, let's advance to a relatively new concept: setters.

8.4.2   Setters

A computed property can act as a *setter* too, meaning that the incoming value is interpreted and properties are modified accordingly. In Listing 8.15, isDead acts as both a getter and a setter simultaneously.

```
struct Monster {
    let name: String
    var health: Int

    var isDead: Bool {
        get { return health <= 0 }
        set(newDead) { health = newDead ? 0 : health + 1 }
    }
}
```

Listing 8.15 Struct Setter Demonstration

Let's check the isDead code block. In the get part, you have exactly what you had before: If health is depleted, return true; otherwise, return false.

The new addition is the set part, which initially imports the newDead Boolean value, symbolizing the incoming isDead value. If isDead is set to true, then health is set to 0. Otherwise, it means that the monster has been resurrected, so you increase health by 1.

Does that sound complex? If so, Listing 8.16 should clarify the situation with the caller syntax.

```
struct Monster {
    let name: String
    var health: Int
```

343

```
    var isDead: Bool {
        get { return health <= 0 }
        set(newDead) { health = newDead ? 0 : health + 1 }
    }
}

var goblin = Monster(name: "Goblin", health: 100)
print("\(goblin.health) - \(goblin.isDead)")    // 100 - false

goblin.isDead = true
print("\(goblin.health) - \(goblin.isDead)")    // 0 - true

goblin.isDead = false
print("\(goblin.health) - \(goblin.isDead)")    // 1 - false
```

Listing 8.16 Accessing Getters and Setters

Initially you declare a goblin with a health value of 100. When goblin.isDead is called, the getter is invoked, returning the value false because goblin.health is greater than 0.

Next you set goblin.isDead = true. Note that isDead is not a real property; it's a "fake" property acting like a real one. When you make the assignment goblin.isDead = true, the setter of isDead is invoked, setting goblin.health = 0.

Finally you set goblin.isDead = false with the intention of resurrecting the goblin in the game. Once again, the setter of isDead is invoked, setting goblin.health = 1 at this time.

Fake Properties

When properly designed, a computed property with a getter and a setter can act as a fake property, relying on underlying real properties to accept or return dynamically calculated values. When you learn about private properties in Section 8.6.1, you'll also discover further benefits of computed properties—like encapsulation.

This was an example of an instance setter, but as you would expect, static setters are also supported in Swift. Listing 8.17 features the static Monster.isActive getter/setter, which acts as the mirror image of the static Monster.isPaused property.

```
struct Monster {
    @MainActor
    static var isPaused: Bool = false

    @MainActor
    static var isActive: Bool {
        get { return !Monster.isPaused }
```

```
        set(newActive) { Monster.isPaused = !newActive }
    }
}
```

```
Monster.isActive = true
print(Monster.isPaused) // false
Monster.isActive = false
print(Monster.isPaused) // true
```

Listing 8.17 Static Getter and Setter Demonstration

In the getter part, if isPaused is true, then isActive returns false; if isPaused is false, then isActive returns true.

In the setter part, if isActive is set as true, then isPaused is changed to false, and if isActive is set as false, then isPaused is set as true.

In summary, computed properties can act like getters and setters, and scope-wise, they can be instance or static types. Although they are already valuable language elements in that sense, they really shine when used in conjunction with private properties, which you'll learn about soon. But before that, let's talk about struct functions.

8.5   Functions

You learned about Swift functions in Chapter 5. Then, in Chapter 7, you also learned that enumerations can contain functions. Likewise, structs can contain functions as well. In this section, we will walk you through different types of struct functions.

8.5.1   Initialization

Let's go in logical order. First and foremost, structs feature a special function called init. As its name implies, init ought to be used to initialize properties. Each struct may contain only one init function (and overloads it if necessary). This function is automatically called only once when the struct is instantiated for the first time.

Let's see init in action for a better understanding. Listing 8.18 contains a simple structure, which is a customer queue. All you have here is a customers property, which is a string array.

```
struct Queue {
    var customers: [String]
}
```

Listing 8.18 Simple Queue Struct

Imagine that your business requirement says that each queue must start with an initial customer. In other words, you are going to need a customer to build the queue. To achieve that, we threw an init function into the mix, as shown in Listing 8.19.

```
struct Queue {
    var customers: [String]

    init(firstCustomer: String) {
        self.customers = [firstCustomer]
    }
}
```

Listing 8.19 Struct with Initializer

Upon instantiation, the init function will automatically execute once. In this example, you import the firstCustomer parameter to init and initialize the customers array with that value.

Note that you have to use the self.customers expression to access the customers property. self is a special keyword, which points to the struct itself. self.customers means "the struct's customer property," preventing ambiguity with local function variables.

Listing 8.20 demonstrates the instantiation of bankQueue, in which bankQueue.init is triggered with the value firstCustomer: "Alice". As a result, bankQueue.customers has been initialized as ["Alice"].

```
struct Queue {
    var customers: [String]

    init(firstCustomer: String) {
        self.customers = [firstCustomer]
    }
}

var bankQueue = Queue(firstCustomer: "Alice")
print(bankQueue.customers)  // Alice
```

Listing 8.20 Triggering init Through Instantiation

Following the initialization state, you can keep using the struct as usual. In Listing 8.21, the example has been enhanced by further operation after the initialization.

```
struct Queue {
    var customers: [String]

    init(firstCustomer: String) {
        self.customers = [firstCustomer]
```

```
    }
}

var bankQueue = Queue(firstCustomer: "Alice")

bankQueue.customers.append("Bob")
bankQueue.customers.append("Charlie")
print(bankQueue.customers) // Alice, Bob, Charlie
```

Listing 8.21 Struct Access After Initialization

Note that the init function in structs is completely optional. None of our structs before this section had an init and they worked just fine. init is used for special requirements, when the instantiation of a structure involves a complex logic beyond basic value assignments.

8.5.2   Nonmutating Functions

Naturally, initialization is not the only spot where a struct may contain functions. You're free to throw in as many functions as you need. *Nonmutating functions* are struct functions that don't mutate any struct property. Listing 8.22 demonstrates two nonmutating functions in the Queue struct.

```
struct Queue {
    var customers: [String]

    init(firstCustomer: String) {
        self.customers = [firstCustomer]
    }

    func printCustomers() {
        for customer in self.customers { print(customer) }
    }

    func containsCustomer(_ subject: String) -> Bool {
        for customer in self.customers {
            if customer == subject { return true }
        }

        return false
    }
}
```

Listing 8.22 Two Nonmutating Struct Functions

printCustomers loops through the queue customers and prints out their names. containsCustomer loops through queue customers and returns true if the given customer is in the queue. Listing 8.23 demonstrates calls to those functions; note that they aren't too different from regular function calls. The only difference is that the variable name needs to be a prefix during the call, as in bankQueue.printCustomers().

```
struct Queue {
    var customers: [String]

    init(firstCustomer: String) {
        self.customers = [firstCustomer]
    }

    func printCustomers() {
        for customer in self.customers { print(customer) }
    }

    func containsCustomer(_ subject: String) -> Bool {
        for customer in self.customers {
            if customer == subject { return true }
        }

        return false
    }
}

var bankQueue = Queue(firstCustomer: "Alice")
bankQueue.customers.append("Bob")
bankQueue.customers.append("Charlie")

bankQueue.printCustomers()              // Alice, Bob, Charlie
bankQueue.containsCustomer("Bob")   // true
bankQueue.containsCustomer("David") // false
```

Listing 8.23 Calling Nonmutating Struct Functions

8.5.3 Mutating Functions

As their name implies, *mutating functions* are struct functions that may change values of struct properties. Struct functions are nonmutating by default; they need to be marked with the mutating prefix to gain that privilege.

Listing 8.24 offers an alternative Queue implementation with two simple mutating functions. addCustomer adds the next customer to the tail of the queue, while addVipCustomer adds them to the head.

```
struct Queue {
    var customers: [String] = []

    mutating func addCustomer(_ customer: String) {
        customers.append(customer)
    }

    mutating func addVipCustomer(_ customer: String) {
        customers.insert(customer, at: 0)
    }
}

var queue = Queue()
queue.addCustomer("Bob")
queue.addVipCustomer("Alice")
print(queue.customers[0]) // Alice
```

Listing 8.24 Struct with Mutating Functions

Without the mutating prefix, none of those functions would be allowed to make changes
to customers. But now they can! Note that queue was declared using var to allow changes
after initialization. If it was declared using let, it would become read-only and you
wouldn't be able to call its mutating functions.

8.5.4   Static Functions

You learned about static properties in Section 8.3. Logically speaking, static functions
follow the exact same principle. They are functions defined on the struct level, not on
the instance level. Because the static concept was explained already, let's move to the
code sample in Listing 8.25.

```
struct Student {
    var name: String
    var grades: [Double]

    var averageGrade: Double {
        guard !grades.isEmpty else { return 0 }
        return grades.reduce(0, +) / Double(grades.count)
    }

    static func findTopStudent(_ students: [Student]) -> Student? {
```

```
        return students.max(by: { $0.averageGrade < $1.averageGrade })
    }
}
```

Listing 8.25 Struct with Static Function

Here you have a struct called Student. Following its mutable properties, name and grades, and computed property, averageGrade, Student.findTopStudent comes up as a static function. This function accepts an array of Student instances and returns the student with the best average grade.

Listing 8.26 demonstrates how to call a static struct function. Just like static properties, you need to provide the struct's name as a prefix, making the call Student.findTopStudent. In this case, Alice is returned as the top student because she has the best average grade.

```
struct Student {
    var name: String
    var grades: [Double]

    var averageGrade: Double {
        guard !grades.isEmpty else { return 0 }
        return grades.reduce(0, +) / Double(grades.count)
    }

    static func findTopStudent(_ students: [Student]) -> Student? {
        return students.max(by: { $0.averageGrade < $1.averageGrade })
    }
}

let alice = Student(name: "Alice", grades: [90, 80, 77])
let bob = Student(name: "Bob", grades: [85, 92, 44])
let marc = Student(name: "Marc", grades: [15, 30, 22])

if let top = Student.findTopStudent([alice, bob, marc]) {print(top.name)} // Al-
ice
```

Listing 8.26 Calling Static Function

Technically speaking, findTopStudent could have been a global function instead, residing outside of the Student struct. But in a large codebase with dozens of structs, having many global functions may be a source of confusion and mayhem. Planting such functions in their related structs keeps things tidy and easy to navigate/understand. Besides, static struct functions have access to private static struct properties too, which isn't an option with global functions.

Private properties? What are those? To answer that question, we need to enter the territory of encapsulation, which is our next topic.

8.6   Encapsulation

Encapsulation is one of the pillars of object-oriented programming, in which programmers can restrict external access to some elements to protect data integrity. Although structs are not classes, they support encapsulation to a certain degree via private properties and functions.

By default, all properties and functions of a struct are *public*; this means that any caller can read them and change them if they are mutable. That's how all former examples in this chapter operated.

In special cases, though, you can mark some struct components as *private*, which makes those properties inaccessible and invisible to the outer world of callers. In many cases, this makes complex structs safe and simple, protecting their inner mechanisms while simplifying their interfaces.

In fact, encapsulation is not a programming concept alone. We encounter encapsulation in our daily lives all the time! Think about a piano, for instance. The white and black keys are its public section; anyone can access them to make some noise. However, the underlying strings and hammers are its private section, hidden away from common folk. Only authorized repair technicians should access them.

If the inner mechanisms were exposed to everyone, then some people might damage the instrument by trying to attack the strings and hammers manually instead of using the safe keys on top.

Table 8.3 features some other daily life examples of encapsulation.

Entity	Public	Private
Piano	Keys, pedals	Strings, hammers
Car	Wheel, radio	Engine, battery
Phone	Screen, buttons	Processor, memory
Bank	Door, clerk	Safe, server

Table 8.3 Encapsulation in Real Life

The same principle applies to structs. In a complex struct holding critical data, it may be a good idea to expose as little as possible, hiding values from sight (like the strings of a piano). Callers would access the values through computed properties or functions that contain defensive programming and safety measures (like the keys of a piano).

Now that the concept is clear, let's discover the syntax and some use cases for private properties and functions.

8.6.1   Private Properties

To introduce the syntax of private properties, we will implement an example centered on products, their prices, and their tax amounts. Listing 8.27 presents the initial definition of the Product struct. Note that the public and private keywords have been thrown in: Although the public keyword is optional, we included it to distinguish public from private properties more clearly.

```
struct Product {
    public var name: String
    private var basePrice: Double = 0
    private var taxAmount: Double = 0

    public var price: Double {
        get {
            return basePrice + taxAmount
        }
        set (newPrice) {
            basePrice = newPrice / 1.2
            taxAmount = newPrice - basePrice
        }
    }
}
```

Listing 8.27 Struct with Private Properties

In this struct, name is a public property, like all properties you have seen before. It can be accessed freely by any caller; that is, its value can be read or changed. Meanwhile, basePrice and taxAmount are private properties. They are invisible and inaccessible to callers; they can only be read and modified by computed properties or functions of the struct.

Why, you might ask? The main reason in this example is data consistency. The base price and tax amount of a product are related; the tax amount ought to be 20% of the base price. So if the base price is 1,000, the tax amount will be 200 and the total price will be 1,200.

If you make these variables public, then a caller might mistakenly apply a tax amount of 300, for instance—and thus cause miscalculations and trouble for the business.

To prevent such risks, you can simply cloak basePrice and taxAmount, keeping them out of sight. Instead of exposing them, you declare a public computed property called price.

Its getter calculates the price by adding the base price to the tax amount as *1,000 + 200 = 1,200*. Its setter breaks down the total price (1,200) into the base price (1,000) and tax amount (200).

Let's see this all in action! In Listing 8.28, we enhanced the example by throwing in an init as well as a printPriceSchema function to print out all the values at hand.

```
struct Product {
    public var name: String
    private var basePrice: Double = 0
    private var taxAmount: Double = 0

    public var price: Double {
        get {
            return basePrice + taxAmount
        }
        set (newPrice) {
            basePrice = newPrice / 1.2
            taxAmount = newPrice - basePrice
        }
    }

    public init(_ name: String, price: Double = 0) {
        self.name = name
        self.price = price
    }

    public func printPriceSchema() {
        print("Base price: \(basePrice)")
        print("Tax amount: \(taxAmount)")
        print("Total price: \(price)")
    }
}

var guitar = Product("Guitar", price: 1200)
guitar.printPriceSchema()
```

Listing 8.28 Caller Example for Product Struct

In the main flow (caller), you declare a new guitar variable of type Product. The provided price = 1200 argument will be broken down as basePrice = 1000 and taxAmount = 200 automatically over the computed price property. The output of guitar.printPriceSchema() is shown in Figure 8.5 as evidence of the calculation's accuracy.

```
Base price: 1000.0
Tax amount: 200.0
Total price: 1200.0
```

Figure 8.5 Output of Guitar's Price

Even if you tried to access them, guitar.basePrice and guitar.taxAmount would not be accessible in the main flow simply because they are marked as private. You could, for instance, set guitar.price = 3000 because guitar.price is public. In that case, the computed price property would set basePrice = 2500 and taxAmount = 500 internally, without leaving any space for called errors.

> **Benefits of Private Properties**
>
> This example should have materialized the benefits of private properties. This feature lets you hide internal implementation details, preventing unintended errors due to external access.
>
> Maintainability is also a significant bonus. In the preceding example, if the tax rate increases from 20% to 25% someday, all you need to do is to change the code of the computed price property. You don't need to make any modifications to the callers, which is a huge benefit in large codebases.

8.6.2   Private Functions

In addition to private properties, a struct can contain private functions, too, which can't be accessed by the outside world of callers. Such functions are typically used for internal purposes and have no value to callers.

Listing 8.29 features the Player struct, containing a private function called produceRandomName. In this struct's init, the caller has the option to provide a player name. If no name has been provided, then a random player name will be generated using produceRandomName.

```
struct Player {
    let name: String

    init(playerName: String? = nil) {
        if let playerName {
            self.name = playerName
        } else {
            self.name = Player.produceRandomName()
        }
    }
}
```

```
    private static func produceRandomName() -> String {
        let names = ["Alice", "Bob", "Charlie", "David"]
        let surnames = ["Smith", "Johnson", "Williams", "Jones"]
        return names.randomElement()! + " " + surnames.randomElement()!
    }
}

let player1 = Player(playerName: "Joe Mall")
print(player1.name) // Joe Mall

let player2 = Player()
print(player2.name) // Bob Johnson (random value)
```

Listing 8.29 Struct with Private Function

Note that produceRandomName is only used internally if needed. Exposing this function to the outside world has no value for a caller. In that case, it is better to hide it by marking it as private, thus simplifying the interface of the struct.

In the main flow, no custom name was provided for player2. In that case, produceRandomName is called internally and a random name is generated on behalf of the caller.

Expose as Little as Possible

Here's some general advice: A struct should expose as little as possible, hiding its inner mechanisms out of sight. Any property or function not needed by callers should be marked as private. That way, a "guest programmer" can understand the scope and purpose of a struct easily without losing time by analyzing its inner mechanisms. Such encapsulation also prevents programming errors due to unintended access to internals—like accessing the engine of a car directly instead of stepping on the pedals.

8.7 Structs as Function Parameters

As you will remember from Chapter 5, functions are able to import, return, or modify variables. Likewise, functions can be used to import, return, or modify struct variables too. Although the syntax is not any different, we will present some examples to keep this chapter self-contained.

Listing 8.30 demonstrates a struct import example. Within the ClassRoom struct, there is a addStudent function, which accepts a Student instance as a parameter to append to its array of students. As stated, there's nothing special about the syntax of the function; we're merely demonstrating an example.

```
struct Student {
    let name: String
}

struct ClassRoom {
    private var students: [Student] = []

    mutating func addStudent(_ student: Student) {
        students.append(student)
    }
}
```

Listing 8.30 Function Importing a Struct

Listing 8.31 contains an enhanced version of the example, in which getFirstStudent()
returns a Student instance—which is a struct.

```
struct Student {
    let name: String
}

struct ClassRoom {
    private var students: [Student] = []

    mutating func addStudent(_ student: Student) {
        students.append(student)
    }

    func getFirstStudent() -> Student? {
        return students.isEmpty ? nil : students[0]
    }
}
```

Listing 8.31 Function Returning a Struct

Finally, Listing 8.32 demonstrates a function that changes a struct. Teacher.assignStudent
accepts a Student instance as an inout parameter to be allowed to change it. You could
have done the same for a simple variable or a collection too; the syntax doesn't change
for structs.

```
struct Teacher {
    let name: String

    mutating func assignStudent(_ student: inout Student) {
        student.advisor = self
    }
```

```
}

struct Student {
    let name: String
    var advisor: Teacher?
}

var bob = Teacher(name: "Bob")
var alice = Student(name: "Alice")
bob.assignStudent(&alice)
print(alice.advisor?.name ?? "")    // Bob
```

Listing 8.32 Function Changing a Struct

Before we get to advanced features, let's look at how another former topic applies to structs. This time, we will inspect generics and their use with structs.

8.8   Generic Structs

In Chapter 5, you learned about generics in functions, and in Chapter 7, you learned that generics can be applied to enumerations too. Generics can be used with structs too! The syntax isn't a far cry from what you've learned so far either, so this should be a smooth ride.

As the platform of our case study, Listing 8.33 features two structs: Student and Teacher. Note that they are not identical; Student has the name and grade properties, while Teacher has the name and subject properties.

```
struct Student {
    let name: String
    let grade: Int
}

struct Teacher {
    let name: String
    let subject: String
}
```

Listing 8.33 Two Simple Structs

The purpose will be to build a third struct, which represents a committee of either students or teachers. For this goal, you can use generics, as demonstrated in Listing 8.34.

```
struct Student {
    let name: String
    let grade: Int
}

struct Teacher {
    let name: String
    let subject: String
}

struct Committee<T> {
    private var people: [T] = []

    mutating func append(_ person: T) {
        people.append(person)
    }

    func printAll() {
        for person in people { print(person) }
    }
}
```

Listing 8.34 Generic Committee Struct

We threw in a generic struct called Committee, which can be used to collect any data type, as symbolized by <T>. Thanks to that flexibility, you can use Committee to build a collection of Student or Teacher instances. Without generics, you would have to declare two distinct structs, StudentCommittee and TeacherCommittee, with identical functionality.

Now, let's see the Committee struct in action! In Listing 8.35, code snippets to build different Committee instances have been added.

```
struct Student {
    let name: String
    let grade: Int
}

struct Teacher {
    let name: String
    let subject: String
}

struct Committee<T> {
    private var people: [T] = []
```

```
    mutating func append(_ person: T) {
        people.append(person)
    }

    func printAll() {
        for person in people { print(person) }
    }
}

var debateClub = Committee<Student>()
debateClub.append(Student(name: "Alice", grade: 80))
debateClub.append(Student(name: "Bob", grade: 81))
debateClub.printAll()

var emergencyTeam = Committee<Teacher>()
emergencyTeam.append(Teacher(name: "Alex", subject: "History"))
emergencyTeam.append(Teacher(name: "Rachel", subject: "Math"))
emergencyTeam.printAll()
```

Listing 8.35 Building Committees of Different Types

In the first part, debateClub has been declared as a Committee to accept Student instances. In the second part, emergencyTeam has been declared as a Committee to accept Teacher instances. Cool, right? If you had other similar structs, such as Janitor, Driver, and the like, you could build Committee instances out of those too.

Other generic features you learned about in Chapter 5, such as type constraints and where clauses, can be applied to structs as well in exactly the same way. We won't repeat that content here, but you are more than welcome to turn it into a small exercise!

8.9   Advanced Features

In this final section, you will learn about some advanced features of structs, which go beyond their daily use cases and are typically used when a special need arises.

8.9.1   Lazy Properties

In Swift structs, *lazy properties* are used to defer the initialization of a property until it is first accessed. Once the lazy property is calculated, it is cached automatically; the calculated result then is returned on later calls. This is useful in terms of performance improvement, especially when the computation of the property is expensive and might not be needed in every instance of the struct.

Let's go over the concrete example in Listing 8.36, where the averageScore lazy property is featured. This property is simply calculating the average value of scores.

```
struct Student {
    let name: String
    let scores: [Int]

    lazy var averageScore: Double = {
        guard !scores.isEmpty else { return 0.0 }
        return Double(scores.reduce(0, +)) / Double(scores.count)
    }()

    init(name: String, scores: [Int]) {
        self.name = name
        self.scores = scores
    }
}

var alice = Student(name: "Alice", scores: [80, 90, 70])
var bob = Student(name: "Bob", scores: [90, 95])
print(alice.averageScore)    // 80 - calculated
print(alice.averageScore)    // 80 - from cache
```

Listing 8.36 Lazy Property Demonstration

The catch is in the lazy prefix, which turns the computed averageScore property into a lazy property. The first time alice.averageScore is called, the calculation is executed and the average value of **80** is displayed. On the second call, the value is called from the cache without rerunning the calculation of averageScore. For bob, no averageScore calculation ever took place because the lazy variable is not accessed at all.

Lazy properties shine in performance-hungry computations, which might be called several times throughout the lifecycle of the struct. Swift ensures that they are executed only once and cached. Some further use cases include the following:

- Thumbnail generation for an image
- Loading large data files
- Database or network queries for stable data
- Setup of optional app features

Despite the performance benefits, lazy properties have a common pitfall to be aware of: Because the property is cached only once, its value may become invalid if its source values change. This pitfall is demonstrated in Listing 8.37, where Student.scores was declared as a mutable property using the var prefix.

```
struct Student {
    let name: String
    var scores: [Int]

    lazy var averageScore: Double = {
        guard !scores.isEmpty else { return 0.0 }
        return Double(scores.reduce(0, +)) / Double(scores.count)
    }()

    init(name: String, scores: [Int]) {
        self.name = name
        self.scores = scores
    }
}

var alice = Student(name: "Alice", scores: [80, 90, 70])
print(alice.averageScore)   // 80 - calculated
alice.scores.append(contentsOf: [100, 100, 100])
print(alice.averageScore)   // 80 - from cache
```

Listing 8.37 Invalid Lazy Property Value

The first time alice.averageScore was accessed, it was calculated with the scores at hand: *(80 + 90 + 70) / 3 = 80* was the obvious result, and *80* was automatically cached for the future.

Later, alice.scores was extended with further values. Alice's average score was supposed to be *(80 + 90 + 70 + 100 + 100 + 100) / 6 = 90* now, but alice.averageScore was cached already; it still returned the former value of *80*, which is obviously incorrect.

The lesson in this example is that lazy properties should be based on immutable values. In Listing 8.36, Student.scores was immutable due to its let prefix, posing no invalidity risk for averageScore. As soon as we made Student.scores mutable, it put Student.averageScore in jeopardy.

This principle doesn't change even if you clone a struct. In Listing 8.38, alice was cloned as bob.

```
struct Student {
    let name: String
    var scores: [Int]

    lazy var averageScore: Double = {
        guard !scores.isEmpty else { return 0.0 }
        return Double(scores.reduce(0, +)) / Double(scores.count)
    }()
```

```
    init(name: String, scores: [Int]) {
        self.name = name
        self.scores = scores
    }
}

var alice = Student(name: "Alice", scores: [80, 90, 70])
print(alice.averageScore)    // 80 - calculated

var bob = alice
bob.scores.append(contentsOf: [100, 100, 100])
print(bob.averageScore)    // 80 - from cache
```

Listing 8.38 Invalid Lazy Property Value in Cloned Struct

Being a clone, bob.averageScore inherited its cached value from alice, preventing a recalculation. This would be a strong feature if scores were immutable, but in this case, the mutability of scores caused the invalidity of the cached value.

As a summary, lazy properties are powerful components of structs in terms of performance—so long as they are based on stable, immutable values.

8.9.2   Property Observers

In Swift, *property observers* allow you to listen to changes in a property's value and respond when the change happens. That way, you have the flexibility to act right before or right after the value change.

In Listing 8.39, you can see that two property observers for Car.price. willSet are executed before the value change takes place, and didSet is executed after the value change takes place.

```
struct Car {
    var make: String
    var model: String

    var price: Double {
        willSet(newValue) {
            print("Changing price from \(price) to \(newValue)")
        }
        didSet(oldValue) {
            print("Changed price from \(oldValue) to \(price)")
        }
    }
}
```

```
var myCar = Car(make: "Toyota", model: "Camry", price: 25000.0)
myCar.price = 30000.0
```

Listing 8.39 Property Observers in Action

In willSet, price carries the current value and newValue carries the incoming value about to be set. In didSet, oldValue carries the former value and price carries the most current value, which has just been set.

The output of this example is shown in Figure 8.6. The first line was produced by willSet and the second line was produced by didSet.

```
Changing price from 25000.0 to 30000.0
Changed price from 25000.0 to 30000.0
```

Figure 8.6 Property Observer Output

In certain cases, property observers can prove to be very useful. Some use cases are listed in Table 8.4.

Use Case	Explanation
User interface updates	A property observer can update the UI whenever the value is changed.
State synchronization	A property observer can post value changes to the server immediately.
Logging	A property observer can log changes to sensitive values.
Triggers	A property observer can trigger notifications or events when a value is changed.

Table 8.4 Sample Use Cases for Property Observers

Avoid Recursive Updates

Be cautious when it comes to recursive updates. If two properties have observers that update each other, then your app might end up caught in an infinite loop, resulting in the app freezing and crashing.

8.9.3   Property Wrappers

In Swift, a *property wrapper* is a special type that "wraps" a value to attach additional logic to it. To understand its use, check out the demonstration in Listing 8.40.

```
@propertyWrapper
struct UpperCased {
    private var text: String

    init(wrappedValue: String) {
        self.text = wrappedValue.uppercased()
    }

    var wrappedValue: String {
        get { self.text }
        set { self.text = newValue.uppercased() }
    }
}

struct Student {
    @UpperCased
    var name: String
}

var joe = Student(name: "joe brave")
print(joe.name) // JOE BRAVE
```

Listing 8.40 Property Wrapper Demonstration

In this example, there is a property wrapper called UpperCased. Its purpose is to wrap a string to ensure that it's always uppercased—that is, set in capital letters. As sketched in Figure 8.7, clients won't be able to access the text value directly. Instead, they must go through the wrapping logic of UpperCased, enforcing the implemented text transformation.

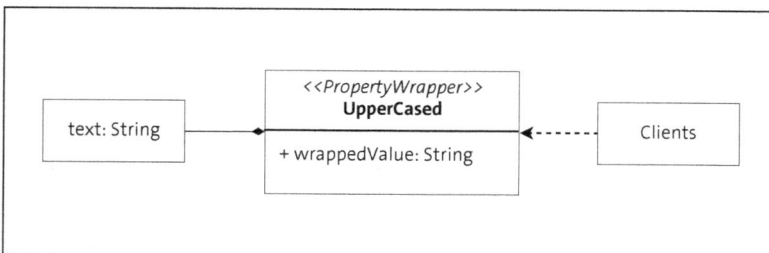

Figure 8.7 Sketch of Wrapper

Although UpperCased is a made-up wrapper name, the @propertyWrapper prefix and the wrappedValue property are requirements for the wrapper to work properly. Swift enforces the wrappedValue name, and you can't rename it to anything else. Its type can vary, though.

Following the declaration, there is a struct called `Student`. Note that `Student.name` was marked with `@UpperCased`. This declaration ensures that `Student.name` is wrapped by `UpperCased`. Any time a student's `name` is read or changed, it will go through the logic defined in `UpperCased`, ensuring that their name is in capitals. This entire chain is sketched in Figure 8.8.

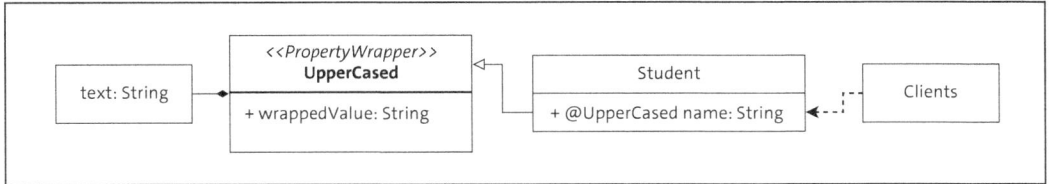

Figure 8.8 Entire Chain Sketched

At this point, the main flow should make sense. Although `joe.name` was set as `"joe brave"`, it went through the wrapper logic and was transformed into `"JOE BRAVE"`.

Naturally, property wrappers offer reusable logic. You can use `UpperCased` for further variables like surname, address, and so on—or even in different structs. Once the wrapper is declared, it can be used anywhere, preventing code duplication elegantly and with high readability. After all, who wouldn't understand the purpose of `UpperCased` at first sight?

8.9.4 Subscripts

In Swift, *subscripts* enable structs to behave like arrays. Once the `subscript` feature is applied, you gain the ability to access a struct using an index, like [1].

Let's examine this feature via the code sample in Listing 8.41. Here there is a struct called `EmployeeTasks`, which has two properties: `responsible` and `tasks`.

```
// Struct with subscript
struct EmployeeTasks {
    let responsible: String
    private var tasks: [String]

    init(responsible: String, tasks: [String]) {
        self.responsible = responsible
        self.tasks = tasks
    }

    subscript(index: Int) -> String {
        get { return tasks[index] }
        set { tasks[index] = newValue }
    }
}
```

```
}

// Adding new tasks
var johnTasks = EmployeeTasks(
    responsible: "John",
    tasks: ["Prepare report", "Send report"])

// Accessing tasks using subscripts
print(johnTasks.responsible)    // John
print(johnTasks[1])             // Send report

// Changing tasks using subscripts
johnTasks[1] = "Send report by email"
print(johnTasks[1])                 // Send report by email
```

Listing 8.41 Simple Struct with Subscript

In the body of the struct, the subscript feature has been implemented. This is not a random name, but a special computed property for index-based access:

- subscript get is triggered when a value is read with an index, such as print(tasks[1]).
- subscript set is triggered when a value is set with an index, such as tasks[1] = "Buy milk".

In the main flow of the example, you start by creating johnTasks with two initial tasks. After this step, print(johnTasks[1]) goes through johnTasks.subscript.get, and ultimately johnTasks.tasks[1] is printed. Finally, johnTasks[1] = "Send report by email" goes through johnTasks.subscript.set, and ultimately johnTasks.tasks[1] is replaced with the new value.

As you can see, subscripts can be considered syntactic sugar in the sense that they offer a concise and expressive syntax but don't add new functionality.

Collections
You saw a similar syntax for collections too, remember? From Chapter 3? Subscripts enable your structs to act like collections.

One cool feature of subscripts is that they can accept multiple parameters, enabling multidimensional structures. Listing 8.42 demonstrates an example, in which AirplaneSeats represents the seats and occupants of a flight.

```
struct AirplaneSeats {
    let rows: Int
    let seatsPerRow: Int
    private var travelers: [String]
```

```
    init(rows: Int, seatsPerRow: Int) {
        self.rows = rows
        self.seatsPerRow = seatsPerRow
        self.travelers = Array(repeating: "", count: rows * seatsPerRow)
    }

    subscript(row: Int, seat: Int) -> String {
        get {
            return self.travelers[(row - 1) * seatsPerRow + (seat - 1)]
        }
        set {
            self.travelers[(row - 1) * seatsPerRow + (seat - 1)] = newValue
        }
    }
}

var seats = AirplaneSeats(rows: 30, seatsPerRow: 6)
seats[1, 3] = "Alice"    // Row 1, Seat 3
seats[2, 5] = "Bob"      // Row 2, Seat 5
seats[4, 1] = "Carol"    // Row 4, Seat 1

print(seats[2, 5])       // Bob
```

Listing 8.42 Multidimensional Subscript Example

In this example, subscript accepts two parameters, row and seat, storing the traveler names in a one-dimensional private array with some minor math magic. That enables the concise syntax seats[2, 5], intuitively representing row and seat numbers.

8.9.5   Failable Initialization

Swift enables structs to have *failable initializers*, which may return nil if the struct instance can't be created. As a demonstration, Listing 8.43 showcases the Pizza struct, in which the small size is not supported; only medium and large pizzas are served.

```
enum FoodSize {
    case small
    case medium
    case large
}

struct Pizza {
    let size: FoodSize

    init?(size: FoodSize) {
```

```
        guard size != .small else { return nil }
        self.size = size
    }
}

let validPizza = Pizza(size: .large)     // OK
let invalidPizza = Pizza(size: .small)  // nil
```

Listing 8.43 Failable Initializer Example

Note the init? expression: The question mark suffix declares it to be a failable initializer, which may return a nil value. The body of the function is as expected: If size is small, then nil is returned instead of creating a Pizza instance.

You see the expected output in the main flow too: validPizza is a realized Pizza instance with large size, whereas invalidPizza is nil because its size has been passed as small.

Although this is a cool feature, it comes with a natural cost: validPizza will be an optional, Pizza?. Every time you access this object, you need to handle it accordingly. Listing 8.44 shows a short corresponding example.

```
enum FoodSize {
    case small
    case medium
    case large
}

struct Pizza {
    let size: FoodSize

    init?(size: FoodSize) {
        guard size != .small else { return nil }
        self.size = size
    }
}

let validPizza = Pizza(size: .large)

if validPizza != nil {
    print("Your \(validPizza!.size) pizza is ready!")
} else {
    print("No pizza for you!")
}
```

Listing 8.44 Handling Optional Pizza Instance

Beyond that simple example, everything you learned about optionals in Chapter 6 will be applicable in this context.

Nil Versus Exception

Some programmers advocate that functions should never return `nil` values and should throw exceptions instead. This makes errors impossible to ignore, reducing the risk of runtime errors.

Because failable initializers return an optional instead of a guaranteed concrete object, Swift is less prone to `nil` access than some other languages. The preceding example had to consider the possibility of a `Pizza` instance to be `nil` every time you accessed it, for example.

However, force-unwrapping `invalidPizza!` might still cause a runtime error. To be on the safe side, you might still consider throwing an exception instead of returning a `nil` value.

You will learn about exceptions and error handling in Chapter 12, at which point this mechanism will become clearer.

8.10 Summary

In this chapter, you learned about structs in Swift, which can be considered data container templates. Typically, structs are used to store and represent the properties of an entity. Beyond simple values, structs can also contain computed properties and functions, which increases their flexibility. Structs even offer encapsulation, via which their components can be marked as private to limit the access of callers. Finally, generics can be applied to structs too.

When all those features are brought together, the borders between structs and classes get blurry—and they certainly overlap in many cases. However, they have significant differences and are typically used for different purposes. As you learn about classes in the next chapter, their similarities and differences should become clear.

Chapter 9
Classes

In Swift, classes are reference types used to define blueprints for objects and methods. Classes are arguably the core entry point for object-oriented Swift, as they support inheritance, initialization, and deinitialization. Classes are ideal for modeling complex and shared elements.

In this chapter, you will learn about *classes*—the conceptual pillars of object-oriented programming. In Swift, classes are very similar to structs, but they have significant differences. While their functionalities overlap, they can't fully replace each other as both offer different mechanisms. As an introduction to classes, let's compare them to structs to understand where each stands in the Swift ecosystem. Table 9.1 acts as an entry point, highlighting the similarities between structs and classes.

Feature	Description
Properties	Both feature concrete and computed properties.
Functions	Both can contain functions, including an initializer.
Encapsulation	Both can mark their components as public or private.
Protocols	Both can adopt protocols to conform to shared interfaces (see Chapter 10).
Extensions	Both can be extended to add functionality (see Chapter 11).
Type Safety	Both support Swift's type system features, like generics and optionals.

Table 9.1 Struct and Class Similarities

If you focus on the similarities alone, you may think that these are nearly the same thing. But wait: Your mind will change when you see the differences, highlighted in Table 9.2.

Feature	Struct	Class
Semantics	Value type; copied on assignment	Reference type; shared on assignment
Mutability	Immutable by default; requires mutating for function mutations	Mutable by default, even with let

Table 9.2 Struct and Class Differences

Feature	Struct	Class
Inheritance	Not possible	Supports single inheritance
Deinitialization	Not possible	Supports deinit for cleanup on deallocation
Performance	Faster; stored in stack	Slower; stored in heap

Table 9.2 Struct and Class Differences (Cont.)

Some of the keywords in Table 9.2 might be foreign to some of you, but as you proceed through this chapter, any knowledge gaps will be filled in through examples. Basically, structs excel as lightweight, immutable data containers, while classes shine in scenarios that require a shared state and complex behavior.

When you need to make a choice between a struct or a class, here is a simplified initial anchor point: You can use structs by default; so long as you can get away with it, you'll be fine. You can "upgrade" to classes when you need their specific differing features. Those differing features are the very subject of this chapter, so let's start inspecting them.

9.1   Declaring Classes

In this section, we will warm you up with classes by declaring some basic classes and deriving objects from them. The examples may look very similar to structs at this point, but in due course we will highlight their key differences.

Struct properties and class properties mostly share the same functionality; therefore, this topic will not be readdressed in detail. What you learned in Chapter 8 can be applied to classes too.

9.1.1   Basic Class Declaration

Let's start with a simple class and go from there. Listing 9.1 features the Player class, which has two properties and a function.

```
class Player {
    var name: String
    var health: Int

    init(name: String) {
        self.name = name
        self.health = 100
    }
```

```
    func greet() {
        print("Hello, my name is \(self.name)")
    }
}

let player1 = Player(name: "Alice")
player1.greet() // Hello, my name is Alice
```

Listing 9.1 Simple Class with Property and Function

The `Player` class features the `name` and `health` properties, as well as an initializer and a greet function—just like a struct could.

So far, the syntax isn't too different from a struct, right? We have merely replaced the struct keyword with the `class` keyword; the rest looks the same. However, looks may be deceiving. Proceed to the next topic to discover the core difference.

9.1.2 Reference Semantics

One of the core features of classes is that they are *reference* types, not *value* types. This potentially confusing sentence deserves an clarifying example, so here we go. Listing 9.2 features a simplified version of the `Player` class.

```
class Player {
    var name: String = "Empty"
    var health: Int = 100
}

var p1 = Player()
var p2 = p1
p1.name = "Bob"
print(p2.name) // Bob
```

Listing 9.2 Reference Semantics Demonstration

In the main flow, you declare p1 as a `Player()`. Before making any changes, you make the p2 = p1 assignment. After this assignment, when you change p1.name to "Bob", p2.name also becomes "Bob".

What? How? Is that a bug?

No, it's not; it's a feature. In fact, this is one of the core features of classes.

Unlike structs, p1 and p2 aren't value containers. They are simply pointers to a `Player` object. The p2 = p1 assignment doesn't create a copy of an existing object; instead, the reference of the object is copied from p1 to p2. As a result, both p1 and p2 point to the same object.

Therefore, a change to p1.name is reflected in p2.name too; both variables are merely mirrors of the name property of the same Player object.

Let's use a visual to understand this mechanism a little better. When you make the p1 = Player() assignment, the computer behaves as illustrated in Figure 9.1. An object instance of the Player class is created, and p1 acts as a reference to that object.

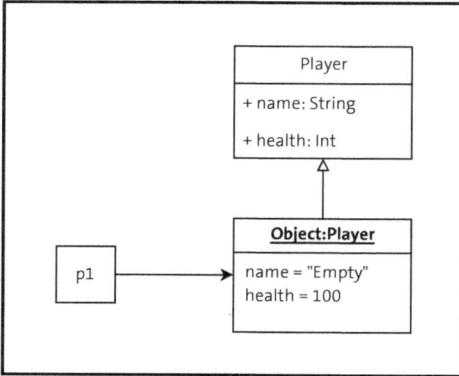

Figure 9.1 Result of p1 Assignment

It's important to understand that p1 doesn't hold any values (unlike a struct). Instead, p1 merely points to the object holding the values.

In the following line, you make the p2 = p1 assignment. As a result, the computer behaves as illustrated in Figure 9.2. Note that the existing object was not cloned (unlike a struct); instead, the p1 reference variable was copied to p2, and both point to the same object now.

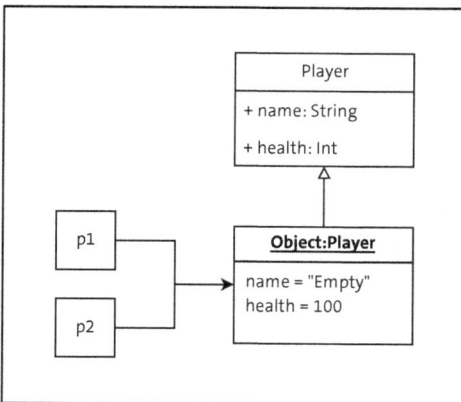

Figure 9.2 Result of p2 Assignment

At this point, setting p1.name = "Bob" or setting p2.name = "Bob" has the exact same effect as both variables point to the same object in the memory.

To contrast this feature with structs, Figure 9.3 demonstrates the behavior you'd see if Player was a struct instead of a class. In that case, p1 and p2 would be value types instead of reference types, so p2 = p1 would produce a copy of the values in p1 as p2 and isolate them forever. Changes to p1.name would not be reflected in p2.name because they are separate containers.

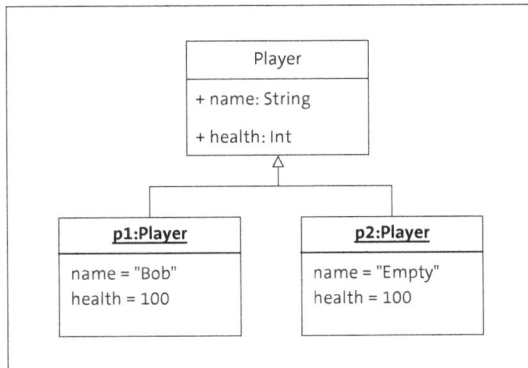

Figure 9.3 Player as Struct

This distinction should make the value/reference difference between structs and classes clear. A struct instance is a value type, acting as a concrete container for properties. A class instance is instead a reference type, acting as pointer to an object.

From that point of view, a struct is appropriate for cases where you want an assignment like p2 = p1 to copy all values from p1 to p2. Imagine a paint app in which you have a circle on the canvas. If the user duplicates the circle, causing the circle2 = circle1 assignment, you will want circle2 to be a new distinct circle, right? For that case, the Circle type would be better as a struct.

Imagine that the same app stores values in a database. A DatabaseConnection managing a single connection shared across multiple components could be declared as a class. Even if you make a copy via dbConn2 = dbConn1, both variables point to the same object, thus preventing unnecessary duplications and connections.

Now that we've gotten this core difference out of the way, let's move forward to other class features.

9.1.3   Composition

Classes support *composition*, which means that a class can hold another class as a property. That sounds simple enough—but combined with the reference-based logic of classes, composition turns into a powerful feature.

Listing 9.3 demonstrates composition through a simple example, using the Customer and Invoice classes. Note that Invoice.customer is of the Customer type.

```
class Customer {
    var name: String

    init(name: String) {
        self.name = name
    }
}

class Invoice {
    let customer: Customer
    let amount: Double

    init(customer: Customer, amount: Double) {
        self.customer = customer
        self.amount = amount
    }
}
```

Listing 9.3 Composition Example

Composition enables objects to exist independently while maintaining a connection by referring to each other, as shown in Figure 9.4. Because the properties of the customer, such as name, birthday, and so on, are already declared in Customer, you don't need to duplicate them in Invoice as customerName, customerBirthday, and so on. Instead, Invoice can simply refer to Customer, and all properties and functions become accessible, as in invoice.customer.name.

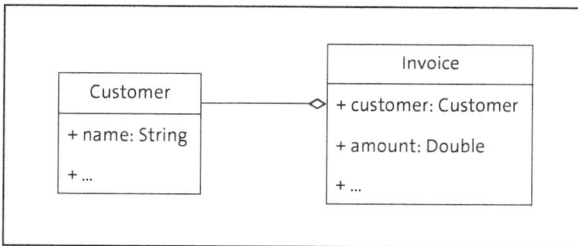

Figure 9.4 Composition Visualized

This access logic is demonstrated in Listing 9.4. Note that invo1.customer.name and invo2.customer.name produce the same output as they point to the same Customer object.

```
class Customer {
    var name: String

    init(name: String) {
```

```
            self.name = name
    }
}

class Invoice {
    let customer: Customer
    let amount: Double

    init(customer: Customer, amount: Double) {
        self.customer = customer
        self.amount = amount
    }
}

let cust = Customer(name: "Bob")
let invo1 = Invoice(customer: cust, amount: 100.0)
let invo2 = Invoice(customer: cust, amount: 200.0)
print(invo1.customer.name)  // Bob
print(invo2.customer.name)  // Bob
```

Listing 9.4 Accessing Invoice Customer

This is important to understand: `invo1` and `invo2` are separate `Invoice` instances with different amounts. However, both point to `cust`, a common `Customer` instance. This means that both invoices were issued for the same customer. They don't contain distinct `Customer` instances; they point to the same `Customer` instance, which is Bob in this case. Check Figure 9.5 for a visualization.

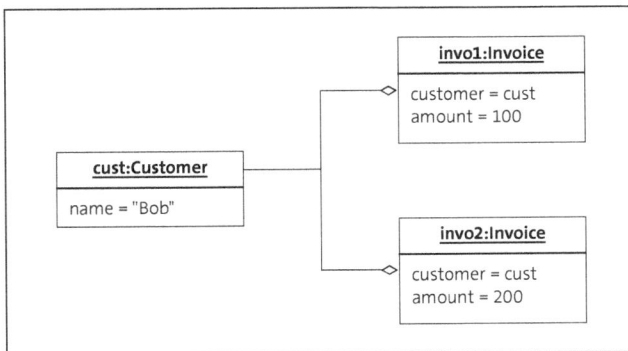

Figure 9.5 Invoice Objects Referencing Same Customer Object

This means that if you change `cust.name` centrally, then `invo1.customer.name` and `invo2.customer.name` will change too as their `Invoice.customer` property is merely a reference to, mirror of, or pointer to the same `Customer` instance. Check Listing 9.5 for a demonstration.

```
class Customer {
    var name: String

    init(name: String) {
        self.name = name
    }
}

class Invoice {
    let customer: Customer
    let amount: Double

    init(customer: Customer, amount: Double) {
        self.customer = customer
        self.amount = amount
    }
}

let cust = Customer(name: "Bob")
let invo1 = Invoice(customer: cust, amount: 100.0)
let invo2 = Invoice(customer: cust, amount: 200.0)

cust.name = "Jane"
print(invo1.customer.name)   // Jane
print(invo2.customer.name)   // Jane
```

Listing 9.5 Renaming Customer's Name

Here is where things get interesting: What if Customer was a struct instead of a class? There's only one way to find out! Listing 9.6 features nearly the same code example, except that Customer is a struct now.

```
struct Customer {
    var name: String

    init(name: String) {
        self.name = name
    }
}

class Invoice {
    let customer: Customer
    let amount: Double
```

```
    init(customer: Customer, amount: Double) {
        self.customer = customer
        self.amount = amount
    }
}

var cust = Customer(name: "Bob")
let invo1 = Invoice(customer: cust, amount: 100.0)
cust.name = "Michael"
let invo2 = Invoice(customer: cust, amount: 200.0)

cust.name = "Jane"
print(invo1.customer.name)   // Bob
print(invo2.customer.name)   // Michael
```

Listing 9.6 Customer as Struct

Remember: Structs are value types, while classes are reference types. Every time you make a struct assignment, the struct instance is cloned as an independent instance—unlike class instances, where only a reference is passed around.

In Listing 9.6, when cust is passed to invo1, it is cloned into invo1.customer, an independent instance. Likewise, when cust is passed to invo2, it is cloned into invo2.customer. This mechanism is visualized in Figure 9.6.

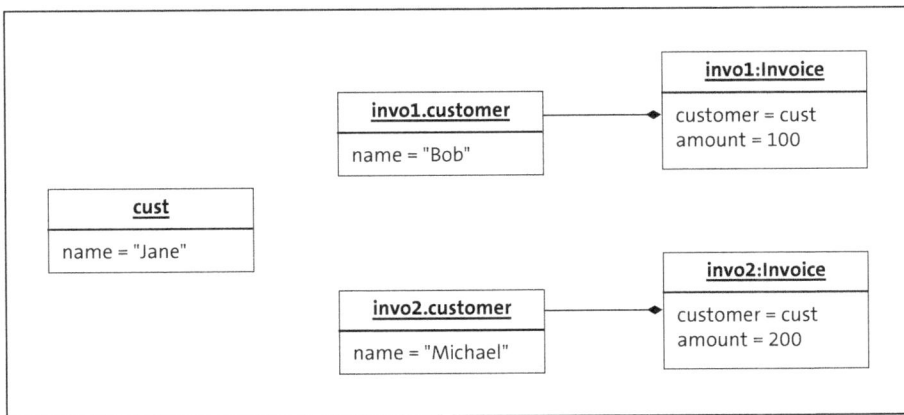

Figure 9.6 Final Form of Customer Structs

At this point, cust, invo1.customer, and invo2.customer are completely isolated, individual instances—unlike with classes, where they would point to the same object.

For this example, declaring Customer as a class makes more sense. If you do some object-oriented thinking, you'll consider that the customer for both invoices is literally the same singular person walking the earth. Both physical invoices point to the same human being. The program structure should reflect this natural phenomenon by making Customer a class so that each Invoice instance can point to the same object. Otherwise, it would look like you are cloning your customers to create separate human beings.

This makes sense, right? Reflecting the natural phenomenon in your code structure is a good compass in many cases if you need to make such a design choice.

Now that we've solidified your understanding of the reference mechanism of classes, we can move forward to functions and see what classes have to offer beyond structs.

9.2   Functions

In Chapter 8, you learned a lot about functions of structs. Nearly all this knowledge is applicable to classes too. There are only a few notable differences, which are covered in the following sections.

9.2.1   Mutability

For structs, methods that modify properties must be marked with the mutating keyword as Swift assumes immutability otherwise. Classes are reference types, however, and their instances are mutable when modified through a reference. In that sense, class functions don't require the mutating keyword to change property values. This difference is highlighted via an otherwise identical struct and class in Listing 9.7.

```
struct PersonStruct {
    var age: Int = 0

    mutating func makeOlder() {
        self.age += 1
    }
}

class PersonClass {
    var age: Int = 0

    func makeOlder() {
        self.age += 1
    }
}
```

Listing 9.7 Mutating Function in Struct and Class

As you can see, even though PersonClass.makeOlder was not marked as mutating, it can change self.age.

9.2.2 Deinitialization

In Swift, *deinitialization* is the process of cleaning up resources or performing final tasks right before a class instance (an object) is destroyed and removed from memory.

Listing 9.8 features a deinitialization example. The sample LogFile class performs a simulation of opening and closing a file. Note that this class has an init function as well as a deinit function—the latter of which is the current focus.

```
class LogFile {
    let fileName: String

    init(_ fileName: String) {
        self.fileName = fileName
        print("File \(fileName) opened")
    }

    deinit {
        print("File \(fileName) closed")
    }
}

var myFile: LogFile? = LogFile("sample.txt")
// Some operations with myFile
myFile = nil
```

Listing 9.8 Deinitialization Example

As you know by now, creation of the myFile object triggers the LogFile.init function. The code in LogFile.init is executed automatically at the point of object creation.

Likewise, LogFile.deinit is executed automatically right before the object is destroyed, as in myFile = nil. If LogFile was a real log file handler, deinit would be a great place to save and close the file. This simple example merely prints out simulation texts, as shown in Figure 9.7.

```
File sample.txt opened
File sample.txt closed
```

Figure 9.7 Initialization and Deinitialization Outputs

Beyond the myFile = nil approach, there are many other ways to destroy an object, and deinit would be triggered in all those cases. Listing 9.9 contains a modified version of the example, in which LogFile objects are put into an array.

```
class LogFile {
    let fileName: String
```

```
    init(_ fileName: String) {
        self.fileName = fileName
        print("File \(fileName) opened")
    }

    deinit {
        print("File \(fileName) closed")
    }
}

var myFiles = [LogFile]()
myFiles.append(LogFile("file1.txt"))
myFiles.append(LogFile("file2.txt"))
myFiles.removeAll()
```

Listing 9.9 Deinitialization via Array Reset

When `myFiles.removeAll()` is called to reset the array, all `LogFile` objects within the array are destroyed, triggering their `deinit` code for each. The output of this sample is shown in Figure 9.8.

```
File file1.txt opened
File file2.txt opened
File file1.txt closed
File file2.txt closed
```

Figure 9.8 Array Reset Output

This is a good time to remind you that class instances/objects operate as reference types. The `myFile = LogFile(…)` statement creates a new `LogFile` object in memory, and `myFile` merely acts like a pointer to that object. Likewise, `myFiles.append(LogFile(…))` creates a new `LogFile` object in memory, and `myFiles[1]` acts like a pointer to that object. Both cases are sketched in Figure 9.9.

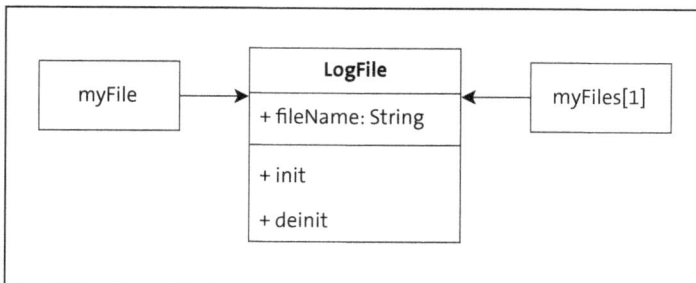

Figure 9.9 LogFile Memory Sketch

When an object has no pointers left, Swift decides that it's time to destroy that object, freeing up memory. That's the exact moment when `deinit` is triggered. So long as an object has a pointer, it won't be deinitialized. Listing 9.10 contains a solid example of this situation.

```swift
class LogFile {
    let fileName: String

    init(_ fileName: String) {
        self.fileName = fileName
        print("File \(fileName) opened")
    }

    deinit {
        print("File \(fileName) closed")
    }
}

var file1 : LogFile? = LogFile("sample.txt")
var file2 = file1
file1 = nil
```

Listing 9.10 Object with Multiple Pointers

In this example, both `file1` and `file2` point to the same object in memory. Even if `file1` is set as `nil`, `file2` still points to the object, preventing deinitialization. Check the sketch in Figure 9.10: To trigger deinitialization, you would have to deallocate `file2` as well.

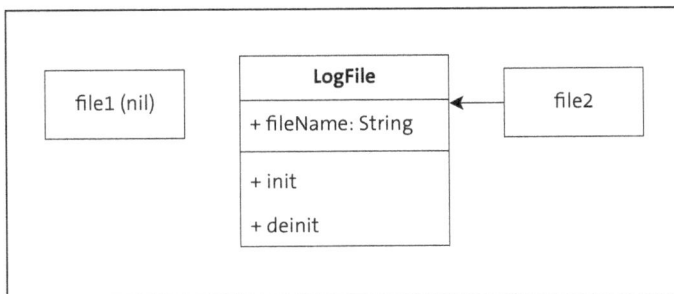

Figure 9.10 Object Sketch for Multiple Pointers

In the examples so far, you have used `deinit` to simulate saving and closing a file. However, `deinit` may serve many other purposes. Some possible use cases are listed in Table 9.3.

Scenario	Use Case
Networking	`deinit` closes the network socket.
Database	`deinit` logs out and closes the database connection.
Camera	`deinit` stops the session and releases the camera.
Image processor	`deinit` deletes temporary files to free up storage.

Table 9.3 Some Use Cases for Deinitialization

9.2.3   Convenience Initialization

In Swift classes, you can implement *convenience initializers*. Such initializers give clients the option to create objects easier with fewer parameters. Listing 9.11 features the sample `CoffeeOrder` class to showcase that mechanism.

```
class CoffeeOrder {
    let drinkType: String
    let size: Int
    let customerName: String

    init(drinkType: String, size: Int, customerName: String) {
        self.drinkType = drinkType
        self.size = size
        self.customerName = customerName
    }

    convenience init (drinkType: String, size: Int) {
        self.init(drinkType: drinkType,
                size: size,
                customerName: "Stranger")
    }

    convenience init (drinkType: String) {
        self.init(drinkType: drinkType,
                size: 1,
                customerName: "Stranger")
    }
}

let myOrder = CoffeeOrder(drinkType: "Coffee")
print(myOrder.drinkType + " for " + myOrder.customerName) // Coffee for
Stranger
```

Listing 9.11 Class with Convenience Initializers

The default initializer of CoffeeOrder expects all necessary parameters to realize the order. If the customer's name is unknown, then the first convenience init is in place to help; it lacks the customerName parameter and fills the name with a default value. The second convenience init expects the drink type only; it initializes the object with a default drink size and customer name.

That way, the caller can place an order merely by providing the drink's name. Default value handling operates centrally within CoffeeOrder. Cool, right?

In this section, we went over class-specific function features. In the next section, you'll learn about inheritance, which is a class-specific behavior.

9.3   Inheritance

In this section, you'll learn about inheritance, which is a controversial characteristic of classes. Although the mechanism is solid and clear, some programmers believe that inheritance is an antipattern and should be mostly avoided in favor of composition. In fact, relatively young languages like Go and Rust don't support inheritance at all.

Swift supports single inheritance in classes—but just because you can doesn't mean that you should. Let's start by exploring what inheritance has to offer and then discuss the potential drawbacks in due course.

9.3.1   Superclasses and Subclasses

Basically, *inheritance* is the process of deriving a child class from an existing parent class. Typically, the child class inherits all properties, methods, and so on of the parent class, but it may also add its own features on top.

As a starting point, Figure 9.11 features a draft structure for an Instrument class. This basic class has the self-explanatory brand, model, and price properties, as well as a method called isInStock() that checks the database for stock availability.

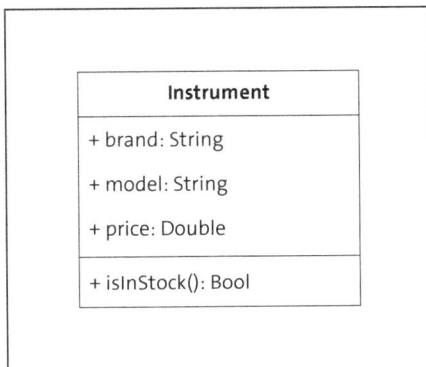

Instrument
+ brand: String
+ model: String
+ price: Double
+ isInStock(): Bool

Figure 9.11 Instrument Class Structure

So far, so good: Instrument covers the basic functionality required for any instrument. However, different instruments may have different additional features, as we'll discuss in the next section.

Single-Level Inheritance

Consider a Guitar class: It would need all the features of Instrument, but would have some additional properties of its own that are important for a buyer. String count and electric/acoustic distinction, for example, would be some basic features for Guitar missing from Instrument.

Likewise, a Keyboard class would need all the features of Instrument, but it would need some additional properties too, like key count and the number of sounds.

The first thing that might come to mind would be to clone the Instrument class twice, once as Guitar and once as Keyboard, as sketched in Figure 9.12. Each clone can then add its special properties.

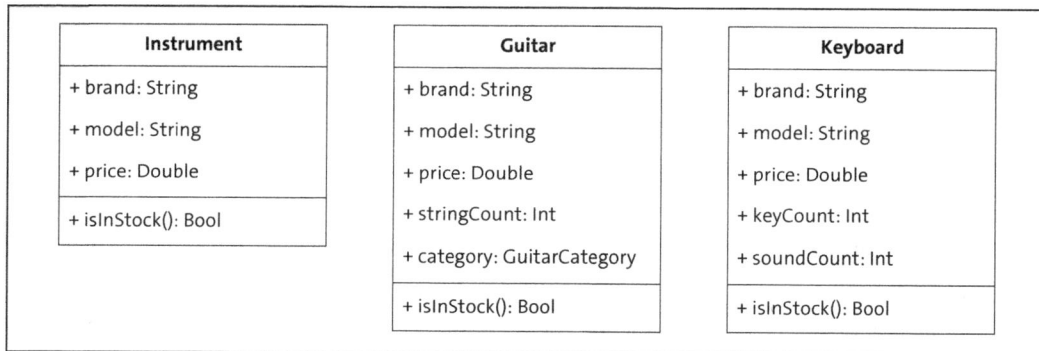

Instrument	Guitar	Keyboard
+ brand: String	+ brand: String	+ brand: String
+ model: String	+ model: String	+ model: String
+ price: Double	+ price: Double	+ price: Double
+ isInStock(): Bool	+ stringCount: Int	+ keyCount: Int
	+ category: GuitarCategory	+ soundCount: Int
	+ isInStock(): Bool	+ isInStock(): Bool

Figure 9.12 Feature Duplication Between Classes

However, the *don't repeat yourself* (DRY) principle in programming is good advice. Some exceptions aside, such duplication is not advised. For example, consider the following cases:

- Imagine that you add a new property called Instrument.color. In that case, Guitar and Keyboard won't automatically gain this property; instead, you have to add it to them manually. That might seem easy with two classes, but when you have dozens of dependent objects, it becomes a nightmare.

- Imagine that the code in Instrument.isInStock() has to be updated due to a database change. In that case, you would have to modify Guitar.isInStock() and Keyboard.isInStock() too, which is a redundant task prone to coding errors.

To mitigate such pitfalls, inheritance offers a mechanism by which you can incrementally develop class structures by inheriting from more generic classes (like Instrument) in more specialized classes (like Guitar and Keyboard).

Figure 9.13 shows this approach visually. If you derive `Guitar` from `Instrument`, then `Guitar` will inherit all properties and functions of `Instrument` and add its own "stuff" on top. The inherited `Guitar.brand` property and the special `Guitar.stringCount` property will both be valid and accessible. The same applies to its sibling class: `Keyboard.model` (inherited) and `Keyboard.soundCount` (special) are equally valid and usable.

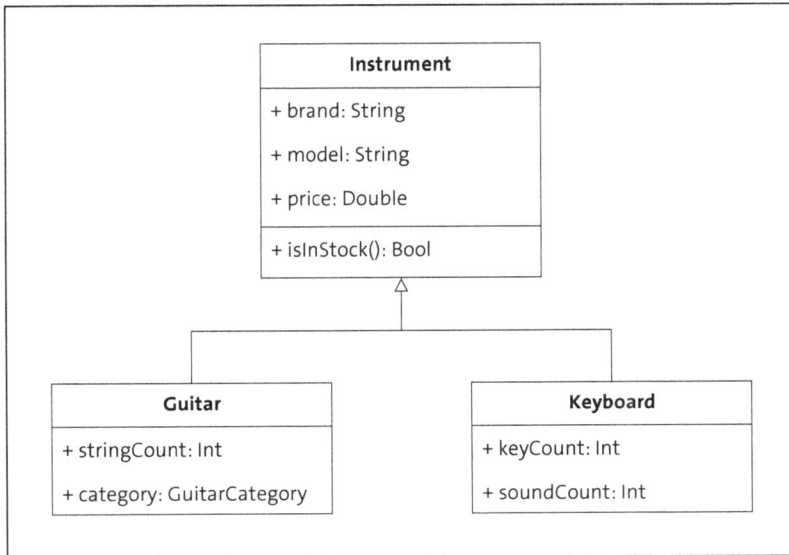

Figure 9.13 Instrument Inheritance Plan

And the best part? If `Instrument.isInStock()` is changed in the future, this change is automatically reflected in `Guitar.isInStock()` and `Keyboard.isInStock()` too! The same applies to `Instrument` properties: If `Instrument.color` is added later, it would automatically be added to `Guitar` and `Keyboard` too.

In that sense, you can think of a child class (`Guitar`) as a mirror of its parent class (`Instrument`). The structure of `Instrument` is completely reflected by `Guitar`—future changes included. To that mirror, you can add `Guitar`-specific properties and functions. Such child-level additions have no effect on the parent class.

Single Parent Principle
Unlike some other languages, Swift only supports single-level inheritance at this time. A child class may have only one parent class, no more.

Now that we've covered the topic logically, let's see it in action. Listing 9.12 demonstrates inheritance between `Instrument` and `Keyboard`.

```
class Instrument {
    var brand: String = ""
    var model: String = ""
    var price: Double = 0

    func isInStock() -> Bool {
        return true // Simulation
    }
}

class Keyboard: Instrument {
    var keyCount = 88
    var soundCount = 100
}

let myKeys = Keyboard()
myKeys.brand = "Nord"
myKeys.model = "A1"
myKeys.soundCount = 300
print(myKeys.isInStock()) // True
```

Listing 9.12 Inheritance Code Example

Syntax-wise, adding the :Instrument suffix is enough to declare that Keyboard is a child class of Instrument. As a result, Keyboard will mirror/inherit everything in Instrument and add whatever else is necessary.

myKeys.brand (inherited) and myKeys.soundCount (special) are equally valid; it doesn't matter that brand was inherited from Instrument. One thing that does matter is that all parent changes are reflected in children. If Instrument.isInStock() was modified to return false, then myKeys.isInStock() would also return false, reflecting its parent, just like a mirror.

Multilevel Inheritance

Inheritance can also be implemented as a multilevel structure. Check the diagram in Figure 9.14, where you see that Instrument is the parent of Guitar, and Guitar is the parent of BassGuitar.

In that case, BassGuitar inherits all features of Instrument and Guitar simultaneously, while adding its own hasBattery and scaleLength properties. Naturally, any child class can add its special functions too.

Implementation of this structure is shown in Listing 9.13. Note that you can access both the Instrument and Guitar features of myBass, as well as its own special features.

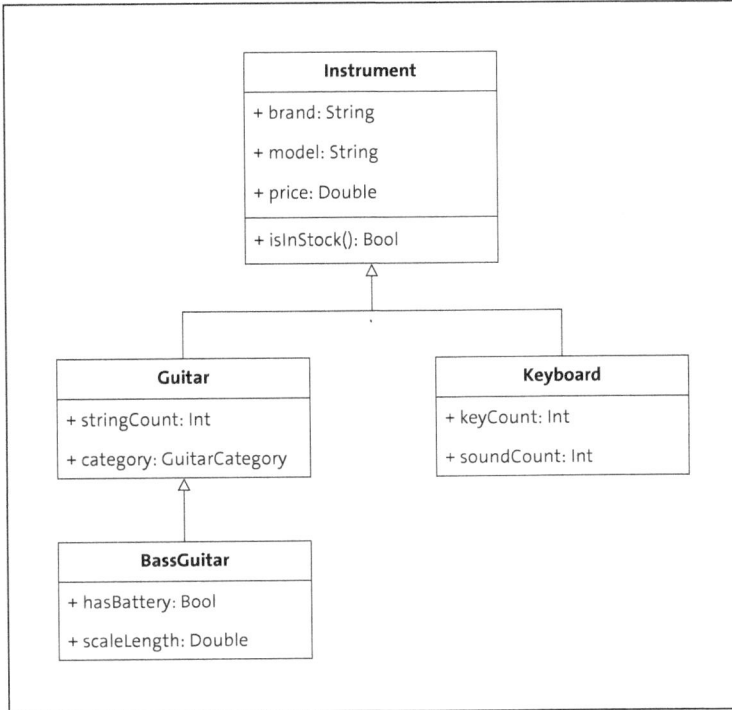

Figure 9.14 Multilevel Inheritance Plan

```
class Instrument {
    var brand: String = ""
    var model: String = ""
    var price: Double = 0

    func isInStock() -> Bool {
        return true // Simulation
    }
}

enum GuitarCategory {
    case acoustic
    case electric
}

class Guitar: Instrument {
    var stringCount = 4
    var category: GuitarCategory = .electric
}
```

```
class BassGuitar: Guitar {
    var hasBattery = false
    var scaleLength: Double = 34
}

let myBass = BassGuitar()
myBass.brand = "Fender"
myBass.model = "Jazz Bass"
myBass.price = 1000
myBass.stringCount = 5
myBass.hasBattery = true
```

Listing 9.13 Multilevel Inheritance Implementation

What's Wrong with Inheritance?

Technically speaking, there is nothing wrong with inheritance. It's a ready-to-use mechanism you can implement or choose to ignore. Some benefits are shown in Table 9.4.

Benefit	Explanation
Code reuse	Common code is shared in a hierarchy, reducing duplication.
Extensibility	Parent features can be extended in child classes.
Polymorphism	Subclasses can be treated like their parents (upcoming topic).
Organized structure	Model hierarchies can be logically understood.

Table 9.4 Some Benefits of Inheritance

However, you should be aware of inheritance detriments, too, some of which are described in Table 9.5.

Detriment	Explanation
Tight coupling	Parent and child classes are tightly bound, making changes risky.
Fragile parents	Changes to a parent class can break subclasses.
Complexity	Deeply nested inheritance structures can be hard to understand and maintain.
Misuse	Some programmers use simple inheritance instead of composition or protocols.

Table 9.5 Some Detriments of Inheritance

If that's the case, how could inheritance have been avoided elegantly in the previous example? Let's see an alternative approach via composition. Figure 9.15 illustrates a model in which Keyboard *contains* an instance of CoreFeatures (formerly known as Instrument) instead of inheriting from it.

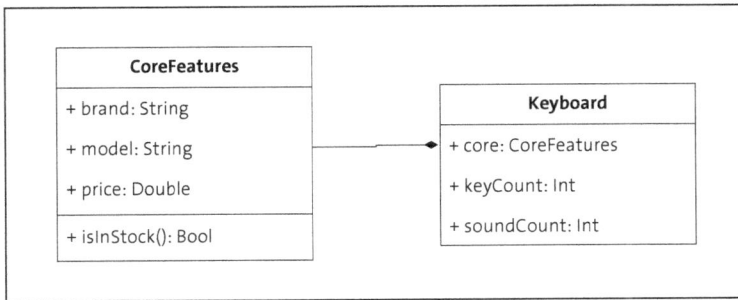

Figure 9.15 Composition Instead of Inheritance

As you can see, Keyboard doesn't have an inheritance flow from a parent any longer, preventing inheritance completely. A code sample for this approach is shown in Listing 9.14.

```
struct CoreFeatures {
    var brand: String = ""
    var model: String = ""
    var price: Double = 0

    func isInStock() -> Bool {
        return true // Simulation
    }
}

class Keyboard {
    var core = CoreFeatures()
    var keyCount: Int = 88
    var soundCount = 100
}

let myKeys = Keyboard()
myKeys.core.brand = "Nord"
myKeys.core.model = "A1"
myKeys.soundCount = 300
print(myKeys.core.isInStock()) // True
```

Listing 9.14 Composition Code Sample

This is merely a simple showcase of composition; things can be more complex in a real-world app. The main idea will be the same, though: Instead of inheriting from a parent, contain another object.

To Inherit or Not?

Should you avoid inheritance altogether, even though it's available? Our advice is to use inheritance sparingly at best—only when its benefits clearly outweigh the risks, or technical obligations arise.

Inheritance shouldn't be used as the default approach for class derivation and code reusability. There are better tools like protocols and dependency injection for that kind of situation. When you learn about protocols in Chapter 10, you'll see what we mean.

Now that you've learned the basics of inheritance, let's discuss how to empower child classes with more autonomy via overrides.

9.3.2   Overriding Functions

Although a child class inherits nearly everything from its parent class, it still has the freedom to replace the code of an inherited function with its own specialized code. That's a technique called *overriding*, which is present in many object-oriented languages. In this section, you will discover various override techniques that you can apply to your classes in Swift.

Basic Override

Figure 9.16 illustrates the upcoming example. There is a (parent) class called Dog, which can bark and react to a stranger. The derived (child) class, GuardDog, will inherit all features of Dog, but it will react to a stranger differently. Therefore, GuardDog.reactTo-Stranger() will override Dog.reactToStranger() with its own code.

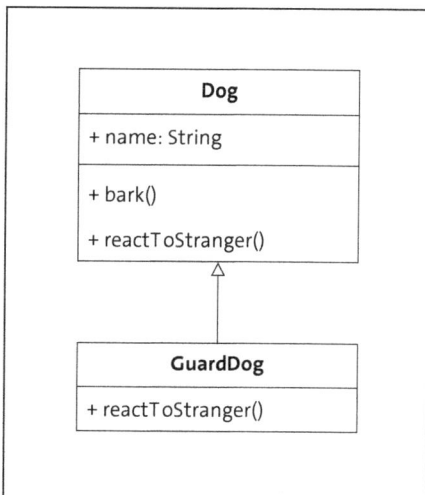

Figure 9.16 Diagram for Function Override Example

The syntax for a common override operation is very simple, as shown in Listing 9.15. All you have to do is add the override prefix to GuardDog's reactToStranger function. This tells Swift that the override is intentional and not a mistake.

```
class Dog {
    var name: String

    init(_ name: String) { self.name = name }

    func bark() { print("Woof!") }

    func reactToStranger() {
        print("\(self.name) sniffs the stranger")
    }
}

class GuardDog: Dog {
    override func reactToStranger() {
        print("\(self.name) growls at the stranger")
    }
}

let smallDog = Dog("Muffin")
smallDog.reactToStranger()  // Muffin sniffs the stranger

let bigDog = GuardDog("Rex")
bigDog.reactToStranger()    // Rex growls at the stranger
```

Listing 9.15 Function Override Example

At this point, Dog.bark and GuardDog.bark are completely identical because bark was inherited but not overridden. On the other hand, Dog.reactToStranger and Guard-Dog.reactToStranger are different now. As you can see in the main flow, they produce different outputs.

Super Keyword

In a child class, super is a keyword that points to its parent. Child classes may access their parent features via the super keyword. Listing 9.16 offers an example in which this keyword is used.

```
class Dog {
    var name: String

    init(_ name: String) { self.name = name }
```

```
    func reactToStranger() {
        print("\(self.name) sniffs the stranger")
    }
}

class GuardDog: Dog {
    override func reactToStranger() {
        super.reactToStranger()
        print("\(self.name) growls at the stranger")
    }
}

let bigDog = GuardDog("Rex")
bigDog.reactToStranger()
```

Listing 9.16 Parent Function Call via super Keyword

Let's focus on GuardDog.reactToStranger. This function is overridden again here, which means that GuardDog will implement its own special Swift code to replace the parent's code.

However, it starts with the expression super.reactToStranger(). This tells Swift to run the original code in Dog.reactToStranger. Once this function is executed, the remaining code in GuardDog.reactToStranger will resume.

The output of this sample code is shown in Figure 9.17: super.reactToStranger() produced the first line, and the following print statement produced the second line.

```
Rex sniffs the stranger
Rex growls at the stranger
```

Figure 9.17 Output of reactToStranger

Obviously, the first line was printed out by Dog.reactToStranger because that's what super.reactToStranger pointed to.

Final Functions

Although child classes can override parent functions by default, a parent class may use its initiative to prevent function overrides. Such functions can be protected using the final prefix. In Listing 9.17, bark has been marked as final. Therefore, a child class like GuardDog can't override this function; it will have to inherit its code as is.

```
class Dog {
    var name: String

    init(_ name: String) { self.name = name }
```

```
    final func bark() { print("Woof!") }

    func reactToStranger() {
        print("\(self.name) sniffs the stranger")
    }
}
```

Listing 9.17 Usage of final Prefix

In a parent-child relationship, functions aren't the only components to be overridden. Let's look at some other options.

9.3.3   Overriding Initializers

Now that you know about overriding functions, overriding an initializer should be a breeze. In that scope, we'll address two initializer override methods: basic and required.

Basic Initializer Override

Check the example in Listing 9.18, where GuardDog has overridden the initializer of its parent, Dog.

```
class Dog {
    var name: String

    init(_ name: String) { self.name = name }

    func bark() { print("Woof!") }
}

class GuardDog: Dog {
    override init(_ name: String) {
        super.init(name)
        self.name += " the guard"
    }
}

let bigDog = GuardDog("Rex")
print(bigDog.name)   // Rex the guard
```

Listing 9.18 Overriding Initializer

GuardDog.init begins by invoking super.init (meaning Dog.init) to run the default initialization code. Once that's completed, GuardDog adds its own logic by appending "the guard" to the end of the dog's name. Easy, right?

Required Initializer

In Swift, you have the option to add the required prefix to the initializer of a parent class. This enforces that each and every child class must implement that initializer. This option can be used when you want to enforce a standard way of initializing objects. In Listing 9.19, Dog has a required initializer. This means that child classes such as GuardDog must implement this exact initializer.

```
class Dog {
    var name: String

    required init(_ name: String) { self.name = name }

    func bark() { print("Woof!") }
}

class GuardDog: Dog {
    var yearsOfService: Int

    required init(_ name: String) {
        self.yearsOfService = 0
        super.init(name)
        self.name += " the guard"
    }

    init (_ name: String, yearsOfService: Int) {
        self.yearsOfService = yearsOfService
        super.init(name)
        self.name += " the guard"
    }
}

let bigDog = GuardDog("Rex", yearsOfService: 3)
print(bigDog.name)              // Rex the guard
print(bigDog.yearsOfService)  // 3
```

Listing 9.19 Required Initializer

In this example, GuardDog couldn't have been implemented without the required init in the form enforced by its parent. However, after fulfilling this basic requirement, you can include custom code in the initializer, as well as throw in an alternative initializer with two parameters. Those are allowed, valid cases. The parent, Dog, merely enforces the existence of a certain initializer signature.

9.3.4 Overriding Computed Properties

Similar to overriding functions, child classes can override computed properties too. The syntax doesn't change dramatically either: All you need to do is add the `override` prefix before the property's name. Let's look at some alternative applications of property overrides.

Basic Override

Listing 9.20 demonstrates an example, in which `SpecialBeverage` overrides the parent computed property, `Beverage.salesPrice`. As you can see, a different calculation took place in the child class.

```
class Beverage {
    var name: String = ""
    var rawPrice: Double = 0

    var salesPrice: Double {
        get { return self.rawPrice * 1.2 }
        set(newValue) { self.rawPrice = newValue / 1.2 }
    }
}

class SpecialBeverage: Beverage {
    override var salesPrice: Double {
        get { return (self.rawPrice * 1.4) + 10 }
        set(newValue) { self.rawPrice = (newValue - 10) / 1.4 }
    }
}

let shake = SpecialBeverage()
shake.salesPrice = 100
print(shake.rawPrice)    // 64.28
```

Listing 9.20 Overriding Computed Property

Super Keyword

Remember the `super` keyword from function overrides? You used it to invoke the parent's version of a function. The same keyword can be used for computed properties too. Listing 9.21 features an enhanced version of the `SpecialBeverage` example in which the `super` keyword is demonstrated.

```
class Beverage {
    var name: String = ""
    var rawPrice: Double = 0
```

```
    var salesPrice: Double {
        get { return self.rawPrice * 1.2 }
        set(newValue) { self.rawPrice = newValue / 1.2 }
    }
}

class SpecialBeverage: Beverage {
    override var salesPrice: Double {
        get { return super.salesPrice + 10 }
        set(newValue) { super.salesPrice = newValue - 10 }
    }
}

let shake = SpecialBeverage()
shake.salesPrice = 100
print(shake.rawPrice)    // 75
```

Listing 9.21 Using super Keyword for Computed Properties

In the getter of `SpecialBeverage.salesPrice`, you call `super.salesPrice` first, which points to the getter of `Beverage.salesPrice`. Once the parent returns a value, Special-Beverage adds 10 on top. Assuming that the `rawPrice` is 75, then

- `super.salesPrice` invokes the code in `Beverage.salesPrice`, which calculates *75 * 1.2 = 90*; and

- 10 is added on top, making the final result *90 + 10 = 100*.

Likewise, in the setter of `SpecialBeverage.salesPrice`, you reduce by 10 from the new value and call the `super.salesPrice` setter after that, letting it handle further calculations. Assuming that the incoming value is 100, then

- 10 is taken away from the value, calculating it as *100 – 10 = 90*; and

- `super.salesPrice` is called as a setter, which calculates `rawPrice` as *90 / 1.2 = 75*.

As you can see, calculated properties can invoke their parents as needed, like functions — which increases your flexibility and reduces redundant code.

Final Properties

For special computed properties that shouldn't be overridden by children, you can use the `final` prefix. You can see an example of this in Listing 9.22, where `Beverage.sales-Price` was protected against overrides.

```
class Beverage {
    var name: String = ""
```

```
    var rawPrice: Double = 0

    final var salesPrice: Double {
        get { return self.rawPrice * 1.2 }
        set(newValue) { self.rawPrice = newValue / 1.2 }
    }
}
```

Listing 9.22 Using final Keyword for Computed Properties

Even if you derive a child class like SpecialBeverage, it still has to reflect salesPrice as is without overriding it.

9.3.5   Overriding Subscripts

A subscript is another class component that can be overridden by children. In this section, we'll address three methods of overriding subscripts in Swift classes.

Basic Override

In Listing 9.23, a basic subscript override is demonstrated via VendingMachine. The syntax is exactly what you'd expect: In the child LoggingVendingMachine class, you add the override prefix before the subscript.

```
class VendingMachine {
    var items: [String: String]

    init() {
        // Database simulation
        items = ["A1": "Soda",
                 "B2": "Chips",
                 "C3": "Chocolate"]
    }

    subscript(code: String) -> String? {
        return self.items[code]
    }
}

class LoggingVendingMachine: VendingMachine {
    override subscript(code: String) -> String? {
        print("Accessing item \(code)")
        return self.items[code]
    }
```

```
}
```

```
LoggingVendingMachine()["C3"]    // Chocolate
```

Listing 9.23 Subscript Override Example

LoggingVendingMachine.subscript added a print statement before returning the appropriate item. Although this example is basic, it demonstrates a proper use case for subscript overriding.

Super Keyword

Like in other override topics, subscript overriding supports the super keyword too. As a demonstration, Listing 9.24 shows an enhanced version of the previous example.

```
class VendingMachine {
    private var items: [String: String]

    init() {
        // Database simulation
        items = ["A1": "Soda",
                 "B2": "Chips",
                 "C3": "Chocolate"]
    }

    subscript(code: String) -> String? {
        return self.items[code]
    }
}

class LoggingVendingMachine: VendingMachine {
    override subscript(code: String) -> String? {
        print("Accessing item \(code)")
        return super[code]
    }
}
```

```
LoggingVendingMachine()["C3"]    // Chocolate
```

Listing 9.24 Accessing Parent Subscript Using super Keyword

In this version of the example, LoggingVendingMachine.subscript called super[code] to invoke VendingMachine.subscript, which is the parent subscript. It's easy and intuitive, right?

Final Subscripts

Finally, the `final` keyword can be used to mark a subscript as not overridable by children. In Listing 9.25, `VendingMachine.subscript` is protected against overrides by child classes because it bears the `final` prefix.

```
class VendingMachine {
    private var items: [String: String]

    init() {
        // Database simulation
        items = ["A1": "Soda",
                 "B2": "Chips",
                 "C3": "Chocolate"]
    }

    final subscript(code: String) -> String? {
        return self.items[code]
    }
}
```

Listing 9.25 Marking Subscript as final

With this example, we conclude our content on overrides. Now that you understand the core logic of inheritance and overriding, let's move on to the topic of casting, which unlocks polymorphism for parent/child classes.

9.3.6 Casting

In programming terms, *polymorphism* is the ability to treat subclasses as if they were instances of their common superclass. To achieve polymorphism in Swift, you have to process objects through an operation called *casting*. There are two main options for casting in Swift, upcasting and downcasting, which we'll cover in this section. You'll also learn how to ensure type consistency when addressing a cast object.

Upcasting

In a nutshell, upcasting is the act of converting a subclass instance to its superclass; like casting a `Dog` as an `Animal`.

Let's take a step back and explore this concept logically. As a framework for this topic, Figure 9.18 features a simple hierarchy of animals. In this hierarchy, `Animal` is the parent class, and the rest are child classes derived from it.

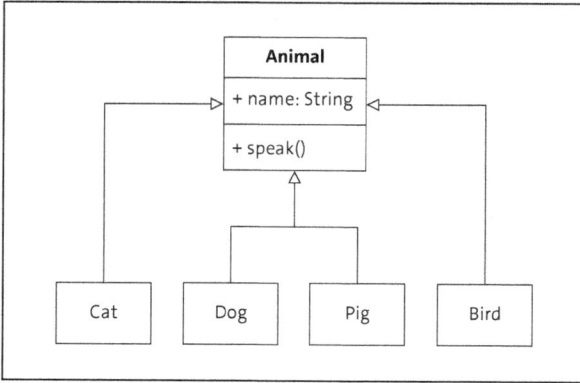

Figure 9.18 Simple Animal Hierarchy

Naturally, Animal.speak would be implemented differently by each child class. As shown in Listing 9.26, each class prints out a different sound.

```
class Animal {
    var name: String = ""

    func speak() { print("Silence") }
}

class Cat: Animal { override func speak() { print("Meow!") } }
class Dog: Animal { override func speak() { print("Woof!") } }
class Pig: Animal { override func speak() { print("Oink!") } }
class Bird: Animal { override func speak() { print("Tweet!") } }
```

Listing 9.26 Different Implementations of Animal Class

So far, so good: Although the example is very simple, it demonstrates a typical case in which a parent class might have multiple implementations. Now let's extend the goal of the application: Let's ask the user about their favorite animal and print out its sound.

For this purpose, the example would be enhanced as shown in Listing 9.27, where casting is introduced.

```
// Classes
class Animal {
    var name: String = ""

    func speak() { print("Silence") }
}

class Cat: Animal { override func speak() { print("Meow!") } }
class Dog: Animal { override func speak() { print("Woof!") } }
```

```
class Pig: Animal { override func speak() { print("Oink!") } }
class Bird: Animal { override func speak() { print("Tweet!") } }

// Main flow
var userAnimal: Animal?

print("Enter your favorite animal (1-4): ", terminator: "")
let animalCode = Int(readLine() ?? "") ?? 0

switch animalCode {
case 1: userAnimal = Cat()
case 2: userAnimal = Dog()
case 3: userAnimal = Pig()
case 4: userAnimal = Bird()
default: userAnimal = nil
}

userAnimal?.speak()
```

Listing 9.27 Casting Example

Let's go through the main flow step by step. You begin by declaring userAnimal of the Animal parent type. Note that it is not of a child type like Cat, Dog, Pig, or Bird; instead, it has the parent type.

Next, you ask the user to enter a number between 1 and 4. Obviously, each number will correspond to a favorite animal: enter "1" for Cat, "2" for Dog, "3" for Pig, and "4" for Bird.

The following switch statement is where casting occurs. userAnimal, which is of the Animal parent type, can be assigned any of its child objects! In other words, userAnimal can act like a pointer to a Cat, Dog, Pig, or Bird object. This mechanism is sketched in Figure 9.19 for a clearer understanding.

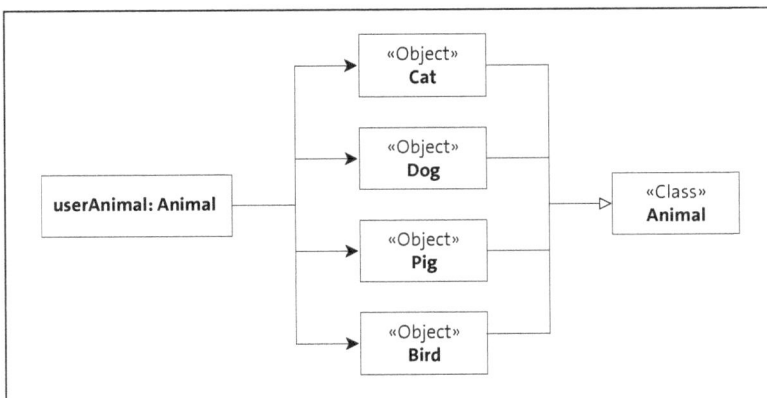

Figure 9.19 Casting Sketched Schematically

Finally, userAnimal?.speak() is invoked. How will this statement behave? Let's take a look:

- If userAnimal points to a Cat object, userAnimal.speak will invoke Cat.speak, printing "Meow".

- If userAnimal points to a Dog object, userAnimal.speak will invoke Dog.speak, printing "Woof".

- If userAnimal points to a Pig object, userAnimal.speak will invoke Pig.speak, printing "Oink".

- If userAnimal points to a Bird object, userAnimal.speak will invoke Bird.speak, printing "Tweet".

A sample output of this small app is shown in Figure 9.20. When the user entered "2" to pick their favorite animal, userAnimal = Dog() was executed, making userAnimal a pointer to the created Dog object. Then userAnimal?.speak invoked Dog.speak as expected, and the output was **"Woof!"**.

```
Enter your favorite animal (1-4): 2
Woof!
```

Figure 9.20 Sample Output of Casting App

This is the power of casting. Once a child object is cast to a parent variable as userAnimal, you can refer to userAnimal throughout the rest of the program without worrying about what kind of object lies behind it. That approach is sure to simplify a large codebase, making your programs easier to understand and maintain.

Checking Types

Casting is a solid mechanism that improves flexibility and simplicity in your programs. However, it comes at the potential cost of ambiguity. Within a complex codebase, how can you know if an encountered object was inherited from a parent class or not?

Check the inheritance hierarchy in Figure 9.21. If you encounter a bird object, how can you check if it was inherited from Mammal or from Oviparous?

Luckily, Swift empowers you with an easy and intuitive keyword for such cases. All you need to do is use the is keyword, as demonstrated in Listing 9.28.

```
// Classes
class Animal {
    var name: String = ""
    func speak() { print("Silence") }
}

class Mammal: Animal {
    func giveBirth() { print("A baby!") }
```

```
}

class Oviparous: Animal {
    func layEgg() { print("An egg") }
}

// Casting
let myDog: Animal = Mammal()
let myParrot: Animal = Oviparous()

// Type check
if myDog is Mammal {
    print("My dog is a mammal")
}

if !(myParrot is Mammal) {
    print("My parrot is not a mammal")
}
```

Listing 9.28 Type Checking via is Keyword

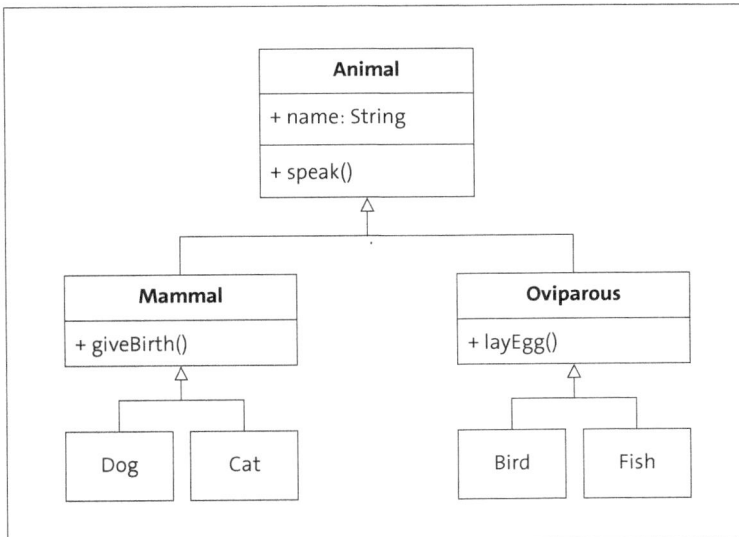

Figure 9.21 Complex Animal Class Hierarchy

This code sample is based on the animal hierarchy shown in Figure 9.21. In this case, myDog is a Mammal instance, but it has been cast as an Animal. Likewise, myParrot is an Oviparous instance, but it has also been cast as an Animal.

In such a case, myDog is Mammal returns true. In fact, myDog is Animal would return true as well. Any parent class included in the hierarchy of myDog would return true.

On the other hand, myParrot is Mammal returns false. Mammal is not part of myParrot's inheritance hierarchy. Easy, right?

Downcasting

Another common technique is downcasting. In contrast to upcasting, downcasting is the act of converting a superclass instance to a subclass; like casting an Animal as a Dog.

Downcasting typically works through optional bindings. You might remember optional bindings from Chapter 6, and the syntax won't change much here. Check the demonstration in Listing 9.29, where myDog goes through an optional binding to fulfill a type check and downcasting.

```
// Classes
class Animal {
    var name: String = ""
    func speak() { print("Silence") }
}

class Mammal: Animal {
    func giveBirth() { print("A baby!") }
}

// Casting
let myDog: Animal = Mammal()

// Type check
if let mammal = myDog as? Mammal {
    mammal.giveBirth() // A baby!
}
```

Listing 9.29 Type Check by Optional Binding

In this case, myDog was really inherited from Mammal. Therefore, the optional binding will succeed and mammal.giveBirth() will be executed, invoking myDog.giveBirth() behind the veil. Although myDog was cast as an Animal instance, you are able to *downcast* it as a Mammal too.

This is an important mechanism to understand. Ultimately, myDog is a Mammal object in memory. It can be temporarily cast as an Animal object, in which case giveBirth() becomes temporarily unavailable for the sake of standardization. But such a casting doesn't make giveBirth()disappear altogether; myDog can be accessed as a Mammal anytime you need to do so, as demonstrated previously.

For the ultimate flexibility, though, the AnyObject type can be used as a universal place-holder for any object type—which is the next topic.

9.3.7   AnyObject Type

In Swift, AnyObject is a protocol to which all classes conform. In other words, you can use the special AnyObject type as a placeholder for any class. This is particularly useful if you are working with a collection of different classes as any class can be cast as AnyObject.

Listing 9.30 features a typical use case. The mixedAnimals: [AnyObject] array declaration implies that mixedAnimals can be used to store objects of any type. You gain this flexibility via AnyObject.

```
// Classes
class Animal {
    var name: String = ""
    func speak() { print("Silence") }
}

class Mammal: Animal {
    func giveBirth() { print("A baby!") }
}

class Oviparous: Animal {
    func layEgg() { print("An egg") }
}

// Definition
let myDog = Mammal()
let myParrot = Oviparous()
let mixedAnimals: [AnyObject] = [myDog, myParrot]
```
Listing 9.30 Consolidating Different Objects into Array via AnyObject

In this particular example, a declaration of mixedAnimals: [Animal] would also be valid because both myDog and myParrot are child classes of Animal—but let's stick to the topic. In Listing 9.31, you can see how an AnyObject can be downcast to the original types/classes. The downcasting syntax is the same as in the previous section.

```
// Classes
class Animal {
    var name: String = ""
    func speak() { print("Silence") }
}
```

```
class Mammal: Animal {
    func giveBirth() { print("A baby!") }
}

class Oviparous: Animal {
    func layEgg() { print("An egg") }
}

// Definition
let myDog = Mammal()
let myParrot = Oviparous()
let mixedAnimals: [AnyObject] = [myDog, myParrot]

// Loop
for animal in mixedAnimals {
    if let mammal = animal as? Mammal {
        mammal.giveBirth()
    }
    if let oviparous = animal as? Oviparous {
        oviparous.layEgg()
    }
}
```

Listing 9.31 Downcasting AnyObject

Another possible use case for AnyObject is to make it a function parameter type to increase the flexibility of a function. In Listing 9.32, the reproduce function takes an Any-Object argument and operates via downcasting.

```
// Classes
class Animal {
    var name: String = ""
    func speak() { print("Silence") }
}

class Mammal: Animal {
    func giveBirth() { print("A baby!") }
}

class Oviparous: Animal {
    func layEgg() { print("An egg") }
}

// Function
func reproduce(_ animal: AnyObject) {
```

```
    if let mammal = animal as? Mammal {
        mammal.giveBirth()
        return
    }
    if let oviparous = animal as? Oviparous {
        oviparous.layEgg()
        return
    }
    print("Can't reproduce")
}

// Definition
let myDog = Mammal()
reproduce(myDog) // A baby!
```

Listing 9.32 Using AnyObject as Function Parameter

Although `AnyObject` increases the flexibility of your codebase, we recommend keeping its usage to a minimum and avoiding it unless it's absolutely justified. Some costs that come with `AnyObject` are shown in Table 9.6.

Cost	Explanation
Reduced readability	The code becomes harder to follow and maintain because the actual type isn't immediately obvious.
Increased boilerplate	Downcasting is nearly unavoidable, cluttering the codebase.
Decreased performance	Casting operations add runtime overhead.
Technical debt	Overusing `AnyObject` is a shortcut that may grow your technical debt.

Table 9.6 Some Costs of Using AnyObject

In the last example, you saw that `AnyObject` can be passed as a function parameter. However, a specific class type can be used as a function parameter too. In the next section, you'll see how.

9.4   Classes as Function Parameters

Just like structs, classes can also be used as function parameters. Listing 9.33 demonstrates an example, in which `areNameSake` accepts two `Person` instances (objects) to determine if they are namesakes.

```
class Person {
    var name: String
    init(name: String) { self.name = name }
}

func areNameSake(_ p1: Person, _ p2: Person) -> Bool {
    return p1.name == p2.name
}

let student1 = Person(name: "Alice")
let student2 = Person(name: "Bob")
let student3 = Person(name: "Alice")

print(areNameSake(student1, student2)) // false
print(areNameSake(student1, student3)) // true
```

Listing 9.33 Classes as Function Parameters

So far, so good: If Person was a struct, you could proceed with the same syntax. However, a logical difference occurs when you want to mutate the argument.

In Chapter 8, you learned that if a function needs to be able to modify a struct instance, it must be passed with the inout prefix. You can check Listing 9.34 as a memory refresher.

```
struct Person {
    var name: String
}

func rename(_ person: inout Person) {
    person.name += "X"
}

var student = Person(name: "Alice")
rename(&student)
print(student.name) // AliceX
```

Listing 9.34 Modifying Struct Instance in a Function

However, you don't need to specify the inout prefix for class parameters. Check the syntax in Listing 9.35, which is nearly the same as Listing 9.34 except that a class is used instead of a struct.

```
class Person {
    var name: String
    init(name: String) { self.name = name }
```

```
}

func rename(_ person: Person) {
    person.name += "X"
}

var student = Person(name: "Alice")
rename(student)
print(student.name) // AliceX
```

Listing 9.35 Modifying Class Instance in a Function

The reason for this syntax difference should be obvious. Structs are value types, while classes are reference types. When you pass a struct to a function, you are passing real values. When the function modifies person.name, it is modifying the very name value held by person. Therefore, you need extra measures to enable mutability. However, when you pass a class instance to a function, you are merely passing the reference/pointer. When the function modifies person.name, it is modifying the object referenced or pointed to by person, and that object is already mutable by default.

This differentiation is sketched in Figure 9.22 for clarity:

- On the left side, student is a struct instance, acting as a concrete component holding its own values. When this student is passed to a function, real values are passed and so inout is needed for mutation.
- On the right side, student is a class instance, merely acting as a reference/pointer to the concrete object in memory. When this student is passed to a function, only the reference is passed (not the concrete object) and so inout is not needed for mutation. The object referenced or pointed to in memory is already mutable.

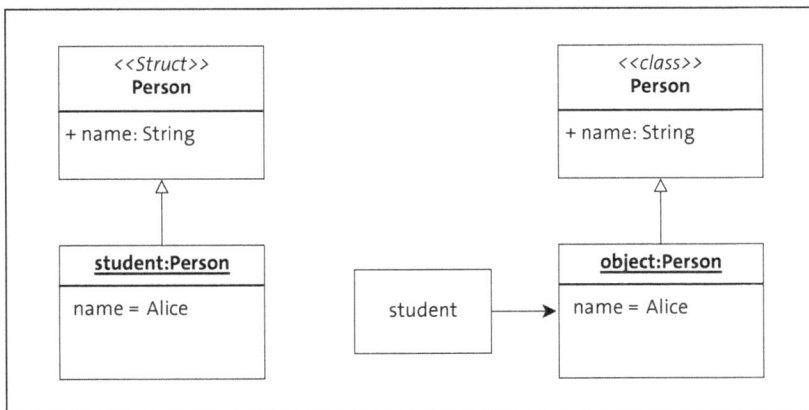

Figure 9.22 Struct and Class Instance Comparison of Student

This case study also acted as a memory refresher for value versus reference types, which is an important distinction to be aware of. The timing is also perfect as this distinction is a prerequisite for the next topic: memory management.

9.5   Memory Management

To wrap up the topic of classes in Swift, we'll address memory management now. This topic revolves around the lifecycle of classes—that is, how they live and die. In this section, you will learn how Automatic Reference Counting (ARC) works and how different types of references affect this lifecycle.

Because classes are reference types, multiple variables can point to the same object. If there are no more variables pointing to an object, then it can be safely removed from memory. Swift uses ARC to govern this mechanism.

Basically, Swift keeps count of the number of variables pointing to an object in memory. As soon as this count drops to zero, the object becomes obsolete and is destroyed, triggering its deinit.

9.5.1   Strong References

Let's see ARC in action via Listing 9.36, which features two variables pointing to the same object in memory.

```
// Class
class Person {
    let name: String

    init(name: String) {
        self.name = name
        print("\(name) is initialized")
    }

    deinit {
        print("\(name) is deinitialized")
    }
}

// Main flow
var p1: Person? = Person(name: "Bob")   // Reference count = 1
var p2: Person? = p1                     // Reference count = 2

p1 = nil                                 // Reference count = 1
print("p1 is now nil")
```

```
p2 = nil                                    // Reference count = 0 (deinit)
print("p2 is now nil")
```

Listing 9.36 ARC Demonstration

Once p1 is initialized, a new Person instance is created and its init is executed, as you've seen before. After that, the memory and ARC looks like Figure 9.23. The fresh Person object is only referenced by p1 at this point, so ARC has the reference count 1.

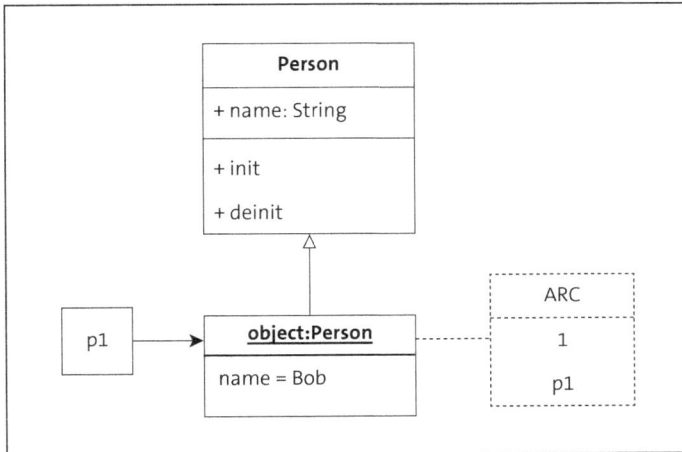

Figure 9.23 ARC After p1 Initialization

Once p2 = p1 is executed, the pointer/reference of the Person instance is copied to p2. Now this object has two references, and ARC looks like Figure 9.24.

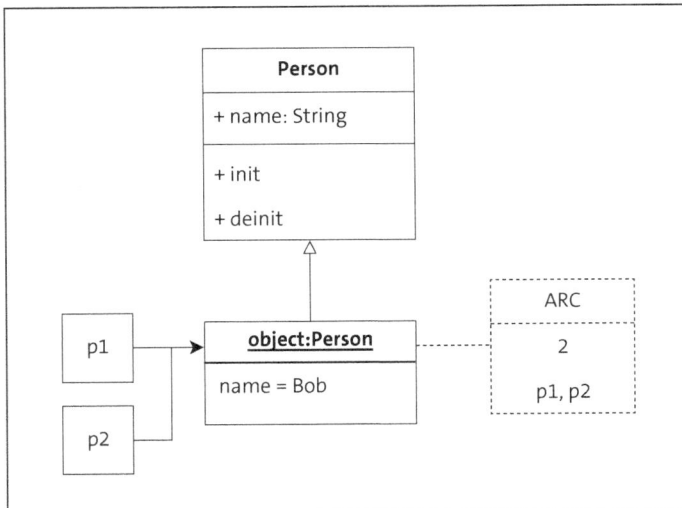

Figure 9.24 ARC After p2 Initialization

At this point, both p1 and p2 are *strong references*; this means that they can hold the Person instance in memory. So long as at least one strong reference exists, the object stays in memory.

Once p2 = nil is executed, the reference count for the object drops to 1, as p1 is now the sole pointer. ARC should look like Figure 9.25 at that point.

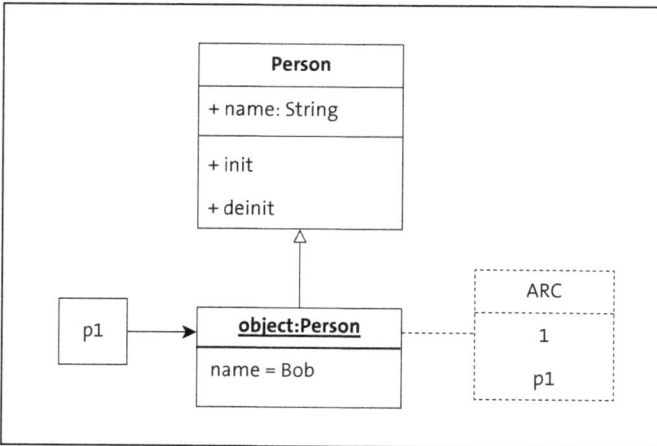

Figure 9.25 ARC After p2 Deinitialization

Finally, after p1 = nil is executed, the reference count for the object drops to 0, and ARC should look like Figure 9.26.

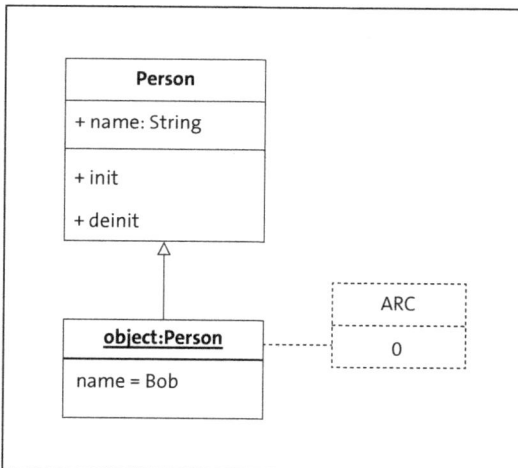

Figure 9.26 ARC After p1 Deinitialization

As soon as the reference count drops to 0, ARC realizes that the object is no longer needed and destroys it, freeing up memory. Right before that, the deinit of the object is triggered to conduct any cleanup operations.

As you can see, ARC is an automatic mechanism that gracefully handles objects in memory, removing them when they are no longer needed. ARC *mostly* frees you from the responsibility of dealing with object memory handling.

We say *mostly* because a certain degree of manual care is required to ensure that ARC runs smoothly. In situations in which two objects are referring to each other with strong references, ARC may fall into a deadlock, unable to remove either object from the memory. That's a case you want to avoid.

To understand and handle such potential problems, let's go through alternative types of object references at our disposal and see how they might support ARC.

9.5.2   Weak References

To help you understand weak references and their usefulness, we'll implement an example centered on an apartment and its tenant, as sketched in Figure 9.27. In this model, there is an Apartment class, which references its Tenant. On the other side, there is a Tenant class, which references its Apartment.

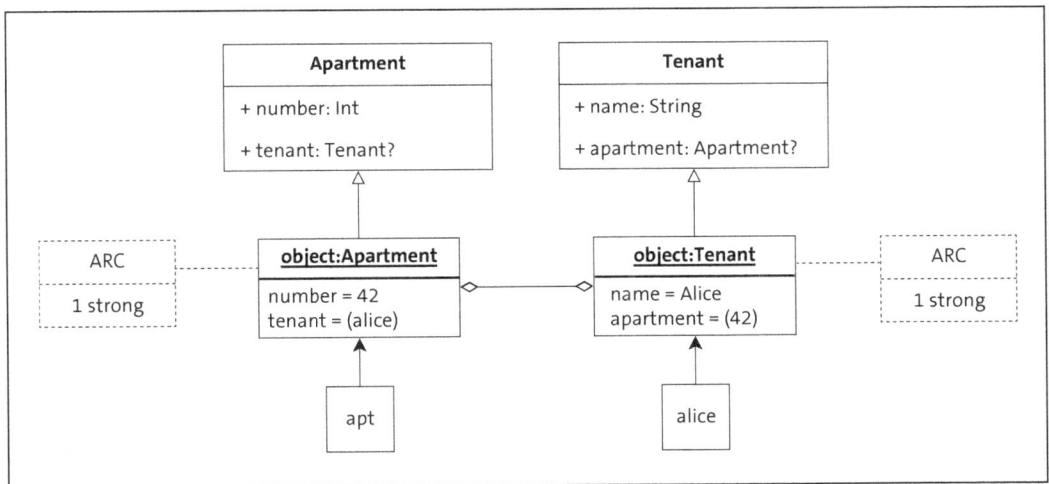

Figure 9.27 Apartment-Tenant Model with Strong Reference

Although the example seems sensible and logical, there is a pitfall: When two objects hold strong references to each other, neither can be deallocated. This is called a *retain cycle*. To avoid this situation, one of the references must be declared weak.

In Swift, a *weak reference* is a special type of reference that doesn't stop ARC from removing the object from memory.

In this example, Apartment's reference to Tenant should be declared weak. Logically, an apartment can exist without a tenant; that's why marking Tenant as weak makes sense.

Listing 9.37 features the code implementation of this example. The most notable part is in the beginning of the Apartment class, where the tenant variable has been marked with the weak prefix.

```
class Apartment {
    let number: Int
    weak var tenant: Tenant?

    init(number: Int) {
        self.number = number
        print("Apartment \(number) is created.")
    }
    deinit { print("Apartment \(number) is demolished.") }
}

class Tenant {
    let name: String
    var apartment: Apartment?

    init(name: String) {
        self.name = name
        print("\(name) moves in.")
    }
    deinit { print("\(name) moves out.") }
}

var apt: Apartment? = Apartment(number: 42)
var alice: Tenant? = Tenant(name: "Alice")

alice?.apartment = apt
apt?.tenant = alice // Weak reference
alice = nil         // Alice moves out
apt = nil           // Apartment is removed
```

Listing 9.37 Weak Reference Implementation

As soon as apt and alice refer to each other, the memory would logically look like Figure 9.28. ARC is certainly aware of the mutual reference, but it is also aware that apt refers to alice weakly.

This allows ARC to remove the Tenant object from the memory when alice = nil is executed. When references to the Tenant object are checked, only one weak reference is found. This isn't significant for ARC and the memory removal takes place. The output of the code example is shown in Figure 9.29.

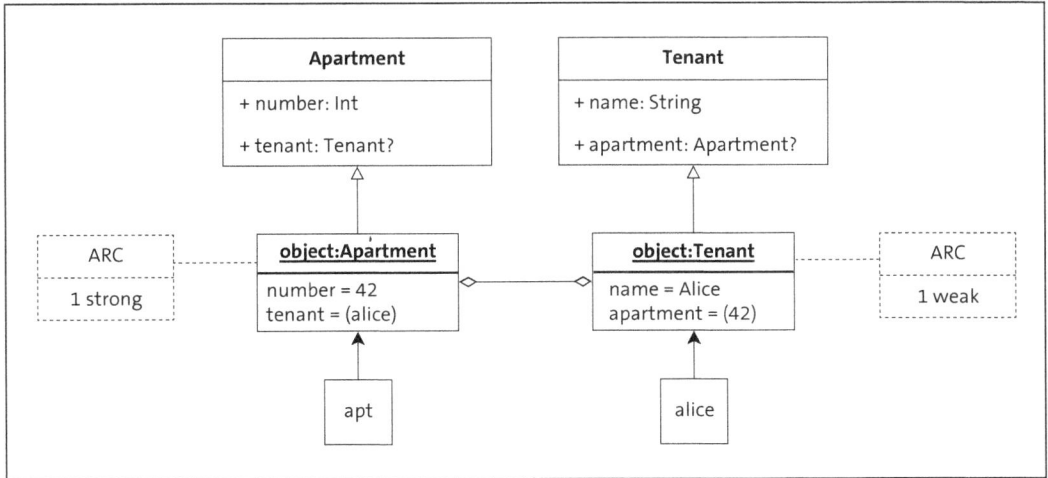

Figure 9.28 Apartment-Tenant Model with Weak Reference

```
Apartment 42 is created.
Alice moves in.
Alice moves out.
Apartment 42 is demolished.
```

Figure 9.29 Output of Weak Reference Implementation

If Apartment.tenant was a strong reference, the output of the code example would look like Figure 9.30. Even when alice = nil is executed, the strong reference count for the Tenant object would still be 1 because the Apartment object is still pointing to the Tenant object strongly. Likewise, apt = nil won't trigger Apartment.deinit because the Tenant object is still pointing to the Apartment object strongly.

```
Apartment 42 is created.
Alice moves in.
```

Figure 9.30 Output of Strong Reference Implementation

Mutual References

Strong mutual references between objects may cause retain cycles, in which neither object can be released from the memory. Whenever you create two mutually referring classes, it's wise to consider the correct reference type.

Although weak references can break unwanted retain cycles, they are not the only option. Let's examine the other one.

9.5.3 Unowned References

Like the weak reference prefix, Swift features another reference prefix, called unowned. These two prefixes are very similar in the sense that both allow object deallocation. Using either of them, you can break retain cycles caused by mutual references.

But there must be a difference, right? Right! The difference lies in the optionality of the referred object.

In the preceding example, Apartment.tenant was a weak *and optional* reference. That's a feature of weak variables: They are optional and can become nil. Weak references are typically used when one object can exist without another, like an Apartment can exist without a Tenant.

On the other hand, unowned references are not optional and won't become nil when the referenced object is released. These references are typically used when one object can't exist without another, like a Book can't exist without an Author.

Speaking of books and authors, look at Listing 9.38, which implements an unowned reference between Book and Author.

```swift
class Author {
    let name: String
    var books: [Book] = []

    init(name: String) {
        self.name = name
        print("Author \(name) is created.")
    }

    deinit { print("Author \(name) is removed.") }
}

class Book {
    let title: String
    unowned let author: Author // Book doesn't keep Author alive

    init(title: String, author: Author) {
        self.title = title
        self.author = author
        print("Book '\(title)' by \(author.name) is published.")
    }

    deinit { print("Book '\(title)' is removed.") }
}

var orwell: Author? = Author(name: "George Orwell")
```

```
orwell!.books.append(Book(title: "1984", author: orwell!))
orwell!.books.append(Book(title: "Animal Farm", author: orwell!))

orwell = nil  // Deallocates both the author and the books
```

Listing 9.38 Unowned Reference Between Author and Their Books

In this example, the Book class has two significant points to note:

- The reference to Author starts with the unowned prefix.
- The reference to Author is declared as let, making it immutable.

This standard leads Swift to assume that a Book must have an Author reference—which makes sense, right? No book writes itself. However, as soon as orwell = nil is executed, all Book objects in orwell.books will be released from memory too; no retain cycle will occur.

The proof is shown in Listing 9.38: orwell = nil released all books, and their deinit code was executed.

```
Author George Orwell is created.
Book '1984' by George Orwell is published.
Book 'Animal Farm' by George Orwell is published.
Author George Orwell is removed.
Book '1984' is removed.
Book 'Animal Farm' is removed.
```

Figure 9.31 Output of Unowned Reference Example

Weak or Unowned?

In general, use weak when the reference is an optional and might become nil. Use unowned when the reference is not an optional and will never become nil so long as the referrer lives. Both prefixes, weak and unowned, prevent retain cycles.

This final distinction marks the end of class memory management, as well as our chapter on classes. Hopefully, the differences between structs and classes are clear at this point as both offer different features and have different use cases.

9.6 Summary

In this chapter, you learned about classes in Swift. We started with the distinction between structs and classes. The most significant difference is that structs are value types, while classes are reference types. Classes also offer additional function types, like deinitialization and convenience initialization.

One option available for classes is inheritance, via which a child class can inherit properties and functions of a parent class as a starting point and add its own logic to that. Although the usage of inheritance is a controversial topic, the facility is there for when you need it. Finally, you learned about Swift's object memory mechanism called ARC and how to deal with mutual referring objects via weak and unowned references.

Now that we have structs and classes covered, the next chapter will be about protocols, which can be used to standardize the behavior of similar structs or classes.

Chapter 10
Protocols

In Swift, protocols define a common blueprint of classes, structs, or enums. They enable polymorphism and abstractions, allowing for development of interchangeable objects. Protocols are key elements of design patterns and object thinking.

In this chapter, you will learn about protocols in Swift, which enable an enormous degree of flexibility in codebases by providing blueprints for classes, structs, or enums.

In many object-oriented languages, protocols are referred to as *interfaces*. So if you're familiar with interfaces in another language, know that Swift calls them *protocols*. But even for experienced programmers, this chapter is worth going through. Swift protocols might offer different merits or shortcomings than those of your former language.

Without further ado, let's discover what protocols have to offer!

10.1 Purpose of Protocols

In a nutshell, *protocols* enable programmers to define blueprints for classes, and classes conforming to the protocol can be used interchangeably, just like changing from one computer keyboard to another that conforms to the same USB protocol. Note that protocols are also applicable to structs and enums.

Because we've mentioned USB keyboards already, let's approach protocols from that angle. Your computer surely has some USB ports, right? USB is a universal port type, able to accept many kinds of hardware devices—but we'll focus on keyboards for now. So long as a keyboard conforms to USB standards, it can be used with any computer with a USB port.

In this approach, you can imagine the USB port as a Swift protocol and each keyboard as a concrete class—as sketched in Figure 10.1.

The USB protocol defines the standards for how keyboards must operate, and your operating system and motherboard only listen to the USB port—not directly to the device behind it. As soon as a keyboard's key is tapped, the keyboard sends a signal to the USB port in a format defined by the protocol. Because the operating system knows what to expect from the USB, it understands that key-tap signal and ultimately displays the corresponding character on the screen.

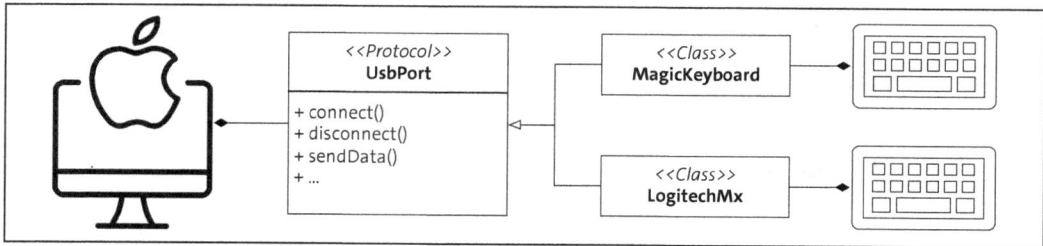

Figure 10.1 USB Modeled as Protocol

This approach frees computer manufacturers from the need to know about the technicalities of each keyboard on the market. It is the keyboard manufacturer's job to conform to the USB protocol, and once that's done, the keyboard can work with any USB-equipped computer. Neat, right?

Now, let's approach protocols through a printer example. Programs on your computer don't care what kind of printer you have, right? So long as your operating system has the appropriate drivers for your printer, any program can print outputs by communicating with the printer API. Whatever printer driver is installed behind the API listens to API commands and translates them to signals for the target printer.

Therefore, the printer API would be a protocol in Swift's terms, and each driver would be a concrete class implementing that protocol. Check the sketch in Figure 10.2.

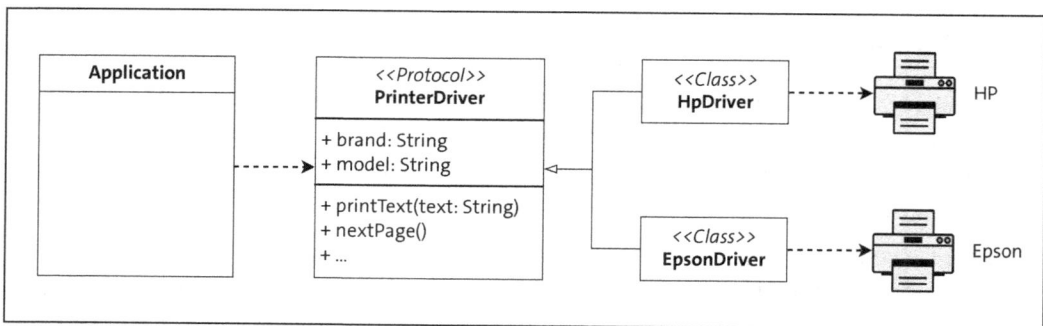

Figure 10.2 Printing Architecture

If the operating system didn't provide a common protocol to hide all printer drivers behind, then each application you install would need to know about each and every printer in the world and how they operate internally. Instead, an application merely calls the protocol's `printText` method, and each printer's driver translates that request to the signal that its target printer understands.

This approach also frees applications from worrying about printers to be produced in the future. So long as the printer comes with a driver that supports the `PrinterDriver` protocol, all applications will be able to work with that new printer too.

Finally, let's go over a software protocol example, moving closer to our real use cases. Suppose you are developing an app that should have AI capabilities. Your big plan is to query a large language model (LLM) engine with prompt inputs and use its replies to generate app content. In this scenario, you may want to use LLM providers interchangeably. Some users may prefer ChatGPT, while others prefer Gemini or Grok. Each provider may have different authentication and API models, as demonstrated in Figure 10.3.

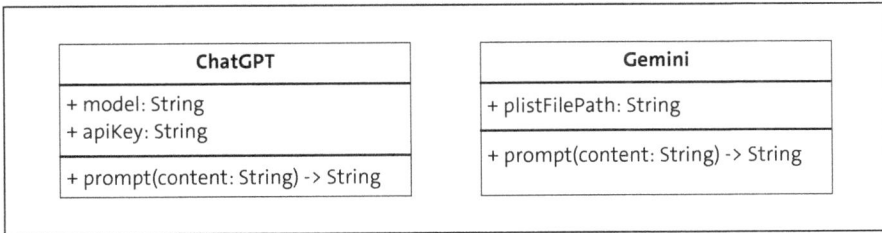

Figure 10.3 Sample LLM Classes

Instead of coding the internal details of each LLM into the heart of your application's codebase, you can hide those classes behind a protocol, as if each LLM engine was a different USB keyboard or printer. Check the architecture in Figure 10.4.

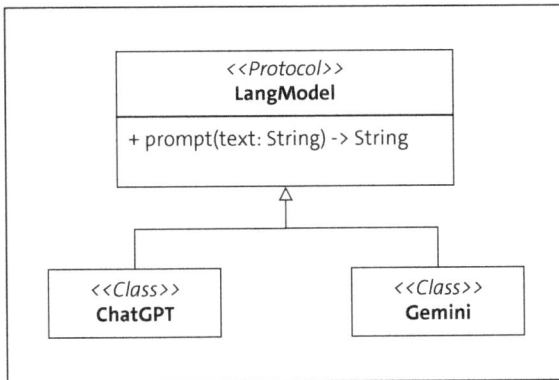

Figure 10.4 LLM-Based Protocol Model

In this architecture, the main application code will only deal with the LangModel protocol, without caring about the concrete LLM class behind it. Even if you switch to a different LLM in the future, you won't have to change a single line of code in the main codebase; all you need to do is to plug in a different LLM class, just like plugging a different keyboard or printer into your computer. Check the extended sketch in Figure 10.5.

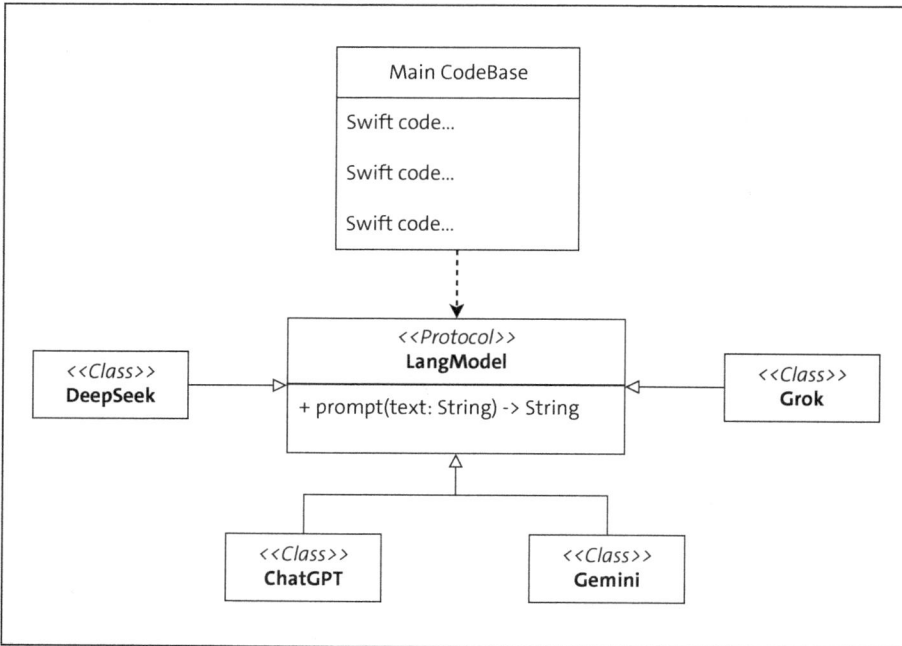

Figure 10.5 Main Codebase with Different LLM Models

As such, the ability to use classes (and other types) interchangeably is one of the major benefits of using protocols. But it isn't the only benefit; check Table 10.1 for an overview.

Benefit	Explanation
Polymorphism	Different classes, structs, or enums implementing the same protocol can be swapped.
Unit test mocking	Mock objects can be injected into unit tests, enabling isolated testing.
Decoupling	The code will depend on protocols instead of concrete types, reducing tight coupling and making component replacements easier.
Multi-inheritance	If needed, a type can conform to multiple protocols, overcoming the limitation of single inheritance.
Plug-ins	You can load external modules that conform to a protocol, enabling third-party extendibility in your apps.

Table 10.1 Some Benefits of Protocols

Protocol Encapsulation

Because protocols are meant to be public interfaces, all components of protocols are assumed to be public.

Now that the overall logic and purpose of protocols is clear, let's move to hands-on examples and see the various features that protocols offer.

10.2   Function Requirements

First and foremost, Swift protocols can enforce concrete classes to implement certain functions. That way, protocol clients can rest assured that the underlying concrete class will support that function without knowing the implementation details. In this section, we will go over different case studies to discover how. This will also be the hands-on debut of protocol syntax.

10.2.1   Nonmutating Function

To learn about nonmutating functions, let's implement the LLM case study introduced earlier. The sample application will feature the ability to talk to multiple LLMs—or at least pretend to do so.

For that purpose, Listing 10.1 features a simple protocol called LangModel, with a single function.

```
protocol LangModel {
    func prompt(text: String) -> String
}
```

Listing 10.1 Protocol with Single Nonmutating Method

Note that the protocol declaration syntax is very similar to that of structs and classes. The obvious difference is that protocols don't have function bodies. Protocols only declare the signature of functions and leave the implementation to concrete structs or classes. Speaking of which, Listing 10.2 demonstrates two sample classes that implement the LangModel protocol: ChatGPT and Gemini.

```
protocol LangModel {
    func prompt(text: String) -> String
}

class ChatGPT: LangModel {
    func prompt(text: String) -> String {
        return ("ChatGPT heard your prompt: \(text).")
    }
}

class Gemini: LangModel {
    func prompt(text: String) -> String {
```

```
        return ("Gemini heard your prompt: \(text).")
    }
}
```

Listing 10.2 Two Classes Implementing LangModel Protocol

Obviously, those sample classes don't actually communicate with LLMs at this time; we're merely focusing on the architecture.

Note that the syntax to implement a protocol is nearly the same syntax as that for class inheritance, which you learned about in the previous chapter. ChatGPT: LangModel means that ChatGPT conforms to the LangModel protocol and guarantees that it will implement all required functions with the same signature. And it does: The ChatGPT. prompt function has exactly the same signature as LangModel.prompt. The Swift compiler doesn't let your code behave otherwise—but you're free to try it if you want!

Gemini is another class implementing the LangModel protocol. Like ChatGPT, Gemini must also implement all functions imposed by LangModel with the exact same signature. How it does so is a free choice for each concrete class. As you can see, ChatGPT and Gemini behave differently within the function.

Now, let's spice things up, see how those classes can be used interchangeably, and enter the world of polymorphism. For this demonstration, shown in Listing 10.3, you let the user pick an LLM engine and send their prompt to that engine.

```
protocol LangModel {
    func prompt(text: String) -> String
}
class ChatGPT: LangModel {
    func prompt(text: String) -> String {
        return ("ChatGPT heard your prompt: \(text).")
    }
}
class Gemini: LangModel {
    func prompt(text: String) -> String {
        return ("Gemini heard your prompt: \(text).")
    }
}

// Let user pick an engine
print("Pick an engine (C/G): ", terminator: "")
let userEngine = readLine() ?? "C"
let engine: LangModel
if userEngine == "C" { engine = ChatGPT() } else { engine = Gemini() }

// Send a prompt
print("Prompt: ", terminator: "")
```

```
let userPrompt = readLine() ?? ""
let output = engine.prompt(text: userPrompt)
print(output)
```

Listing 10.3 Polymorphism Demonstration

Let's break down this code for a better understanding. In the first part, you let the user pick an LLM engine; they enter either "C" for ChatGPT or "G" for Gemini.

Right after, there is a variable called engine, which is of the LangModel type. This is very important: Note that engine is neither of type ChatGPT nor of type Gemini. Instead, engine has been declared type LangModel. In this case, engine will be our "USB port," so to speak, to access plugged-in classes/objects—in this case, either ChatGPT or Gemini.

Luckily, the syntax to do this is very easy! If the user entered "C", then engine = ChatGPT() is executed; otherwise, engine = Gemini() is executed. Because both ChatGPT and Gemini are compatible with LangModel, an object of either type can be plugged into engine, just like plugging a USB-compliant keyboard into a USB port!

Once you have the desired object plugged into engine, invoking engine functions will invoke the function of the desired object—in other words,

- if engine = ChatGPT() was executed, engine.prompt() will invoke ChatGPT.prompt(); and
- if engine = Gemini() was executed, engine.prompt() will invoke Gemini.prompt().

This is also apparent in the rest of the example. After setting up engine, you ask the user for a prompt and query the chosen LLM engine with that prompt.

Figure 10.6 shows a sample output, where ChatGPT was picked as the target LLM. In that case, engine.prompt() invoked ChatGPT.prompt(), and that code produced the output that you see.

```
Pick an engine (C/G): C
Prompt: Hello
ChatGPT heard your prompt: Hello.
```

Figure 10.6 Prompt Example with ChatGPT

Likewise, Figure 10.7 shows a sample output in which Gemini was picked as the target LLM. In that case, engine.prompt() invoked Gemini.prompt(), and that code produced the output that you see.

```
Pick an engine (C/G): G
Prompt: Hola
Gemini heard your prompt: Hola.
```

Figure 10.7 Prompt Example with Gemini

Cool, right? It's just like plugging in USB keyboards, as mentioned in the introduction! Now, let's advance to an example involving a protocol with a mutating function.

10.2.2   Mutating Function

A protocol imposing a mutating function is not too different from a protocol imposing a nonmutating function. All you need to do is add the `mutating` prefix to the function signature. As a demonstration, let's implement such a protocol for a struct.

The core architecture is illustrated in Figure 10.8, centered on a video game example. There is a protocol called `Damageable`, which represents any individual (player or NPC) that can be damaged. This protocol includes a mutable function called `takeHit`, which will reduce the health of the individual.

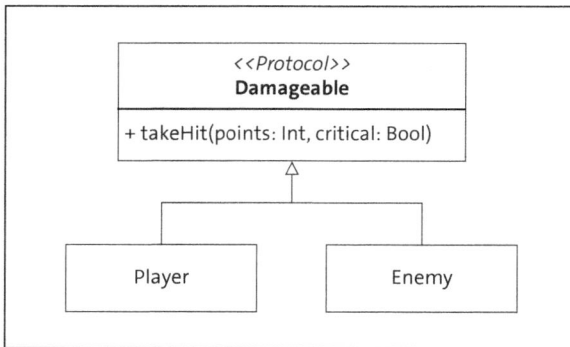

Figure 10.8 Architecture of Mutating Function Example

This protocol will be implemented for two structs: `Player` and `Enemy`. Because both are damageable individuals in the game, it makes sense to consolidate the function signature into the protocol, right? If further individual types are required later, then new structs can be declared and have `Damageable` implemented.

Now check out the implementation in Listing 10.4. As expected, `Damagable.takeHit` was declared as usual, merely adding the `mutating` prefix to enable mutations. Note that `Player` and `Enemy` have similar but different implementations for `takeHit`: Can you spot the difference?

```
protocol Damageable {
    mutating func takeHit(points: Int, critical: Bool)
}

struct Player: Damageable {
    var name: String
    var health: Int

    mutating func takeHit(points: Int, critical: Bool) {
        let damage = critical ? points * 2 : points
        self.health -= damage
```

```
        }
    }
}

struct Enemy: Damageable {
    var name: String
    var health: Int

    mutating func takeHit(points: Int, critical: Bool) {
        let damage = critical ? points * 3 : points
        self.health -= damage
    }
}

var hero = Player(name: "Arthur", health: 100)
hero.takeHit(points: 10, critical: true) // Down to 80

var orc = Enemy(name: "Orc", health: 80)
orc.takeHit(points: 10, critical: true) // Down to 50
```

Listing 10.4 Mutating Protocol Function Implementation

The main flow should be intuitive enough: Both the Player and Enemy instances have a function called takeHit with the exact same signature, as imposed by the protocol. That's not too different from class implementations, right?

10.2.3 Static Function

Swift protocols support static functions too. You can add a nonmutating static function or mutating static function to a protocol, enforcing subtypes to implement this function.

In the example for this topic, we'll implement a protocol with a static function for enumerations. That way, both static functions and protocol-to-enumeration implementations will be demonstrated simultaneously. Keep in mind that static functions can be implemented for classes and structs equally well.

The example is sketched in Figure 10.9. The Describable protocol will have a static function called getTypeDescription, implemented by the Direction and Season enumerations.

The code implementation of this case study is supplied in Listing 10.5. Like any static function, it should be accessed by using the type name—so Direction.getTypeDescription or Season.getTypeDescription.

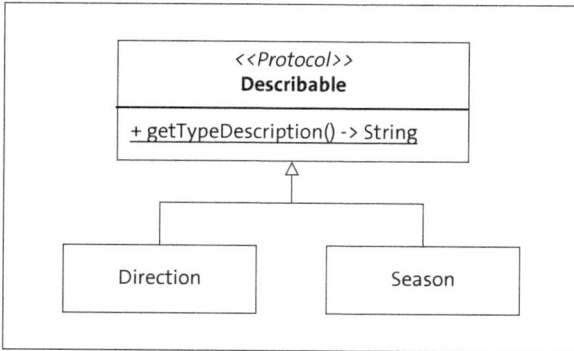

Figure 10.9 Architecture of Static Protocol Example

```
protocol Describable {
    static func getTypeDescription() -> String
}

enum Direction: Describable {
    case north
    case south
    case east
    case west

    static func getTypeDescription() -> String { return "Direction Enum" }
}

enum Season: Describable {
    case spring
    case summer
    case autumn
    case winter

    static func getTypeDescription() -> String { return "Season Enum" }
}

print(Direction.getTypeDescription())   // Direction Enum
print(Season.getTypeDescription())      // Season Enum
```

Listing 10.5 Protocol with Static Function Implemented for enums

In this section, you learned about the function requirements of Swift protocols. But that's not all: You can make protocols impose certain initializer signatures too.

10.3   Initializer Requirements

Just like imposing certain functions, protocols also can impose certain initializer signatures. All implementing data types will have to conform to that signature.

Listing 10.6 demonstrates such an example. The Notification protocol has an initializer requirement, by which init needs to be implemented with two parameters: recipient and message.

```
protocol Notification {
    init(recipient: String, message: String)
    func send()
}

struct Email: Notification {
    var target_address: String
    var mail_body: String

    init(recipient: String, message: String) {
        self.target_address = recipient
        self.mail_body = message
    }

    func send() { print("Email sent to \(self.target_address)") }
}

struct Sms: Notification {
    var target_phone: String
    var short_msg: String

    init(recipient: String, message: String) {
        self.target_phone = recipient
        self.short_msg = message
    }

    func send() { print("SMS sent to \(self.target_phone)") }
}

let email = Email(recipient: "hello@hello.com", message: "Hello!")
email.send() // Email sent to hello@hello.com
```

Listing 10.6 Initializer Requirement Example

In this example, both the Email and Sms structs implement the Notification protocol. To conform to the protocol, both structs have implemented init the way it was imposed. `Email` interprets `recipient` as `target_address` and message as `mail_body`, whereas `Sms` interprets `recipient` as `target_phone` and `message` as `short_msg`.

When it comes to creating a struct instance, the good old syntax can still be used, like `let email = Email()`. Just because `Email.init` was imposed by the protocol doesn't mean that the client's syntax needs to change.

There is a cool trick, though, that's useful for increasing your code's flexibility. Check the enhanced code sample in Listing 10.7.

```
protocol Notification {
    init(recipient: String, message: String)
    func send()
}

struct Email: Notification {
    var target_address: String
    var mail_body: String

    init(recipient: String, message: String) {
        self.target_address = recipient
        self.mail_body = message
    }

    func send() { print("Email sent to \(self.target_address)") }
}

struct Sms: Notification {
    var target_phone: String
    var short_msg: String

    init(recipient: String, message: String) {
        self.target_phone = recipient
        self.short_msg = message
    }

    func send() { print("SMS sent to \(self.target_phone)") }
}

func notify(_ type: Notification.Type, recipient: String, message: String) {
    let notification = type.init(recipient: recipient, message: message)
    notification.send()
}
```

```
notify(Email.self, recipient: "hello@hello.com", message: "Hello!")
notify(Sms.self, recipient: "+1234567890", message: "Hi!")
```

Listing 10.7 Type-Based Instantiation Example

This enhanced version takes advantage of type-based instantiation by throwing in the `notify` function. Like the `type` parameter, this function accepts a `Notification.Type` value, which can be `Email.self` or `Sms.self`, which both implement the `Notification` protocol. Within the function's body, you can create an object instance via the `type.init` syntax:

- If `Email.self` is passed, `type.init` will invoke `Email.init`.
- If `Sms.self` is passed, `type.init` will invoke `Sms.init`.

This technique allows you to consolidate the construction of the object into a function instead of asking clients to create and pass an object instance. This isn't always necessary, but it's useful in some cases.

Now, let's inspect property requirements in protocols.

10.4   Property Requirements

As you learned in previous chapters, enumerations, structs, and classes can have properties, and different flavors of properties are available, like the following:

- Immutable properties (`let`)
- Mutable properties (`var`)
- Computed properties (`get`/`set`)
- Static properties (type-based)

Because protocols standardize such data types, it's only natural for protocols to impose property requirements too. Implementing data types will have to provide such properties of the same name and type.

In the following sections, you'll discover different types of property requirements and how they are implemented.

10.4.1   Get Requirement

A *get requirement* is a property requirement in which the protocol imposes an immutable/read-only property on implementing data types. Listing 10.8 shows the corresponding syntax, in which the `Product` protocol has a `get` requirement for `price`.

```
protocol Product {
    var price: Double { get }
}
```

```
struct Instrument: Product {
    let price: Double
}
```

```
var guitar: Product = Instrument(price: 1000.0)
print(guitar.price) // 1000
```

Listing 10.8 get Requirement Example

The get requirement is coded like a regular variable declaration, but it ends with the {
get } suffix, which implies that the property is immutable. In correspondence, the
implementing Instrument struct had to declare a namesake immutable property, let
price: Double. That perfectly conforms to the protocol's requirement, and the rest of the
code flows smoothly.

However, implementing an immutable variable is not the only way to fulfill the protocol's requirement. A namesake computed property works equally well, as shown in Listing 10.9.

```
protocol Product {
    var price: Double { get }
}
```

```
struct Instrument: Product {
    var price: Double {
        get {
            // Simulating internal calculations
            return 1000
        }
    }
}
```

```
var guitar: Product = Instrument()
print(guitar.price) // 1000
```

Listing 10.9 Fulfilling get Requirement with Computed Property

In this example, Instrument fulfilled the price requirement using a computed property.
In a live application, this code would probably connect to a database to fetch the latest
price, but we simulated that with a hard-coded price value instead. In the end, the protocol's expectations were satisfied.

10.4.2   Get Set Requirement

To impose a mutable property, a protocol may contain a get set requirement for a property. That's similar to the previous syntax; all you have to do is to add a set keyword to the property's suffix. Listing 10.10 shows how.

```
protocol Payable {
    var amount: Double { get }
    var isPaid: Bool { get set }
}

class Invoice: Payable {
    let amount: Double
    var isPaid: Bool = false

    init(amount: Double) {
        self.amount = amount
    }
}

var invo: Payable = Invoice(amount: 100.0)
invo.isPaid = true
invo.isPaid = false
```

Listing 10.10 get set Requirement Example

Pay attention to the Payable protocol, which contains both immutable and mutable property requirements. Payable.amount is immutable because it has the { get } suffix, whereas Payable.isPaid is mutable because it has the { get set } suffix. The difference is that easy!

The implementing Invoice class fulfills those requirements as expected: Invoice. amount is an immutable let property, whereas Invoice.isPaid is a mutable var property.

Like get requirements, get set requirements can be fulfilled with computed properties too. Listing 10.11 demonstrates the same example converted to a computed property.

```
protocol Payable {
    var amount: Double { get }
    var isPaid: Bool { get set }
}

class Invoice: Payable {
    let amount: Double
```

```
    init(amount: Double) {
        self.amount = amount
    }

    var isPaid: Bool {
        get {
            // Read payment status from the database
            return true
        }
        set (newVal) {
            // Record payment status to the database
        }
    }
}

var invo: Payable = Invoice(amount: 100.0)
invo.isPaid = true
invo.isPaid = false
```

Listing 10.11 Fulfilling get set Requirement with Computed Property

In this case, `Invoice.isPaid` was implemented as a computed property. Its get section could read the payment status dynamically from the database and return it—which was simulated by a hard-coded `true` value in this case. Meanwhile, its set section could update the database with the passed payment status.

Different implementations are also possible, of course. But the key takeaway is the ability to fulfill get set requirements with computed properties.

10.4.3 Static Requirement

In the preceding examples, instance get set requirements were implemented. However, you are not limited to instance properties; static property requirements can also be a part of protocols.

As a demonstration, let's reimplement the previous example in Listing 10.5. This time, we'll use a static property requirement instead of a static function. Check the sample code in Listing 10.12, focusing on `Describable.typeDescription`.

```
protocol Describable {
    static var typeDescription: String { get }
}
```

```
enum Direction: Describable {
    static let typeDescription: String = "Direction Enum"

    case north
    case south
    case east
    case west
}

enum Season: Describable {
    case spring
    case summer
    case autumn
    case winter

    static var typeDescription: String { get { return "Season Enum" } }
}

print(Direction.typeDescription)    // Direction Enum
print(Season.typeDescription)       // Season Enum
```

Listing 10.12 Static Property Requirement Example

In this example, the Describable protocol has two enumeration implementations. Direction.typeDescription was implemented as a static let constant, whereas Season. typeDescription was implemented as a static computed property. Both approaches are equally valid.

In the examples so far, we have covered data types that implement a single protocol. But that is not a limitation by any means! Let's walk through some other options in the next section.

10.5   Implementing Multiple Protocols

In Swift, protocol-supporting data types are free to implement multiple protocols. So long as all imposed properties and functions are implemented, there is no restriction against that. Figure 10.10 illustrates a sample architecture in which different classes implement different sets of protocols.

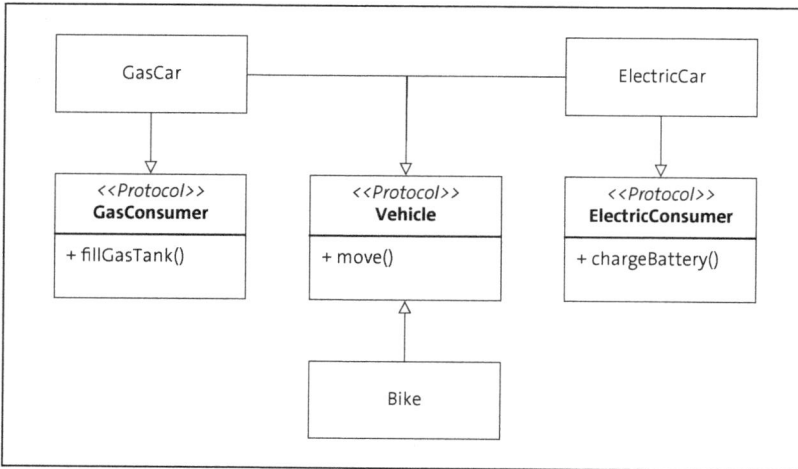

Figure 10.10 Sample Architecture for Multiprotocol Implementations

The protocol implementation plan is summarized in Table 10.2. As seen here, GasCar and ElectricCar both implement multiple protocols rather than a single one.

Class	Implemented Protocols	Supported Functions
Bike	Vehicle	move
GasCar	Vehicle, GasConsumer	move, fillGasTank
ElectricCar	Vehicle, ElectricConsumer	move, chargeBattery

Table 10.2 Classes with Implemented Protocols

The code implementation of this architecture is shown in Listing 10.13. Note that classes implementing multiple protocols need to mention the protocol names separated by commas, such as in Vehicle, GasConsumer.

```
protocol Vehicle { func move() }
protocol GasConsumer { func fillGasTank() }
protocol ElectricConsumer { func chargeBattery() }

class Bike: Vehicle {
    func move() { print("Bike moved") }
}

class GasCar: Vehicle, GasConsumer {
    func move() { print("Gas car moved") }
    func fillGasTank() { print("Car filled up") }
}
```

```
class ElectricCar: Vehicle, ElectricConsumer {
    func move() { print("Electric car moved") }
    func chargeBattery() { print("Car charged") }
}

var myRide: GasConsumer = GasCar()
myRide.fillGasTank() // Car filled up

var herRide: ElectricConsumer = ElectricCar()
herRide.chargeBattery() // Car charged
```

Listing 10.13 Multiprotocol Implementation Example

Interface Segregation Principle

The *interface segregation principle* (ISP) says that classes should not be forced to depend on methods they don't use. According to ISP, having large interfaces (protocols) with many methods may be a bad idea. Instead, they should be split into separate interfaces (protocols). That's exactly what we did in the latest example: It wouldn't make sense for Bike to have a method called fillGasTank, for instance.

Another approach for multiprotocol implementation is protocol inheritance, as sketched in Figure 10.11. Because each gas vehicle or electric vehicle is already a vehicle, you can inherit GasConsumer and ElectricConsumer from the Vehicle protocol.

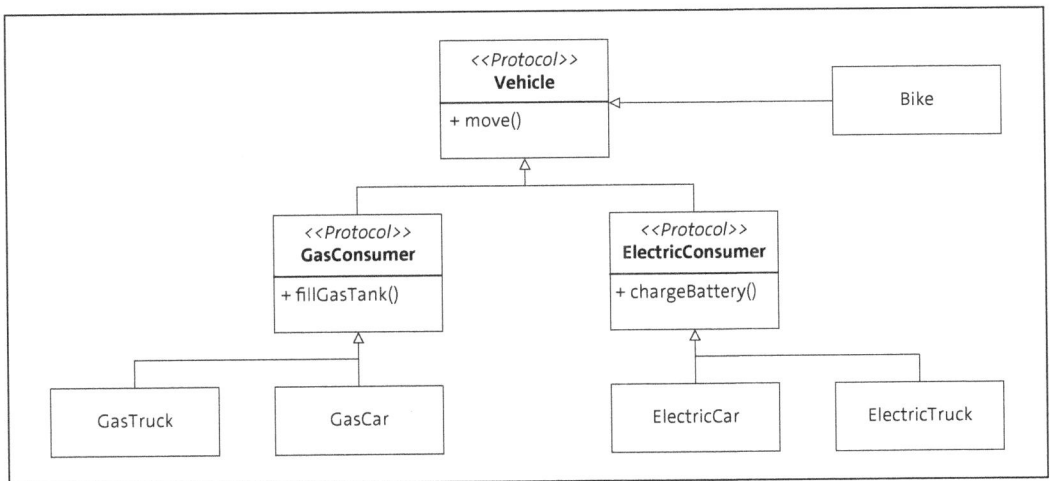

Figure 10.11 Protocol Inheritance Architecture

Protocol inheritance works similarly to class inheritance, which you learned about in Chapter 9. Each child protocol inherits properties and functions from the parent protocol. In this example, GasConsumer will end up with the move (from Vehicle) and fillGasTank (its own) methods. ElectricConsumer will end up with the move (from Vehicle) and charge-Battery (its own) methods.

As such, it's possible to build incrementally growing protocol structures with a similar logic as that for class inheritance. An implementation of this approach is shown in Listing 10.14.

```
protocol Vehicle { func move() }
protocol GasConsumer: Vehicle { func fillGasTank() }
protocol ElectricConsumer: Vehicle { func chargeBattery() }

class Bike: Vehicle {
    func move() { print("Bike moved") }
}

class GasCar: GasConsumer {
    func move() { print("Gas car moved") }
    func fillGasTank() { print("Car filled up") }
}

class ElectricCar: ElectricConsumer {
    func move() { print("Electric car moved") }
    func chargeBattery() { print("Car charged") }
}

var myRide: GasConsumer = GasCar()
myRide.fillGasTank() // Car filled up

var herRide: Vehicle = ElectricCar()
herRide.move() // Electric car moved
```

Listing 10.14 Protocol Inheritance Implementation

As you can see, you can cast an object to any protocol in its hierarchy. For myRide, GasCar was cast as GasConsumer (mid-level protocol). For herRide, ElectricCar was cast as Vehicle (top-level protocol).

Now, let's explain how to find out if a data type has implemented a protocol or not.

10.6   Checking for Protocol Conformance

In Chapter 9, as you were learning about class inheritance, we highlighted that class conformance can be checked using the is and as keywords. Because protocols also act like parent types, the same keywords can be used to check protocol conformance—and the syntax won't change much either.

Listing 10.15 demonstrates the usage of the is keyword. if obj is Food will return true for Burger instances but false for Computer instances, because Burger implements the Food protocol.

```
protocol Food { var calories: Int { get } }

struct Burger: Food { let calories = 300 }

struct Computer {
    var brand: String
    var model: String
}

let objects: [Any] = [Burger(),
                    Computer(brand: "Apple", model: "MacBook Pro")]

for obj in objects {
    if obj is Food { print("\(obj) is a food") }
}
```

Listing 10.15 Protocol Conformance Check Using is Keyword

The is keyword is useful for quick checks, but it doesn't allow you to access protocol elements in this example. After all, obj is a generic object. Even when obj points to a Food-compatible object (like Burger), obj needs to be cast as a Food instance before Food elements can be accessed.

An easy way to achieve that is to use the as keyword. Check the demonstration in Listing 10.16.

```
protocol Food { var calories: Int { get } }

struct Burger: Food { let calories = 300 }

struct Computer {
    var brand: String
    var model: String
}

let objects: [Any] = [Burger(),
                    Computer(brand: "Apple", model: "MacBook Pro")]

for obj in objects {
    if let food = obj as? Food {
```

10

```
        print("\(food) is a food with \(food.calories) calories")
    }
}
```

Listing 10.16 Protocol Conformance Check Using as Keyword

The familiar if let food = obj as? Food expression is the key point of this example. This will only return true for Food-compliant instances (like a Burger instance), and obj will be cast as Food, allowing you to access Food features, like calories.

Easy, right? Just like checking parent-child class conformances!

10.7   Protocols as Function Parameters

In Swift, a function can import or return a value that conforms to a protocol, such as a class or struct instance. Functions returning a protocol can effectively act like a factory or builder. Although this sounds simple enough, there are different ways to return protocol-conforming types, which will be explored in this section.

10.7.1   Protocol Type

A function can use a protocol type directly. For instance, returning a strict protocol type via casting is the most basic approach. This approach is demonstrated in Listing 10.17, where you should focus on the makeRandomShape function.

```
protocol Shape {
    var area: Double { get }
}

class Circle: Shape {
    let radius: Double
    init(radius: Double) { self.radius = radius }
    var area: Double { return .pi * radius * radius }
}

func makeRandomShape() -> Shape {
    return Circle(radius: 10) as Shape
}

let myShape = makeRandomShape()
print(myShape.area) // 314.15
if myShape is Circle { print("It's a Circle!") }
if let c = myShape as? Circle { print(c.radius)  }
```

Listing 10.17 Returning Protocol Type from a Function

In this example, makeRandomShape is returning an object of the Shape protocol type. The underlying object is Circle in this case, but it could have been Triangle or Square; the client doesn't necessarily need to know about that. The client would mostly be interested in Shape.area, which is a common property for all Shape types.

However, the client can still use the if myShape is Circle expression, which will return true if that's the case. Likewise, the following if let c = myShape as? Circle expression is equally valid; if myShape is really Circle, then you can downcast it to c and access Circle-only features, such as Circle.radius.

Likewise, a function could also import a protocol type as a parameter, such as func drawShape(s: Shape).

Using a strict type is the most basic, intuitive, and straightforward method to return a protocol from a function. However, this syntax is discouraged in newer versions of Swift. To achieve the same functionality, existential types can be used—which is our next topic.

10.7.2   Existential Type

For existential types, the syntax changes slightly, but you basically gain the same functionality as with the syntax for returning a protocol type. Check the enhanced demonstration in Listing 10.18, where makeRandomShape randomly returns a Circle or Square instance, fulfilling its purpose more realistically.

```
protocol Shape {
    var area: Double { get }
}

class Circle: Shape {
    let radius: Double
    init(radius: Double) { self.radius = radius }
    var area: Double { return .pi * radius * radius }
}

class Square: Shape {
    let side: Double
    init(side: Double) { self.side = side }
    var area: Double { return side * side }
}

func makeRandomShape() -> any Shape {
    return Bool.random() ? Square(side: 10) : Circle(radius: 10)
}

let myShape = makeRandomShape()
```

```
print(myShape.area) // 314.15
if myShape is Circle { print("It's a Circle!") }
if let c = myShape as? Circle { print(c.radius)  }
```

Listing 10.18 Returning Existential Type from a Function

Careful readers will have already noticed that the function's signature has changed slightly, becoming `func makeRandomShape() -> any Shape`. This means that `makeRandomShape` can return *any* object that conforms to the Shape protocol: `Circle` instances, `Square` instances, whatever.

If the function returns a `Square` instance, the last two lines won't print anything, simply because the object is not a `Circle` instance.

Boxed Types

Existential types are also called *boxed types*. If you see the term *boxed type* anywhere, you can assume that it is a type returned via the any keyword.

Adding the any keyword makes the purpose and intent of the function more transparent. A programmer seeing the any keyword will understand that this is a flexible function able to return many object types and will prepare their client code accordingly.

A function could also use the any keyword when importing parameters. The `func paintShapes(shapes: [any Shape])` signature, for example, means that a `Circle` and a `Square` instance can be sent to `paintShapes` simultaneously in the same array.

As an alternative, let's discuss opaque types.

10.7.3   Opaque Type

Syntax-wise, opaque types are very similar to existential types. As shown in Listing 10.19, all you need to do is replace any Shape with some Shape in the function's signature.

```
protocol Shape {
    var area: Double { get }
}

class Circle: Shape {
    let radius: Double
    init(radius: Double) { self.radius = radius }
    var area: Double { return .pi * radius * radius }
}

func makeRandomShape() -> some Shape {
    return Circle(radius: 10)
```

```
}

let myShape = makeRandomShape()
print(myShape.area) // 314.15
if myShape is Circle { print("It's a Circle!") }
if let c = myShape as? Circle { print(c.radius)  }
```

Listing 10.19 Returning Opaque Type from a Function

Although the syntax change is minor, the underlying effect is not all that small. With any Shape, the function gains the ability to return any Shape type (Circle, Square, etc.). However, with some Shape, the function is limited to return only one Shape subtype.

In this instance, makeRandomShape can't freely return a Square or Circle. Instead, it must return either a Square or a Circle in all cases, reducing the flexibility of the function but simplifying things for the client.

A function could also use the some keyword when importing parameters. The func paintShapes(shapes: [some Shape]) signature means that a Circle and a Square instance *cannot* be sent to paintShapes simultaneously. All shapes must either be Circle instances or Square instances: No mix-ups are allowed.

Existential Versus Opaque Types

Existential types are more flexible, allowing any protocol implementer to pass. But this might cost the protocol consumer more effort for defensive programming and management.

Opaque types are simpler, allowing only one protocol implementer to pass. Once the protocol consumer knows about the concrete type, it can proceed accordingly.

Now, let's talk about generics in protocols.

10.8   Associated Types

In previous chapters, you learned how generics increase the flexibility of any data type. Protocols are no exception; generics can increase the abstraction degree of any protocol. In this section, you'll learn how generics apply to protocols.

So far, protocol examples have contained concrete data types, such as String or Double. However, fixating on data types might be a limiting factor, leading to the creation of multiple similar protocols.

Check the sample in Listing 10.20. StringReversible and IntReversible are very similar protocols, with a slight data type difference: StringReversible.reverse is expected to turn "ABC" into "CBA", while IntReversible.reverse is expected to turn 123 into 321.

```
protocol StringReversible {
    func reverse(_ input: String) -> String
}

protocol IntReversible {
    func reverse(_ input: Int) -> Int
}
```

Listing 10.20 Two Very Similar Protocols

As you can see, both protocols are identical except for the data type difference. In such cases, you can merge those protocols into a single protocol with an associated type. Listing 10.21 shows the merged protocol.

```
protocol Reversible {
    associatedtype AT
    func reverse(_ input: AT) -> AT
}
```

Listing 10.21 Merged Protocols with Associated Type

Note the associatedtype AT expression, which acts as a placeholder for the actual data type. The implementing struct or class is responsible for replacing the made-up AT token with an actual data type. Listing 10.22 demonstrates how to fulfill this responsibility.

```
protocol Reversible {
    associatedtype AT
    func reverse(_ input: AT) -> AT
}

struct TextReverser: Reversible {
    func reverse(_ value: String) -> String {
        return String(value.reversed())
    }
}

struct NumberReverser: Reversible {
    func reverse(_ value: Int) -> Int {
        let reversedString = String(String(value).reversed())
        return Int(reversedString) ?? value
    }
}
```

```
print(TextReverser().reverse("ABC"))     // BCA
print(NumberReverser().reverse(123))     // 321
```

Listing 10.22 Associated Type Protocol Implementation

In this implementation, TextReverser used the String type to replace AT, while Number-Reverser used the Int type to replace AT. Both are perfectly valid implementations of the Reversible protocol, taking advantage of its data type flexibility.

Naturally, a protocol can host multiple associated types too. In Listing 10.23, KeyValStore hosts two associated types: K (key type) and V (value type). The implementing TextKey-ValStore class has used String for both placeholder associated types, but other implementations are also possible—like String for K and Int for V.

```
protocol KeyValStore {
    associatedtype K
    associatedtype V

    func store(_ key: K, _ val: V)
    func get(_ key: K) -> V?
    func remove(_ key: K)
}

class TextKeyValStore: KeyValStore {
    private var db: [String: String] = [:]

    func store(_ key: String, _ val: String) {
        self.db[key] = val
    }

    func get(_ key: String) -> String? {
        return self.db[key]
    }

    func remove(_ key: String) {
        self.db[key] = nil
    }
}
```

Listing 10.23 Protocol with Multiple Associated Types

Another topic for protocol generics is type constraints, which will be covered in the next section. You will learn how to apply standard protocols to your custom objects, and you'll see application examples of type constraints as well.

10.9   Using Standard Protocols

Swift comes with a bag full of built-in protocols ready to be used for your convenience in your developments. And guess what: You may have already used them throughout the book, without us necessarily naming them as protocols yet. For example, the == token corresponds to the Equatable protocol, while the <= token corresponds to the Comparable protocol.

If you're a little confused, don't worry; we'll cover those two protocols in this section, followed by some other common ones.

10.9.1   Equatable

In Swift, Equatable is a protocol that allows you to compare two values using the == operator. Any type conforming to Equatable can be compared, as in a == b. In this section, we'll go over class and struct examples as those are the typical implementors.

In Listing 10.24, the Equatable protocol is implemented for the Book class. As a prerequisite of the protocol, Book has to contain a function called ==, which determines if two Book objects are equal.

```
class Book: Equatable {
    let title: String
    let author: String
    let pages: Int = 0

    init(title: String, author: String) {
        self.title = title
        self.author = author
    }

    static func == (b1: Book, b2: Book) -> Bool {
        return b1.title == b2.title && b1.author == b2.author
    }
}

let b1 = Book(title: "1984", author: "George Orwell")
let b2 = Book(title: "1984", author: "George Orwell")
let b3 = Book(title: "Animal Farm", author: "George Orwell")

b1 == b2 // true
b1 == b3 // false
```

Listing 10.24 Equatable Class Implementation

During the comparison, you only compare the key properties: title and author. If both are equal, those books should be equal too, so there's no need to waste time on further properties.

Once Equatable is properly implemented, books can be compared with the usual simple syntax—that is, b1 == b2. Cool, right?

If Book was a struct instead of a class, the overall syntax wouldn't change much. Check the enhanced example in Listing 10.25, where Equatable was implemented for a struct.

```
struct Book: Equatable {
    let title: String
    let author: String
    let pages: Int = 0

    static func == (b1: Book, b2: Book) -> Bool {
        return b1.title == b2.title && b1.author == b2.author
    }
}


let b1 = Book(title: "1984", author: "George Orwell")
let b2 = Book(title: "1984", author: "George Orwell")
let b3 = Book(title: "Animal Farm", author: "George Orwell")


b1 == b2 // true
b1 == b3 // false
```

Listing 10.25 Equatable Struct Implementation

Here's a cool trick for structs: If you're going to compare all struct properties for equality, don't bother; Swift can do that automatically. You don't need to explicitly implement the == function in that case, as demonstrated in Listing 10.26.

```
struct Book: Equatable {
    let title: String
    let author: String
}


let b1 = Book(title: "1984", author: "George Orwell")
let b2 = Book(title: "1984", author: "George Orwell")
let b3 = Book(title: "Animal Farm", author: "George Orwell")


b1 == b2 // true
b1 == b3 // false
```

Listing 10.26 Synthesized Implementation for Equatable

This automation technique is called *synthesized implementation*, and it's a time-saving feature of Swift. Note that even if you add new properties to Book in the future, all of them will be automatically included in the == comparison—meaning one less maintenance task to worry about.

Another empowerment from standard protocols like Equatable is that they open doors for generic functions with type constraints. In Listing 10.27, the allEqual function can accept both Book and Album arrays, simply because allEqual can work with any array containing some Equatable objects.

```
struct Book: Equatable {
    let title: String
    let author: String
}

struct Album: Equatable {
    let name: String
    let band: String
}

func allEqual<T: Equatable>(_ elements: T...) -> Bool {
    for element in elements[1...] {
        if element != elements[0] { return false }
    }
    return true
}

let b1 = Book(title: "1984", author: "George Orwell")
let b2 = Book(title: "1984", author: "George Orwell")
let b3 = Book(title: "Animal Farm", author: "George Orwell")
allEqual(b1, b2)        // True
allEqual(b1, b2, b3)    // False

let a1 = Album(name: "Nevermind", band: "Nirvana")
let a2 = Album(name: "Let It Be", band: "Beatles")
allEqual(a1, a2)    // False
```

Listing 10.27 Equatable as Type Constraint

Naturally, allEqual(a1, b2) would produce an error because such generic functions expect a single concrete type, like some Equatable (opaque logic). Logically, it doesn't make sense to compare a book to an album anyway; Swift merely reflects that.

Now, let's cover some other similar protocols.

10.9.2   Comparable

Now that we have Equatable under control, Comparable should be a breeze. In Swift, Comparable is a protocol that allows you to compare two values using comparison operators, such as <= and >=. Any type conforming to Comparable can be compared, like a > b.

Listing 10.28 demonstrates a Comparable-compliant class: Person. The purpose here is to compare people based on their ages.

```swift
class Person: Comparable {
    let age: Int
    let name: String

    init(age: Int, name: String) {
        self.age = age
        self.name = name
    }

    static func == (lhs: Person, rhs: Person) -> Bool {
        return lhs.age == rhs.age
    }

    static func < (lhs: Person, rhs: Person) -> Bool {
        return lhs.age < rhs.age
    }
}

let bob = Person(age: 30, name: "Bob")
let alice = Person(age: 25, name: "Alice")
let david = Person(age: 25, name: "David")

alice == david  // true
alice > bob     // false
bob >= alice    // true
```

Listing 10.28 Comparable Class Implementation

As you can see, you have to implement two functions in Person: == and <. Using those two base functions, Swift automatically takes care of other comparison options, such as > or <=. In the following lines, alice, david, and bob can be compared based on their ages.

If Person was a struct, you could take advantage of synthesized implementation for the == function given that we accept all properties to be compared for equality. But you still need to implement the < function manually. Check the demonstration in Listing 10.29.

```
struct Person: Comparable {
    let age: Int
    let name: String

    static func < (lhs: Person, rhs: Person) -> Bool {
        return lhs.age < rhs.age
    }
}

let bob = Person(age: 30, name: "Bob")
let alice = Person(age: 25, name: "Alice")
let david = Person(age: 25, name: "David")

alice == david  // false
alice > bob     // false
bob >= alice    // true
```

Listing 10.29 Synthesized Comparable Implementation in Struct

Note that alice == david returns false this time. Synthesized implementation checks if all properties are the same. Although alice.age and bob.age are equal, alice.name and bob.name are not; therefore, alice == david is returned as false. To compare ages only, you have to implement the == function manually, as you did for the Person class.

However, a more autonomous synthesized implementation mechanism is available for enumerations. Swift can handle their Comparable operations automatically, given that the enumeration doesn't have associated values and that values are in ascending order.

In Listing 10.30, an enumeration called Priority fulfills all requirements: No associated values are present, and priority values are sorted from lowest to the highest.

```
enum Priority: Comparable {
    case low
    case medium
    case high
    case critical
}

let p1: Priority = .low
let p2: Priority = .medium
let p3: Priority = .high
let p4: Priority = .critical

p1 < p2     // true
p3 > p1     // true
p2 <= p3    // true
```

```
let priorities: [Priority] = [.high, .medium, .critical, .low]
let sortedPriorities = priorities.sorted() // Automatic sort
print(sortedPriorities) // [low, medium, high, critical]
```

Listing 10.30 Synthesized Comparable Implementation in enum

In this case, all comparison and sorting operations can be handled by Swift automatically. If you had to implement ad hoc logic, you could still implement the == and < functions, of course.

Finally, let's look at an example in which Comparable is used as a generic type. Check the code sample in Listing 10.31.

```
struct Person: Comparable {
    let age: Int
    let name: String

    static func < (lhs: Person, rhs: Person) -> Bool {
        return lhs.age < rhs.age
    }
}

func getOldest<T: Comparable>(_ elements: T...) -> T {
    var oldest = elements[0]

    for element in elements[1...] {
        if element > oldest { oldest = element }
    }

    return oldest
}



let bob = Person(age: 30, name: "Bob")
let alice = Person(age: 25, name: "Alice")
let david = Person(age: 25, name: "David")

getOldest(alice, david, bob) // bob
```

Listing 10.31 Comparable as Generic Type

This isn't much different than the usage of Equatable as a generic type, right? The only significant difference is that you can use the > token because it's supported by Equatable.

10.9.3   Identifiable

The Identifiable protocol is typically used to label objects with unique identifiers, similar to primary keys in databases.

In real life, car plate numbers, student numbers, employee email addresses, and bank account numbers are unique IDs, right? No two cars can share the same plate number, and no two employees can have the same email address. Likewise, for Identifiable, no two objects may share the same ID, ensuring that each object can be recognized uniquely.

Listing 10.32 demonstrates a manual Identifiable implementation. This protocol needs the implementing type to have a property called id, which should be a Hashable type. In this case, Car.id was implemented as a computed property of the String type.

```
struct Car: Identifiable {
    let licensePlate: String

    var id: String {
        return licensePlate
    }
}

let myCar = Car(licensePlate: "XJ3 7QK")
print(myCar.id) // XJ3 7QK
```

Listing 10.32 Computed Identifiable Implementation

However, a regular stored property would work equally well. So long as the property is called id and the type is Hashable, it should work (see Listing 10.33).

```
struct Student: Identifiable {
    let id: Int
    let name: String
}

let kid = Student(id: 1, name: "John")
```

Listing 10.33 Stored Identifiable Implementation

Hashable Protocol

You'll learn about the Hashable protocol shortly. For now, it's enough to know that most basic types in Swift are automatically Hashable, such as String, Int, and Bool—but you can make your custom types Hashable too.

As a further exercise, Listing 10.34 demonstrates a sample in which Identifiable is used as a generic type constraint.

```
struct Student: Identifiable {
    let id: Int
    let name: String
}

func areSame<T: Identifiable>(_ t1: T, _ t2: T) -> Bool {
    return t1.id == t2.id
}

let kid1 = Student(id: 1, name: "John")
let kid2 = Student(id: 2, name: "Eric")
areSame(kid1, kid2) // false
```

Listing 10.34 Identifiable as Generic

Identifiable in SwiftUI

As you advance in your Swift adventures toward SwiftUI, the real value of Identifiable will surely emerge. As you work with UI components to display your objects, SwiftUI will need a unique ID for each object to keep track of them. That's where Identifiable becomes essential in your objects.

As we just mentioned the Hashable protocol, now is the perfect time to address it.

10.9.4   Hashable

In Chapter 3, you learned that some collection types like Set and Dictionary hash their elements and store them uniquely. And in the previous section, you learned that Identifiable requires the data type to be hashed. In those instances, and many more, you need to apply hashing to your custom data types.

To counter that requirement, Swift politely provides the built-in Hashable protocol for your convenience. When Hashable is implemented properly, that data type can be used nearly anywhere hashing is required.

Listing 10.35 features an example in which Hashable is implemented by the Book struct. That protocol requires the hash function to be implemented. In this instance, the hash value was built by combining Book.title and Book.author as this combination is expected to be a unique value.

```
struct Book: Hashable {
    let title: String
    let author: String
```

```
    let pages: Int

    func hash(into hasher: inout Hasher) {
        hasher.combine(title)
        hasher.combine(author)
    }
}

let book1 = Book(title: "Hobbit", author: "Tolkien", pages: 320)
let book2 = Book(title: "LOTR", author: "Tolkien", pages: 1000)
let books: Set<Book> = [book1, book2]
```

Listing 10.35 Manual Hashable Implementation

Because Book is Hashable now, you can put book1 and book2 into the books set. As an exercise, you can remove the Hashable implementation from Book and see that books won't accept those objects any longer.

A good rule of thumb is to ensure that the values fed into hasher.combine are unique. Those are the values you would use for a comparison in case your type implements Equatable. In the preceding example, Book.title and Book.author are the very values you would check to see if two books are identical. Check the sample code in Listing 10.36 for a better understanding.

```
struct Book: Equatable, Hashable {
    let title: String
    let author: String
    let pages: Int

    func hash(into hasher: inout Hasher) {
        hasher.combine(title)
        hasher.combine(author)
    }

    static func == (lhs: Book, rhs: Book) -> Bool {
        return lhs.title == rhs.title && lhs.author == rhs.author
    }
}

let book1 = Book(title: "Hobbit", author: "Tolkien", pages: 320)
let book2 = Book(title: "LOTR", author: "Tolkien", pages: 1000)
let books: Set<Book> = [book1, book2]
```

Listing 10.36 Equatable and Hashable Implemented Simultaneously

Here's a nifty shortcut: If the combination of all stored properties builds a unique value, then you can omit the hash implementation; Swift can automatically handle hashing via synthesized implementation. That version of the preceding example is shown in Listing 10.37.

```
struct Book: Hashable {
    let title: String
    let author: String
    let pages: Int
}

let book1 = Book(title: "Hobbit", author: "Tolkien", pages: 320)
let book2 = Book(title: "LOTR", author: "Tolkien", pages: 1000)
let books: Set<Book> = [book1, book2]
```

Listing 10.37 Synthesized Implementation for Hashable

As a bonus, here's a pro tip: If your data type implements Identifiable, then your id is the perfect candidate for a hash source as id must already be unique. As an exercise, Listing 10.38 features a struct that implements three protocols simultaneously.

```
struct Car: Identifiable, Hashable, Equatable {
    let licensePlate: String
    let owner: String

    var id: String {
        return self.licensePlate
    }

    func hash(into hasher: inout Hasher) {
        hasher.combine(self.id)
    }

    static func == (lhs: Car, rhs: Car) -> Bool {
        return lhs.id == rhs.id
    }
}
```

Listing 10.38 Identifiable, Hashable, and Equatable Implemented Together

10.9.5   CustomStringConvertible

In Swift, the CustomStringConvertible protocol is about producing human-readable strings that describe your object. This protocol merely requires a description property to be implemented. Listing 10.39 shows a simple implementation of this protocol.

```
struct Car: CustomStringConvertible {
    let make: String
    let model: String
    let year: Int

    var description: String {
        return "\(make) \(model), produced in \(year)"
    }
}

let theCar = Car(make: "Toyota", model: "Yaris", year: 2025)
print(theCar) // Toyota Yaris, produced in 2025
```

Listing 10.39 Sample CustomStringConvertible Implementation

Although this example contains a simple print statement, the usefulness of Custom-StringConvertible goes beyond that. Because this is a language-level protocol, Swift automatically uses it in debugger outputs, SwiftUI text views, logging frameworks, and string interpolations. As a second example, a string interpolation demonstration is provided in Listing 10.40.

```
struct BassGuitar: CustomStringConvertible {
    let brand: String
    let model: String
    let strings: Int
    let isActive: Bool

    var description: String {
        return "\(strings)-string \(isActive ? "active" : "") \(brand)
\(model) bass "
    }
}

let bass = BassGuitar(brand: "Yamaha", model: "C400", strings: 4, isActive:
true)
print("I have a \(bass)") // I have a 4-string active Yamaha C400 bass
```

Listing 10.40 CustomStringConvertible in String Interpolation

This concludes our standard protocol examples. Naturally, the Swift language has many more built-in protocols, but the ones we covered here should be a good starting point. Two further protocols, Encodable and Decodable, will be addressed in Chapter 13, where you'll learn more about industry-standard formats like JSON.

10.10   Summary

In this chapter, we covered one of the major pillars of object-oriented Swift.

Protocols act as standardized blueprints for classes, structs, and enums, in which their behavior can be standardized. In return, the client code only needs to know about the components of the protocol, not the implementation details of the underlying object — just like your computer knowing about the USB protocol but not about the technicalities of the connected keyboard.

As part of their standards, protocols can impose function, initializer, and property requirements. Data types conforming to the protocol must implement all these requirements, ensuring a smooth connection to the client code.

Protocols can be used as function parameters in the scope of existential and opaque types. For increased flexibility, protocols can also host associated types, unlocking a generic type of behavior.

10

The Swift language also provides many protocols for your convenience to enable your objects to be used with standard syntax elements.

In the next chapter, you'll learn about extensions, which will enable you to append custom logic to existing data types like protocols, classes, structs, and enumerations.

Chapter 11
Extensions

Swift extensions add new functionality to existing elements without modifying the original source. They can include methods and calculated properties for enumerations, structs, classes, or protocols. Extensions enhance flexibility and code organization.

In the Swift language, extensions offer a unique possibility to enhance existing data types by adding new features using custom code. The best part is that you don't have to modify the original source code for this functionality; extensions work side by side with the extended object.

Extensions are particularly useful when you're working with unmodifiable code, such as a third-party library or sensitive code elements. Without altering the original code, you can add new stuff on top of it. Even Swift's own codebase can be altered using extensions.

Figure 11.1 sketches a simple extension. The original Message class comes with two native functions: sendAsEmail and sendAsSms. To add new functionality to that class in your own app, you can make an extension that includes two new functions: sendToWhatsapp and sendToTelegram.

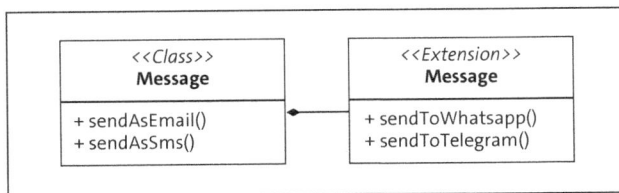

Figure 11.1 Extension Plan

Once this extension is implemented, the codebase will perceive these as native functions. In other words, Message will appear to have four native functions from now on: sendAsEmail, sendAsSms, sendToWhatsapp, and sendToTelegram. Extension elements don't need special treatment; they simply appear as part of the original data type—which is good news for code readability!

Using extensions, you can add the following elements to a data type:

- Computed properties (but not stored properties)
- Functions

461

- Initializers
- Subscripts
- Nested types
- Protocol conformance

Extensions can be applied to enumerations, structs, classes, and protocols. In this chapter, we'll go through examples that address each data type and the elements mentioned previously. Instead of demonstrating each element and data type combination, we will address different extension elements for different data types, covering broad ground economically.

11.1   Extending Enumerations

Let's begin the hands-on exercises by targeting enumerations. In this section, you'll add nested types and computed properties to an enumeration. Note that those elements can be added to structs, classes and protocols too; enumerations stand as a practice platform at this time.

Listing 11.1 demonstrates a simple enumeration called Weather, which features different weather conditions.

```
enum Weather {
    case sunny
    case cloudy
    case rainy
    case stormy
}
```

Listing 11.1 Simple enum

Now, let's create an extension to add a nested type to that enumeration. As you can see in Listing 11.2, the syntax to do so is very simple and straightforward. The extension Weather signature tells Swift that the following code block should be evaluated as an extension of the Weather enumeration.

```
enum Weather {
    case sunny
    case cloudy
    case rainy
    case stormy
}

extension Weather {
    struct ClothingSuggestion {
```

```
        let top: String
        let accessory: String
    }
}

let sugg = Weather.ClothingSuggestion(top: "shirt", accessory: "sunglasses")
```

Listing 11.2 Nested Type Extension

In this code block, you add a nested struct called ClothingSuggestion to Weather, containing the top and accessory properties. As you can see in the following line, this nested struct can be accessed as Weather.ClothingSuggestion, as if it was a native element of Weather. Cool, right?

As a second example, let's add a computed property on top. Listing 11.3 enhances the enumeration even further with suggestion, which returns a clothing suggestion based on the weather condition.

```
enum Weather {
    case sunny
    case cloudy
    case rainy
    case stormy
}

extension Weather {
    struct ClothingSuggestion {
        let top: String
        let accessory: String
    }

    var suggestion: ClothingSuggestion {
        switch self {
        case .sunny:
            return ClothingSuggestion(top: "T-shirt", accessory: "Sunglasses")
        case .cloudy:
            return ClothingSuggestion(top: "Hoody", accessory: "Light jacket")
        case .rainy:
            return ClothingSuggestion(top: "Raincoat", accessory: "Umbrella")
        case .stormy:
            return ClothingSuggestion(top: "Waterproof jacket", accessory:
 "Boots")
        }
    }
}
```

```
let weather = Weather.sunny
print(weather.suggestion.top)        // T-shirt
print(weather.suggestion.accessory)  // Sunglasses
```

Listing 11.3 Nested Type and Computed Property Extension

Once again, you can access the `suggestion` extension property like a native enumeration property, as you can see in `weather.suggestion.top`.

Extensions and Clean Coding

Note that applying an extension to `Weather` made the rest of the code much cleaner. Instead of having dozens of stray code blocks like `getWeatherSuggestion(w: Weather)`, it's tidier to inject them into the corresponding enum as `Weather.suggestion`. As the codebase grows, such orderly approaches add up, keeping the codebase understandable and manageable.

Now, let's move forward to struct extensions and explore some further options.

11.2 Extending Structs

Struct is another data type open to extensions. For a struct example, let's apply a subscript and function extension to an `RGBColor` struct, as shown in Listing 11.4.

```
struct RGBColor {
    let red: Int
    let green: Int
    let blue: Int
}
```

Listing 11.4 Basic Form of RGBColor

What Is RGB Color?

RGB stands for red, green, and blue and is the standard model used to express colors in user interfaces such as web pages. Red, green, and blue are three primary colors of light. By mixing different intensities of those colors, you can produce any other color.

For example, pure red is (255, 0, 0); that is, it has the maximum value for red and the minimum value for other colors. Black is (0, 0, 0), meaning that it has the minimum value for each color. Orange is (255, 165, 0), meaning that it mixes different levels of red and green, but no blue is needed.

Now, let's extend this struct with a numeric subscript, which will allow you to access the colors by index instead of name. Listing 11.5 demonstrates how to do that.

```
struct RGBColor {
    let red: Int
    let green: Int
    let blue: Int
}

extension RGBColor {
    subscript(index: Int) -> Int? {
        switch index {
        case 0: return red
        case 1: return green
        case 2: return blue
        default: return nil
        }
    }
}

let c1 = RGBColor(red: 255, green: 100, blue: 50)
print(c1[1]) // 100
```

Listing 11.5 Subscript Extension for a Struct

The syntax should be very familiar and intuitive. It starts with the `extension RGBColor` signature, expressing the intent to extend the struct. Within the extension, a familiar `subscript` was added, which returns the appropriate color by index. Thanks to that, you can access `c1.green` with the indexed `c1[1]` expression.

As a further example, let's add a function this time. Listing 11.6 extends `RGBColor` with a function called `brightened`. This function brightens the color at hand by the given factor and returns a new `RGBColor` instance containing the brighter color.

```
struct RGBColor {
    let red: Int
    let green: Int
    let blue: Int
}

extension RGBColor {
    subscript(index: Int) -> Int? {
        switch index {
        case 0: return red
        case 1: return green
        case 2: return blue
        default: return nil
        }
```

```
        }

    func brightened(by factor: Double) -> RGBColor {
        func clamp(_ value: Int) -> Int {
            return min(max(value, 0), 255)
        }
        return RGBColor(
            red: clamp(Int(Double(red) * factor)),
            green: clamp(Int(Double(green) * factor)),
            blue: clamp(Int(Double(blue) * factor))
        )
    }
}

let c1 = RGBColor(red: 255, green: 100, blue: 50)
print(c1[1])    // 100
let c2 = c1.brightened(by: 1.1)
print(c2[1])    // 110
```

Listing 11.6 Function Extension for a Struct

As expected, the `brightened` extension function appears to be part of `RGBColor` itself, enabling access via `c1.brightened`. The client code appears to be oblivious to the fact that `RGBColor.brightened` came with an extension, which makes it easy to read and manage.

Now, let's move forward to class extensions.

11.3   Extending Classes

Thanks to your experience with extending enumerations and structs, class extensions should be a breeze. In this section, we'll extend the `User` class, as shown in Listing 11.7.

```
class User {
    let firstName: String
    let lastName: String

    init(firstName: String, lastName: String) {
        self.firstName = firstName
        self.lastName = lastName
    }
}
```

Listing 11.7 User Class in Its Original Form

To demonstrate further extension options, let's extend User with an initializer and protocol conformance. The final code is shown in Listing 11.8.

```
class User {
    let firstName: String
    let lastName: String

    init(firstName: String, lastName: String) {
        self.firstName = firstName
        self.lastName = lastName
    }
}

extension User: CustomStringConvertible {
    convenience init(fullName: String) {
        let parts = fullName.split(separator: " ")
        self.init(firstName: String(parts[0]), lastName: String(parts[1]))
    }

    var description: String { return "User called \(firstName) \(lastName)" }
}

let myUser = User(fullName: "John Doe")
print(myUser) // User called John Doe
```

Listing 11.8 User Extension

In this extension, you add a convenience initializer, which enables clients to create User instances by providing a full name like "John Doe" instead of providing the name and surname separately. Check the client code, where myUser = User(fullName: "John Doe") works via the extension initializer.

You also inject conformance for the CustomStringConvertible protocol. Although the original User class was not compatible with this protocol, the extension adds the required conformance. As a result, print(myUser) prints out a nicely formatted output for myUser.

It's powerful, yet elegant, right? If User was part of an untouchable codebase, such as an external library, you could still add new features via such extensions without modifying the original User code.

Because we just touched on the subject of protocols, now would be a good time to address protocol extensions.

11.4   Extending Protocols

In Swift, protocol extensions have a unique place among extension types. As you should remember from Chapter 10, protocols act as blueprints for other data types. They don't contain implementation code by default.

However, protocol extensions offer the unique capability to add default implementations for protocol requirements! That way, you can transform any protocol into an abstract class, offering a partial implementation. How cool is that?

To begin the next demonstration, check the initial Artwork protocol in Listing 11.9. It's nothing fancy, really—just a protocol with three properties. An Artwork implementation might be a book, painting, or song, for instance.

```
protocol Artwork {
    var name: String { get }
    var artist: String { get }
    var fullDescription: String { get }
}
```

Listing 11.9 Simple Artwork Protocol

Now, let's add an extension. Listing 11.10 implements an extension with two valuable components: a default implementation for fullDescription and a new function called isInStock.

```
protocol Artwork {
    var name: String { get }
    var artist: String { get }
    var fullDescription: String { get }
}

extension Artwork {
    var fullDescription: String { "\(name) by \(artist)" }
    func isInStock() -> Bool { return false }
}
```

Listing 11.10 Artwork Extension

You see? Now you have default code implementations in the protocol. In the end, any class implementing the Artwork protocol will automatically inherit the default fullDescription implementation, as well as the totally new isInStock function. You will be able to access them via object.fullDescription or object.isInStock() as if they were part of the original Artwork protocol. Naturally, the implementing class may override them as necessary.

Without further ado, check the protocol implementation example in Listing 11.11, where the Book class has implemented Artwork in its extended form.

```swift
protocol Artwork {
    var name: String { get }
    var artist: String { get }
    var fullDescription: String { get }
}

extension Artwork {
    var fullDescription: String { "\(name) by \(artist)" }
    func isInStock() -> Bool { return false }
}

class Book: Artwork {
    let name: String
    let artist: String

    init(name: String, artist: String) {
        self.name = name
        self.artist = artist
    }

    func isInStock() -> Bool { return true }
}

let myBook = Book(name: "The Alchemist", artist: "Paulo Coelho")
print(myBook.fullDescription) // The Alchemist by Paulo Coelho
print(myBook.isInStock())     // true
```

Listing 11.11 Extended Protocol Implementation

In this case, you don't override Artwork.fullDescription in Book. The default implementation provided in the extension is inherited by Book. As a result, myBook.fullDescription returns the string built in the extension, displayed as The Alchemist by Paulo Coelho.

In contrast, the default implementation of isInStock was overridden in Book. Instead of returning false, you return true. As a result, myBook.isInStock() will return true. In a real-life scenario, you would check a database for the actual stock situation, but we're keeping these examples simple.

It's worth knowing that Swift requires default implementations for protocol extensions—and for good reason. Imagine the Artwork protocol having two implementations already, as sketched in Figure 11.2.

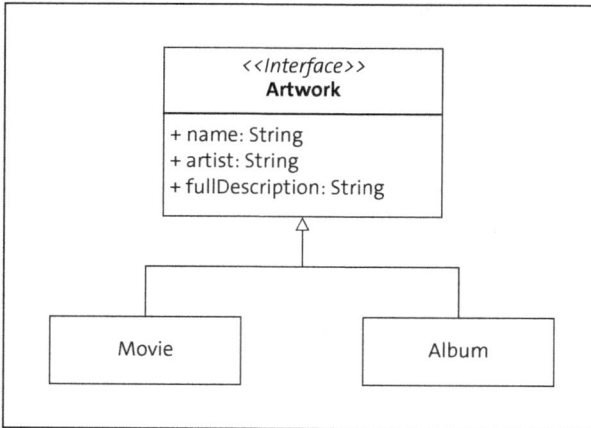

Figure 11.2 Possible Artwork Implementations

If you were to add new protocol requirements via the extension, you would surely break the existing Movie and Album classes because they don't have implementations for the requirements. For instance, if a client called Movie.isInStock(), it would fail because Movie comes from the past, when isInStock() wasn't present yet.

Having a default implementation in the protocol extension mitigates this problem. Instead of crashing the app altogether, Movie.isInStock() would simply execute the default code provided in the extension. That approach ensures the security and consistency of your apps while enabling abstract class–style behavior as a bonus.

Now, let's explore an equally cool feature of extensions, in which you can apply support for operators to existing data types.

11.5   Operator Extensions

Throughout the book, we've introduced many Swift operators. For example, + is used to add two numbers and == is used to check for equality. Using extensions, you can add support for new operators for a data type. In this section, you'll see various examples of operator overloading.

11.5.1   Arithmetic and Boolean Operators

Let's start with a refresher of some of the arithmetic and Boolean operators supported in Swift, as listed in Table 11.1.

Operator	Meaning	Example
+	Addition	a + b
-	Subtraction	a - b
*	Multiplication	a * b
/	Division	a / b
%	Remainder	a % b
&&	AND	a && b
||	OR	a || b

Table 11.1 Some Arithmetic and Boolean Operators

Using extensions, it's possible to add support for any of those operators to an existing data type. In Listing 11.12, the Point struct is extended and empowered with support for the + operator.

```
struct Point {
    var x: Int
    var y: Int
}

extension Point {
    static func + (lhs: Point, rhs: Point) -> Point {
        return Point(x: lhs.x + rhs.x, y: lhs.y + rhs.y)
    }
}

let p1 = Point(x: 3, y: 2)
let p2 = Point(x: 6, y: 1)
let p3 = p1 + p2
print(p3) // Point(x: 9, y: 3)
```

Listing 11.12 Extension with + Operator

In this example, the + operator works by summing the x and y coordinates of two Point instances, creating a new Point.

Now, let's go through a slightly more complex exercise. Listing 11.13 represents the Classroom struct, which contains a set of Student instances. The goal is to apply the - operator via an extension, which returns the students present in one classroom, but not the other.

471

```
import Foundation

struct Student: Hashable {
    let id: UUID
    let name: String
}

struct Classroom {
    var students: Set<Student>
}

extension Classroom {
    static func - (lhs: Classroom, rhs: Classroom) -> Classroom {
        let deltaStudents = lhs.students.subtracting(rhs.students)
        return Classroom(students: deltaStudents)
    }
}

let a = Student(id: UUID(), name: "Alice")
let b = Student(id: UUID(), name: "Bob")
let c = Student(id: UUID(), name: "Charlie")
let d = Student(id: UUID(), name: "David")

let mathClass = Classroom(students: [a, b, c])
let historyClass = Classroom(students: [c, d])
let deltaClassroom = mathClass - historyClass
print(deltaClassroom) // Only Alice
```

Listing 11.13 Extension with - Operator

As you can see, the application of the - operator is easy; you merely take advantage of the subtracting function to find the delta. In the client code, the delta student Alice was detected by the mathClass - historyClass operation. Obviously, when this operation is invoked, the - function of the Classroom extension is executed.

As a final exercise, Listing 11.14 demonstrates an extension with the && operator. In this example, the && operator will return True if both Connection instances are connected.

```
class Connection {
    var isConnected: Bool = false
    func connect() { isConnected = true }
}

extension Connection {
    static func && (lhs: Connection, rhs: Connection) -> Bool {
```

```
            return lhs.isConnected && rhs.isConnected
    }
}

let c1 = Connection()
let c2 = Connection()
c1.connect()
print(c1 && c2) // False
```

Listing 11.14 Extension with && Operator

This extension enabled the easy c1 && c2 syntax instead of the longer c1.isConnected && c2.isConnected syntax. Cool, right?

Now, let's continue with another breed of extension operators.

11.5.2 Equality and Comparison Operators

With a similar approach, extensions can be used to apply equality and comparison operators to an existing data type. Some of those operators are shown in Table 11.2 as a memory refresher.

Operator	Meaning	Example
==	Equal to	a == b
!=	Not equal to	a != b
<	Less than	a < b
<=	Less than or equal to	a <= b
>	Greater than	a > b
>=	Greater than or equal to	a >= b

Table 11.2 Some Equality and Comparison Operators

Without any further ado, let's work through some hands-on examples. In Listing 11.15, the User struct is extended with conformance to the Equatable protocol, adding support for the == operator.

```
struct User {
    let id: Int
    let name: String
}

extension User: Equatable {
    static func == (lhs: User, rhs: User) -> Bool {
```

```
            return lhs.id == rhs.id
    }
}

let u1 = User(id: 1, name: "Jane")
let u2 = User(id: 1, name: "Jane")
let u3 = User(id: 2, name: "Bob")
print(u1 == u2) // true
print(u1 == u3) // false
```

Listing 11.15 Extension with Equatable

You learned about the Equatable protocol in Chapter 10. The only new point in this example is to apply the protocol via an extension. As a result, you gain the ability to compare two User instances for equality via their id properties.

With a similar approach, Listing 11.16 demonstrates how to add support for the Comparable protocol to an existing class, unlocking the option to use comparison operators.

```
class Person {
    let age: Int
    let name: String

    init(age: Int, name: String) {
        self.age = age
        self.name = name
    }
}

extension Person: Comparable {
    static func == (lhs: Person, rhs: Person) -> Bool {
        return lhs.age == rhs.age
    }

    static func < (lhs: Person, rhs: Person) -> Bool {
        return lhs.age < rhs.age
    }
}

let bob = Person(age: 30, name: "Bob")
let alice = Person(age: 25, name: "Alice")
bob >= alice    // true
```

Listing 11.16 Extension with Comparable

In this instance, you gain the ability to apply the simple bob >= alice syntax, instead of the longer bob.age >= alice.age syntax.

11.5.3   Prefix and Postfix Operators

In Swift, *prefix* and *postfix operators* are unary tokens, operating on a single value rather than two. You have seen many of these throughout the book, but Table 11.3 shows some of these operators as a reminder.

Operator	Meaning	Example
-	Prefix minus	b = -a
!	Prefix not	b = !a
++	Postfix plus one	a++
--	Postfix minus one	a--

Table 11.3 Some Prefix and Postfix Operators

Like the previous sections, extensions can be used to add support for such operators in existing data types. Listing 11.17 demonstrates an example in which the – prefix was added to the Vector struct via an extension.

```swift
struct Vector {
    var x: Int
    var y: Int
}

extension Vector {
    static prefix func - (v: Vector) -> Vector {
        return Vector(x: -v.x, y: -v.y)
    }
}

let v1 = Vector(x: 1, y: 2)
let v2 = -v1
print(v2) // Vector(x: -1, y: -2)
```

Listing 11.17 Extension for - Prefix

As a result, v2 = -v1 creates a new Vector instance that is the mirror of v1, negating v1.x and v1.y. The client code turns out to be very clean and readable, don't you think?

Now, let's look at a postfix example. In Listing 11.18, the ++ postfix operator is added to Vector using an extension.

```swift
class Vector {
    var x: Int
    var y: Int
```

```
    init(x: Int, y: Int) {
        self.x = x
        self.y = y
    }
}

extension Vector {
    static postfix func ++ (v: Vector) {
        v.x += 1
        v.y += 1
    }
}

let v1 = Vector(x: 1, y: 2)
v1++
print(v1.x, v1.y) // 2, 3
```

Listing 11.18 Extension for ++ Postfix

This extension allows you to shift the x and y values of the Vector instance by one using the simple v1++ syntax. Sweet!

So far, we've explored code samples that revolve around standard Swift operators. However, Swift is flexible enough to allow custom operators too. Let's explore this option in the next section.

11.5.4   Custom Operators

The engineers behind the Swift language were kind enough to give us the flexibility to add our made-up operators to the codebase, allowing us to simplify the syntax even further. As an example, we'll add the √ operator for square root calculation. Note that this is not a built-in operator in Swift; it's something we made up.

Listing 11.19 shows the target struct called Vector, which will be the target practice data type for the √ operator.

```
import Foundation

struct Vector {
    var x: Double
    var y: Double
}
```

Listing 11.19 Target Struct Vector

First, you need to tell Swift that you're making up a new operator. This will allow the compiler to interpret the syntax accurately. The simple syntax for this declaration is shown in Listing 11.20.

```
import Foundation

struct Vector {
    var x: Double
    var y: Double
}

prefix operator √
```

Listing 11.20 Declaration of New Operator

Note that you use the prefix operator √ expression because the √ operator will be used before the variable—that is, √x. If the custom operator was to be used after the variable, it would need to be declared as postfix operator.

Now that you've introduced the custom operator, you can apply and use the extension. Check the code sample in Listing 11.21, where the extension contains the √ prefix function.

```
import Foundation

struct Vector {
    var x: Double
    var y: Double
}

prefix operator √

extension Vector {
    static prefix func √ (v: Vector) -> Vector {
        return Vector(x: sqrt(v.x), y: sqrt(v.y))
    }
}

let v1 = Vector(x: 16, y: 25)
let v2 = √v1
print(v2) // Vector(x: 4, y: 5)
```

Listing 11.21 Extension for the Custom Operator

Because the made-up √ operator is declared to Swift, you can write code for it in the extension like any Swift-standard prefix (or postfix) operator. In this instance, the code for the √ operator simply calculates the square root of `Vector.x` and `Vector.y`, as expected. The result is returned as a new `Vector` instance.

In the end, all this effort enabled the simple `let v2 = √v1` syntax, which creates `v2` using the square root of `v1` properties.

11.6 Extending Swift Data Types

So far, we've worked through examples that extend custom data types. But here's a surprise: Swift lets you extend its built-in data types too! Just like custom types, you can add computed properties, functions, initializers, protocol conformance, subscripts, and operators to standard types like `String` or `Double`. You can also extend library-specific data types, such as `SpireKit SKScene` or `SwiftUI View`. This section will teach you how to create such extensions and highlight some typical pitfalls.

11.6.1 Sample Usage

As an example, Listing 11.22 features a `String` extension, adding a computed property and a function to the built-in type.

```
extension String {
    var isEmail: Bool {
        return self.contains("@") && self.contains(".")
    }

    func repeated(times: Int) -> String {
        return String(repeating: self, count: times)
    }
}

let a = "hello@hello.com"
a.isEmail // true
let b = "Ho"
b.repeated(times: 3) // HoHoHo
```

Listing 11.22 String Extension with Property and Function

Thanks to the extension, you can check any string via `a.isEmail` and see if it's an email address or not. Likewise, you can build repeated versions of any string via `b.repeated(times: 3)`.

Once you define a `String` extension belonging to your project/framework, any `String` instance in that project/framework will gain the new functionality. Therefore, you don't need to repeat the same extension throughout your project; you only need to do it once. For imported modules, though, the accessibility of such extensions will depend on their access control modifiers. This subject will be covered in Chapter 15 when we talk about modules.

11.6.2   Avoiding Name Conflicts

One pitfall is the risk of name conflict. If you add a function to a standard type that conflicts with a future Swift release, your code might break. In the preceding example, if Swift adds a computed property called `String.isEmail`, then this will conflict with your extension, and you will probably have to make changes. This might be a minor refactoring task on a local project, but in a public Swift library with many clients, it brews trouble because everyone will have to modify their code to comply with your new (fixed) library version.

To avoid this risk, you may consider project-prefixed or domain-prefixed names in your extensions. For example, if you are coding an *Angry Birds* game, your extension elements may begin with the `birds` prefix. Your extension property/function would be called `birdsIsEmail` or `birdsRepeated`, instead of the generic and risky name `isEmail` or `repeated`. It's highly unlikely that Swift will add a standard function called `birdsIsEmail`, right?

An even cleaner approach is to add a static namespace wrapper, if possible. In Listing 11.23, a nested struct `Birds` is added to the `String` extension as a demonstration.

```
extension String {
    struct Birds {
        let base: String

        var isEmail: Bool {
            return base.contains("@") && base.contains(".")
        }

        func repeated(times: Int) -> String {
            return String(repeating: base, count: times)
        }
    }

    var birds: Birds { Birds(base: self) }
}

let a = "hello@hello.com"
a.birds.isEmail // true
let b = "Ho"
b.birds.repeated(times: 3) // HoHoHo
```

Listing 11.23 Static Wrapper in an Extension

In this approach, features you'd like to add to String aren't added to the type directly. Instead, they are added to the Birds extension struct. This gives you the flexibility to have nice, clean names for extension elements (like isEmail) while ensuring that they won't conflict with future releases. After all, it's very unlikely for Apple engineers to add a property called String.birds, right?

The client code also remains clean and simple. The a.birds.isEmail expression speaks for itself; this is obviously a String property introduced as an extension in an *Angry Birds* project.

Before concluding this chapter, let's talk about how generics work with extensions.

11.7 Generic Extensions

On a more advanced note, we'll conclude this chapter with a bunch of demonstrations of using generics in extensions. Naturally, generics can be made part of extensions too; there isn't much specific to it. Still, it should prove useful to go over some concrete syntax examples.

Let's start with a simple example. Listing 11.24 contains a demonstration in which the generic Box struct is extended with the describe function. As expected, the client code can access box.describe() as if it was originally part of the Box struct.

```
struct Box<T> {
    let value: T
}

extension Box {
    func describe() -> String {
        return "Box holds: \(value)"
    }
}

let box = Box(value: "papers")
print(box.describe())  // Box holds: papers
```

Listing 11.24 Extending Generic Struct

Now it's time for a generic where clause. In Listing 11.25, Swift's Array type was extended with the containsGreaterThan function. This function returns true if the array contains an element greater than the passed item.

```
extension Array where Element: Comparable {
    func containsGreaterThan(_ item: Element) -> Bool {
        for x in self { if x > item {return true} }
        return false
```

```
    }
}
```

```
let numbers = [1, 2, 3]
print(numbers.containsGreaterThan(2))  // true
print(numbers.containsGreaterThan(4))  // false
```

Listing 11.25 Extension with Generic where Clause

Naturally, this code would only work for arrays that contain Comparable elements. Otherwise, if x > item won't work because Swift won't know what to do with the > operator. To ensure that constraint, the extension is declared as extension Array where Element: Comparable, meaning that this extension is intended to work in arrays with Comparable elements only.

In the following client code, numbers.containsGreaterThan works just fine because integers are naturally Comparable. As an exercise, you can try calling containsGreaterThan for an array containing non-Comparable elements (like a custom class): The compiler will certainly generate an error. This is runtime safety at its best!

Finally, Listing 11.26 showcases a demonstration in which Swift's Dictionary type is extended with a generic subscript. This implementation allows you to set a default value in case Dictionary doesn't contain the requested key.

```
extension Dictionary {
    subscript<K: CustomStringConvertible>(key: K, defVal: Value) -> Value {
        get {
            return self[key.description as! Key] ?? defVal
        }
    }
}
```

```
let dict = ["name": "Alice"]
print(dict["name", "Unknown"])   // Alice
print(dict["age", "Unknown"])    // Unknown
```

Listing 11.26 Generic Dictionary Subscript Extension

In the client code, dict["name", "Unknown"] returns the value "Alice" because the "name" key is present in dict. Meanwhile, dict["age", "Unknown"] returns the default/fallback value "Unknown" (thanks to the extension) because dict doesn't contain a value for the "age" key.

Beyond the examples in this section, different combinations of generic and extension techniques are applicable. You are more than welcome to mix and match generics with extensions in various examples as a personal exercise.

11.8   Summary

This chapter was about extensions, which empower programmers with the option to add new features to custom data types, as well as to Swift's built-in types, without modifying the original code. Enums, structs, classes, and protocols can be extended with various components, such as computed properties, functions, initializers, subscripts, nested types, and even protocol conformance.

Protocol extensions are especially curious as they enable partial implementations, making protocols behave like abstract classes.

It's possible to add operator support via extensions and even to invent custom operators like √. Finally, generics and extensions also work well together when needed.

This chapter concluded our main content on object-oriented Swift. Although you were taught about the toolbox offered by Swift, strong object-oriented designs require architectural knowledge too. If you're interested in that sort of thing, you may want to research topics like object-oriented thinking, design patterns, and SOLID principles.

In the next chapter, we'll explore error handling in Swift, which is one of the pillars of defensive programming and solid applications. After all, what are all these rich language features worth if your app crashes every other minute, right?

Chapter 12
Error Handling

Defensive programming is a vital part of any programming platform, and Swift is no exception to that. In this chapter, you will learn what Swift has to offer in terms of handling errors.

Errors are unpleasant but unavoidable. As programmers, we don't have the luxury to write a chunk of code and assume that everything will function perfectly just as we've foreseen. This is a false hope, sure to be doomed. Table 12.1 contains some sample scenarios in which errors might occur.

Scenario	Hope	Reality
Accessing a RESTful API	The service will return values in the expected JSON format	The API's JSON format might be changed by an unannounced update
Reading settings from a file	The file is present; settings have proper values	The user might have manually deleted or altered the file
Calculating a division	Numbers are divided	Numbers might be 0
Saving a document	Document is written to the disk	The user might lack the permission to write the file to the target location
Querying a database	Query results are fetched	The administrator might have suspended your database account

Table 12.1 Sample Scenarios with Possible Errors

To ensure the robustness of your apps, you must foresee that such errors can and will occur, eventually—and, of course, implement the necessary safety measures in your code.

No user can be expected to keep using an app that crashes at every oddity. Would you drive a car that shuts the engine down when the wiper fluid runs out? Or would you rather have your car display a polite alert, reminding you to refill? That's how apps should behave too: notice oddities/errors, raise alerts for them, and either handle those alerts automatically or inform the user.

For instance, if the configuration file of your app is missing, you can catch that file error and rebuild the file with default values behind the scenes, handling the problem automatically. If the server behind your weather app is down, you can catch the HTTP error and politely tell the user that the weather forecast is temporarily unavailable, suggesting that they check the weather a little later.

The main point is to be mindful about possible errors and how to handle them. If an unhandled error occurs during runtime, your app will likely shut down, frustrating the users and maybe even causing bigger problems, such as corrupted files or inconsistent database records.

Error handling is a natural requirement in the software industry, no matter which language you use. But the topic here is Swift, so this chapter is dedicated to throwing, catching, and properly handling errors in the Swift language. Let's start with a technical bird's-eye view of Swift errors.

12.1   Understanding Errors in Swift

In Swift, an *error* is an unexpected condition that prevents the app from continuing its normal flow. Some error examples were mentioned earlier.

Primarily, Swift errors are "thrown" from functions or initializers that are explicitly marked with the throws keyword. You'll see the exact syntax soon, but Listing 12.1 demonstrates the signature of such a function. Note the throws suffix, indicating that the function is expected to produce errors in some cases.

```
func login(user: String, pass: String) throws {
    // Check credentials
    // Throw error if invalid
}
```

Listing 12.1 Function That Can Throw an Error

In the end, this function's flowchart should look roughly like Figure 12.1. Note that the function will throw an error for unexpected cases such as a nonexisting user, a locked user, or an invalid password.

A function can throw any built-in Swift error, such as CancellationError, DecodingError, EncodingError, or URLError. If, for instance, your function encounters an error with a URL, you can throw a URLError that's ready to be handled by the client code. How to catch and handle errors is a topic we'll address soon enough, but first let's add the try prefix to the client code, as demonstrated in Listing 12.2. You'll learn about proper error handling a little later; don't worry.

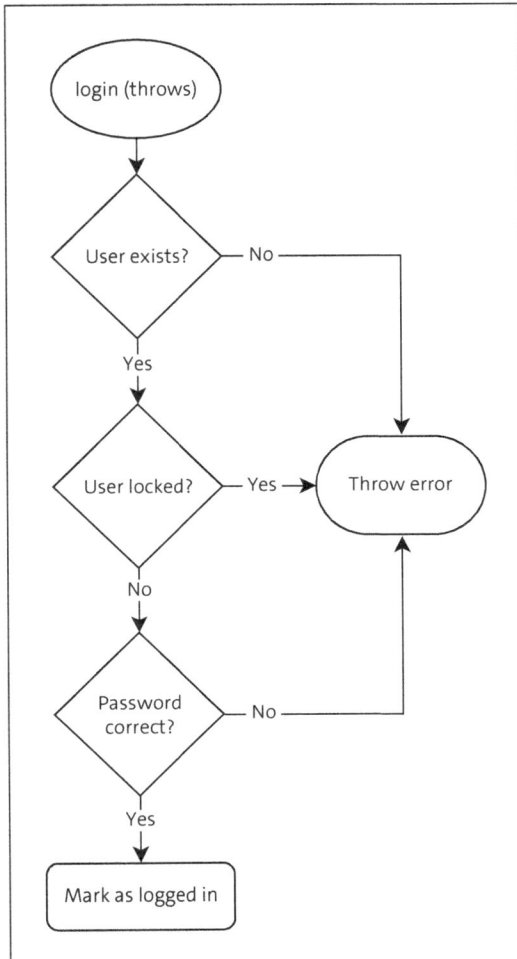

Figure 12.1 Flowchart for Logic Function

```
func login(user: String, pass: String) throws {
    // Check credentials
    // Throw error if invalid
}

try login(user: "alice", pass: "password")
```

Listing 12.2 Calling throws Function

It's also possible to define and throw custom errors. In a social media app, for example, if the user attempts to login with invalid credentials, you may raise your custom-defined LoginError. Typically, Swift errors are represented using enumerations that implement the built-in Error protocol, so the creation of a custom error is nothing but the creation of an appropriate enumeration, as sketched in Figure 12.2.

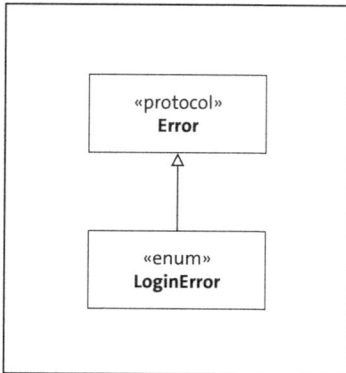

Figure 12.2 Custom Error Enum

A simple Error implementation is shown in Listing 12.3.

```
enum LoginError: Error {
    case noSuchUser
    case userLocked
    case invalidPassword
}
```

Listing 12.3 Declaration of Custom Error

Once you have such an enumeration at hand, you can throw it from your functions and catch it in your client code. You can also use a struct or class for error types, but enumeration is the preferred choice for most use cases.

This should be enough of a general technical overview. Now, it's time to get hands-on with Swift errors via practical examples.

12.2   Throwing Errors

We'll start by exploring how to throw errors. This is the logical first step: If there's no error thrown, then there's nothing to catch and manage in your client code, right? In this section, you'll learn to throw built-in Swift errors and advance from there.

12.2.1   Throwing Built-in Errors

The first hands-on example will be a function that throws an error in the Swift library. For this purpose, Listing 12.4 features the stringToUrl function, which attempts to convert the input string to an URL object. Note that the function signature contains the throws keyword, indicating that it may produce errors.

```swift
import Foundation

func stringToUrl(_ input: String) throws -> URL {
    guard let url = URL(string: input) else {
        throw URLError(.badURL)
    }

    guard url.host == "www.example.com" else {
        throw URLError(.unsupportedURL)
    }

    return url
}

let url1 = try stringToUrl("http://www.example.com/api")    // OK
let url2 = try stringToUrl("http://www.abc.com/api")        // Error
```

Listing 12.4 Function Throwing a URLError

In the first check, if the string isn't a valid URL, an error of type URLError.badURL is thrown using the throw URLError(.badURL) expression. The syntax to throw an error is that simple!

In the second check, the function ensures that the host is *www.example.com*. If that's not the case, then throw URLError(.unsupportedURL) is executed, throwing a different URLError case.

In this example, we're reusing Swift's built-in URLError type. To see all the usable cases under URLError, you can take advantage of the autocomplete feature of Xcode, as shown in Figure 12.3.

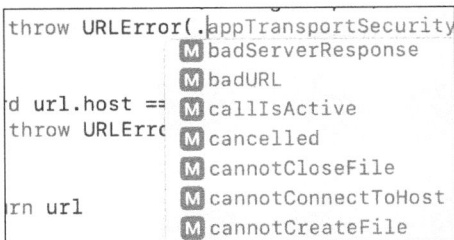

Figure 12.3 Autocomplete Cases for URLError

In the client code, url1 would be assigned successfully because the passed URL is valid and it is under the *www.example.com* domain, passing all checks of the stringToUrl function. The code will generate an error for url2 because the passed URL belongs to a different domain. The playground message about the uncaught error will appear as shown in Figure 12.4.

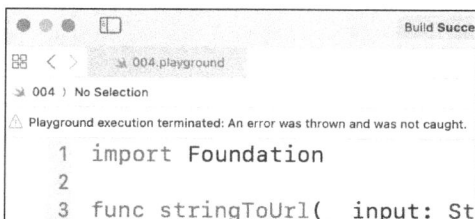

Figure 12.4 Playground Warning About Uncaught Error

If the error was handled properly, the playground wouldn't display such a message of course. This example doesn't contain any error-handling mechanism; we'll address that topic soon enough. For now, we're focusing on throwing errors, not catching them.

Now, let's continue by throwing a custom error.

12.2.2   Throwing Custom Errors

In Swift, throwing a custom error is not too different from throwing a built-in error. The only extra work needed is to declare a custom enumeration implementing the Error protocol. For your convenience, the Error protocol doesn't impose any complex logic or properties; it merely acts as a marker.

You can see such an example in Listing 12.5. AgeValidationError is a custom Error enumeration that includes two cases: negative and tooYoung. Note that the tooYoung case accepts a parameter: tooYoung(minimum: Int) means that you can set a value for minimum to inform the client of the minimum age required.

```
enum AgeValidationError: Error {
    case negative
    case tooYoung(minimum: Int)
}


func validateAge(_ age: Int) throws {
    if age < 0 {
        throw AgeValidationError.negative
    }
    if age < 18 {
        throw AgeValidationError.tooYoung(minimum: 18)
    }
}
```

```
try validateAge(22) // OK
try validateAge(17) // tooYoung
```

Listing 12.5 Function Throwing a Custom Error

In the flow of validateAge, there are two checks:

- If the age is negative, throw AgeValidationError.negative.
- If the age is less than 18, throw AgeValidationError.tooYoung.

In the client code, the second call will fail because the age is less than 18. That's intuitive, right?

In our examples so far, we've worked with functions throwing errors. However, initializers can throw errors too! In Listing 12.6, Member.init is marked with throws. This initializer contains an age-verification routine, and if the age is not OK, it throws an error—similar to the preceding former example.

```
enum AgeValidationError: Error {
    case negative
    case tooYoung(minimum: Int)
}

public class Member {
    var name: String
    var age: Int

    init(name: String, age: Int) throws {
        if age < 0 {
            throw AgeValidationError.negative
        }
        if age < 18 {
            throw AgeValidationError.tooYoung(minimum: 18)
        }

        self.name = name
        self.age = age
    }
}

let m1 = try Member(name: "Alice", age: 22) // OK
let m2 = try Member(name: "Bob", age: 17)   // tooYoung
```

Listing 12.6 Initializer Throwing Errors

Naturally, struct or class functions may throw errors too. Listing 12.7 demonstrates an example in which `Member.setAge` throws an `AgeValidationError` if the passed age is not valid.

```
enum AgeValidationError: Error {
    case negative
    case tooYoung(minimum: Int)
}

struct Member {
    var name: String
    private var _age: Int = 0

    init(name: String) { self.name = name }
    func getAge() -> Int { return self._age }

    mutating func setAge(_ age: Int) throws {
        if age < 0 {
            throw AgeValidationError.negative
        }
        if age < 18 {
            throw AgeValidationError.tooYoung(minimum: 18)
        }
        self._age = age
    }
}

var member = Member(name: "Joe")
try member.setAge(22)    // OK
try member.setAge(17)    // tooYoung
```

Listing 12.7 Struct Function Throwing an Error

Note that throwing an error from a struct/class function has the same syntax as throwing an error from a standalone function; we simply emphasized the option to do so.

12.2.3   Propagating Errors

Imagine that you have an outer function calling an inner function. If the inner function is marked as `throws`, then the outer function must either catch and handle the thrown error, or it must also be marked as `throws` and let the client handle the error.

Did that sound a bit complex? No worries: It will be made crystal clear via Listing 12.8. In this example, the `marryPeople` outer function is internally calling `validateAge`.

```
enum AgeValidationError: Error {
    case negative
    case tooYoung(minimum: Int)
}

struct Person {
    let name: String
    let age: Int
    var spouse: String? = nil
}

func validateAge(_ age: Int) throws {
    if age < 0 {
        throw AgeValidationError.negative
    }
    if age < 18 {
        throw AgeValidationError.tooYoung(minimum: 18)
    }
}

func marryPeople(bride: inout Person, groom: inout Person) throws {
    try validateAge(bride.age)
    try validateAge(groom.age)
    bride.spouse = groom.name
    groom.spouse = bride.name
}
```

Listing 12.8 Outer Function marryPeople Marked as throws

Because the validateAge inner function may throw an error, the marryPeople outer function was also marked as throws. If validateAge throws an error, marryPeople will forward this error to the client, passing the thrown AgeValidationError.

This mechanism frees the programmer from the boilerplate code of catching and rethrowing errors in the outer function. You can simply mark the outer function as throws, making it forward any produced error from its inner mechanism.

Naturally, the outer function may catch and handle the error if necessary; there is no limitation in that regard.

Another mechanism for error propagation works via the rethrows keyword. If you have a function accepting a closure, and the closure may throw an error, then your function ought to be marked as rethrows. Check the corresponding example in Listing 12.9.

```
func backupFileToCloud() throws { }
func saveFileToDisk() throws { }
```

```
func backupAndSave(backupTask: () throws -> Void,
                   saveTask: () throws -> Void) rethrows {
    print("Creating backup...")
    try backupTask()
    print("Backup done. Now saving file...")
    try saveTask()
    print("Operation completed.")
}


try backupAndSave(backupTask: backupFileToCloud,
                  saveTask: saveFileToDisk)
```

Listing 12.9 Error Propagation via Rethrows

In this example, the backupAndSave function accepts two closures able to throw errors. Therefore, backupAndSave was marked as rethrows, indicating that it would forward any closure error it encounters.

Using rethrows is a good practice because it clarifies that the function itself doesn't introduce new error conditions; rather, it relies on the errors from its closure parameters. But if the function throws nonclosure errors too, then it must be marked as throws. Don't worry, the compiler will ensure that anyway.

OK, that's enough about producing errors. Now, let's discuss how to catch and handle them.

12.3   Catching Errors

Throwing errors is all good and fine, but if the client code isn't catching and handling them properly, then the work is only half done. In this section, we'll focus on that second half, completing the groundwork for solid apps.

12.3.1   Direct Pattern Matching

As a framework for the error-catching syntax, Listing 12.10 reintroduces an age-validation example. This time, we are validating the age entered by the user to make it a bit more realistic.

```
enum AgeValidationError: Error {
    case negative
    case tooYoung(minimum: Int)
}


func validateAge(_ age: Int) throws {
    if age < 0 { throw AgeValidationError.negative }
```

```
    if age < 18 { throw AgeValidationError.tooYoung(minimum: 18) }
}

print("Enter your age: ", terminator: "")
let userAge = Int(readLine() ?? "") ?? 0
try validateAge(userAge)
print("Your age is valid")
```

Listing 12.10 User Age Validation Without Catching Errors

In the current state of this example, we are not catching errors (yet). If the user enters a valid age, then the output will appear as shown in Figure 12.5 without any hiccups.

```
Enter your age: 22
Your age is valid
```

Figure 12.5 Valid Age Output

Now let's try entering an invalid age and see the ugly side of Xcode, telling us off. Figure 12.6 shows the result of entering an invalid age and encountering an unmanaged error.

Figure 12.6 Invalid Age Output

Luckily, you can see some important error details in your local environment. Xcode tells you that `AgeValidationError.tooYoung` was thrown but not caught. But if this was an app running on a user's device, they would probably encounter an unwelcome app crash.

So, let's catch those errors! Listing 12.11 contains the same code, enhanced with an error-catching mechanism.

```
enum AgeValidationError: Error {
    case negative
    case tooYoung(minimum: Int)
}

func validateAge(_ age: Int) throws {
```

493

```
    if age < 0 { throw AgeValidationError.negative }
    if age < 18 { throw AgeValidationError.tooYoung(minimum: 18) }
}

print("Enter your age: ", terminator: "")
let userAge = Int(readLine() ?? "") ?? 0

do {
    try validateAge(userAge)
    print("Your age is valid")
} catch AgeValidationError.negative {
    print("Invalid age: Age cannot be negative.")
} catch AgeValidationError.tooYoung(let minimum) {
    print("Invalid age: You must be at least \(minimum) years old.")
} catch {
    print("An unexpected error occurred: \(error)")
}
```

Listing 12.11 Age Validation with Error Handling

First and foremost, "risky" code that's able to produce errors should be placed into a do
{ ... } code block. Basically, this hints to the compiler to be careful about the code
between those brackets. Note that try validateAge(userAge) is between those do brackets.

Following that, there's a catch block for each error you can catch and handle. The catch
AgeValidationError.negative { ... } block will be executed if that error is thrown due to
the user entering a negative age. Here, you simply print out a message. The catch
AgeValidationError.tooYoung(let minimum) { ... } block will be executed if that error is
thrown due to an underage user. Note that you can use the minimum value in your output;
you can throw and catch such helper values along with Error instances in Swift's error
handling.

Finally, there's a generic catch { ... } block for everything else. In other words, if an unhandled error occurs in the do { ... } block, Swift will fall back into that final catch { ... } block,
allowing you to handle errors you couldn't foresee. This block will catch any Error not
caught beforehand. In such generic fallback blocks, Swift automatically provides a local
constant named error, holding the caught Error instance.

Now, let's see the error-handling at work. The enhanced version of the code sample will
produce the proper output shown in Figure 12.7 for underage users. That's much better
than an application crash, don't you think?

```
Enter your age: 15
Invalid age: You must be at least 18 years old.
```

Figure 12.7 Invalid Age Entry Handled Properly

If you're aiming for a lower level of granularity in your error handling, you could simply implement a singular generic catch { ... } block in your code—which is often called *Pokémon error handling* in programming circles. Check the demonstration in Listing 12.12.

```
enum AgeValidationError: Error {
    case negative
    case tooYoung(minimum: Int)
}

func validateAge(_ age: Int) throws {
    if age < 0 { throw AgeValidationError.negative }
    if age < 18 { throw AgeValidationError.tooYoung(minimum: 18) }
}

print("Enter your age: ", terminator: "")
let userAge = Int(readLine() ?? "") ?? 0

do {
    try validateAge(userAge)
    print("Your age is valid")
} catch {
    print("An unexpected error occurred: \(error)")
}
```

Listing 12.12 Pokémon Error Handling

This code is obviously more concise, but it lacks the detailed error-handling mechanism of the former version; you can't produce clear messages for the user this time. Check the output for an underage user shown in Figure 12.8: Although the message is understandable in this case, it surely isn't getting a UX award for clarity.

```
Enter your age: 15
An unexpected error occurred: tooYoung(minimum: 18)
```

Figure 12.8 Pokémon Error Output

Pokémon Error Handling

The classic programming joke about Pokémon error handling ought to be explained at this point. The joke comes from the Pokémon motto, "Gotta catch them all!"—taking a humorous jab at the practice of using a generic catch { ... } block that catches all possible errors.

Some also call this the *diaper antipattern* because it catches, well, all the poop.

> Although those phrases are catchy, they stand as reminders to be intentional with your error handling rather than thoughtlessly catching and handling everything with singular logic.
>
> Still, if you need a safety net for all unforeseeable circumstances, this approach has its place.

Now that you're familiar with basic error handling, we can move on to more advanced topics.

12.3.2   Error Pattern Matching

In the preceding examples, we used a technique called *direct pattern matching*, in which the client code targets Error cases directly, such as catch AgeValidationError.negative. That's a solid and usable approach. However, it comes with a situational disadvantage: Each AgeValidationError case has its own independent code block; there is no shared code among them.

To address that issue, Swift features a second option called *error pattern matching*, which is demonstrated in Listing 12.13.

```
enum AgeValidationError: Error {
    case negative
    case tooYoung(minimum: Int)
}

func validateAge(_ age: Int) throws {
    if age < 0 { throw AgeValidationError.negative }
    if age < 18 { throw AgeValidationError.tooYoung(minimum: 18) }
}

print("Enter your age: ", terminator: "")
let userAge = Int(readLine() ?? "") ?? 0

do {
    try validateAge(userAge)
    print("Your age is valid")
} catch let ave as AgeValidationError {
    print("Invalid age; \(ave)")
} catch {
    print("An unexpected error occurred: \(error)")
}
```

Listing 12.13 Basic Error Pattern Matching

Did you spot the difference? You sure did! Instead of catching each AgeValidationError case in a different catch { … } block, they all were caught in a singular catch let ave as AgeValidationError { … } block. Within that block, only AgeValidationError instances will be caught, and the ave variable is usable as the error object.

But wait—how is that useful? Isn't that similar to Pokémon error handling? Well, yes and no. Although you consolidate all AgeValidationError cases into a single block, you don't catch everything under the sun. Sometimes, that's exactly what you want: to catch all errors of a certain type.

Besides, you haven't seen the full picture yet! Check the enhanced sample in Listing 12.14, where the catch block contains a switch statement to differentiate cases.

```
enum AgeValidationError: Error {
    case negative
    case tooYoung(minimum: Int)
}

func validateAge(_ age: Int) throws {
    if age < 0 { throw AgeValidationError.negative }
    if age < 18 { throw AgeValidationError.tooYoung(minimum: 18) }
}

print("Enter your age: ", terminator: "")
let userAge = Int(readLine() ?? "") ?? 0

do {
    try validateAge(userAge)
    print("Your age is valid")
} catch let ave as AgeValidationError {
    print("Invalid age error!", terminator: " ")

    switch ave {
    case .negative:
        print("Age cannot be negative")
    case .tooYoung(minimum: let minAge):
        print("You must be at least \(minAge) years old")
    default:
        print("Unknown age error")
    }
} catch {
    print("An unexpected error occurred: \(error)")
}
```

Listing 12.14 Error Pattern Matching with switch Statement

In this advanced version, you can cycle through different `AgeValidationError` cases using a `switch` statement. The significant advantage is that all `AgeValidationError` cases share common code, which prints an `"Invalid age error!"` statement. In a real-world app, you could implement a more comprehensive common mechanism, like writing logs or rolling back database updates.

Note that the `switch` statement contains a final `default` case, which is unnecessary in this example because all cases were already covered. In practice, though, the `default` case would be executed for all uncovered `AgeValidationError` cases—like a local Pokémon handler.

In our examples, the `validateAge` function is throwing only one kind of exception, which is `AgeValidationError`. But in real-world apps, a function may throw different types of errors. Listing 12.15 features a code sample in which the imaginary `saveInvoice` function might throw different types of errors. Note that the functions are left empty for the sake of simplicity; we're focused on the architecture here.

```
func saveInvoice() throws {}
func logDecodeError(_ e: Error) {}
func logEncodeError(_ e: Error) {}

do {
    try saveInvoice()
} catch let de as DecodingError {
    logDecodeError(de)

    switch de {
    case .keyNotFound:
        print("Missing key, sorry!")
    default:
        print("Can't decode")
    }
} catch let ee as EncodingError {
    logEncodeError(ee)

    switch ee {
    case .invalidValue:
        print("Invalid value")
    default:
        print("Encoding error")
    }
} catch {
    print("Something went wrong! \(error)")
}
```

Listing 12.15 Calling Function Able to Throw Multiple Types of Errors

In the error-handling part, there are three sections:

- DecodingError instances are caught in de and are logged immediately. Then, the key-NotFound case gets special treatment, while others are handled by the default case.
- EncodingError instances are caught in en and are logged immediately. Then, the invalidValue case gets special treatment, while others are handled by the default case.
- Finally, all other errors are caught with the generic catch block, which acts as a safety net for unforeseeable cases.

Obviously, this is an oversimplified example aimed at demonstrating how multiple errors and cases are handled. In a more realistic program, your error-handling code would do more than log events and print things. But the main architecture and the code structure would remain the same, so you can use this code template as a skeleton structure for the future.

12.3.3   Localized Error Messages

Now let's jump to another topic that will probably be another staple in your error templates. In our examples so far, we have thrown some errors, and the client code was responsible of producing a meaningful error message. In fact, this is not a good practice in many cases because it might violate the *don't repeat yourself* (DRY) principle.

Imagine that you have a central throws function, called from multiple spots in your app, as demonstrated in Figure 12.9.

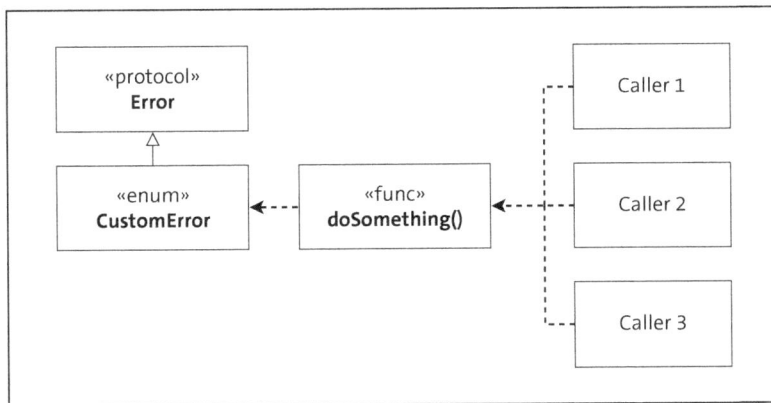

Figure 12.9 Function with Multiple Callers

In this diagram, if you produce the same copy-and-pasted error messages in caller 1, caller 2, and caller 3, you are repeating yourself. Instead, the error messages should come from a centralized repository. That way, you can ensure that when you change the error text centrally, this change will be reflected to all callers.

Apple engineers have already thought of this and made sure that Swift contains a mechanism to centralize error text messages. By applying the LocalizedError protocol instead of the usual Error protocol, you can include error texts centrally in your enumerations. That way, all callers may reuse the central texts in your enumeration.

Don't worry: LocalizedError inherits from the Error protocol, so when your type conforms to LocalizedError, it also conforms to Error automatically. LocalizedError can be seen as a deeper version of Error, enriched by centralized error texts.

Check out the application of LocalizedError in Listing 12.16.

```
import Foundation

enum LoginError: LocalizedError {
    case wrongPassword
    case userNotFound

    var errorDescription: String? {
        switch self {
        case .wrongPassword:
            return "Wrong password, retry!"
        case .userNotFound:
            return "No such user, go register!"
        }
    }
}

do {
    throw LoginError.userNotFound
} catch {
    print(error.localizedDescription) // No such user, go register!
}
```

Listing 12.16 Central Error Texts via LocalizedError

Focusing on the LoginError enumeration, you'll notice that it became a new computed property called errorDescription. This property came as a requirement of LocalizedError. Typically, you would return a different text for each case as errorDescription. That's exactly what we did in this example; wrongPassword and userNotFound returned different texts.

Now, any caller encountering LoginError can access and consume those centralized texts. There's no need to copy and paste error texts any longer! In the example, the caller accessed the central text via error.localizedDescription.

Neat, right? If you change the `LoginError.errorDescription` values in the future, then all callers will automatically reflect this change, using the new text values for error descriptions.

To make the example a bit more realistic, Listing 12.17 contains an enhancement. This time, the localized error is thrown by the `login` function instead of being thrown artificially in the main flow. But as you can see, the main approach doesn't change at all.

```
import Foundation

enum LoginError: LocalizedError {
    case wrongPassword
    case userNotFound

    var errorDescription: String? {
        switch self {
        case .wrongPassword:
            return "Wrong password, retry!"
        case .userNotFound:
            return "No such user, go register!"
        }
    }
}

func login(username: String, password: String) throws {
    if username != "alice" || password != "123" {
        throw LoginError.wrongPassword
    }
}

do {
    try login(username: "alice", password: "456")
    print("Login successful!")
} catch {
    print(error.localizedDescription) // Wrong password, retry!
}
```

Listing 12.17 Function Throwing Localized Error

You've now learned many techniques to catch and handle an error. However, there might be cases in which you don't care about the error details and want to ignore the error. Swift offers a neat bit of syntactic sugar for that, which we'll discuss now.

12.3.4   Optional Try

In many examples in this chapter, you have seen that try is the essential part of calling a function able to throw errors. At the very least, it tells the compiler that you are aware of the error risk and made the function call despite it.

On top of that framework, Swift gives you the option to use try statements as optionals. When used as such, you don't need to write full do-catch blocks. Instead, the function result will be converted to nil and the error will be discarded silently.

For a better understanding, inspect the "regular" code sample in Listing 12.18 first. This example features a function call without an optional try.

```
enum MathError: Error {
    case divByZero
}


func divide(_ a: Double, _ b: Double) throws -> Double {
    guard b != 0 else { throw MathError.divByZero }
    return a / b
}


var division: Double?
do { division = try divide(10, 0) } catch {}
print(division) // nil
```

Listing 12.18 Regular Usage of Try

In this example, we're OK with leaving division as nil if an error occurs. Because the catch { ... } block is empty and has no error-handling mechanism, the entire do-catch block looks like an empty boilerplate section.

For such cases, you can treat the try keyword as an optional, shortening the syntax as shown in Listing 12.19. This code snippet is the equivalent of the previous one.

```
enum MathError: Error {
    case divByZero
}


func divide(_ a: Double, _ b: Double) throws -> Double {
    guard b != 0 else { throw MathError.divByZero }
    return a / b
}


var division: Double?
division = try? divide(10, 0)
print(division) // nil
```

Listing 12.19 Optional Usage of Try

Instead of coding an entire do-catch block without any real value, you can shorten the expression to division = try? divide(10, 0). If you're not interested in any error details and you're OK with leaving division as nil in case of an error, that shortened syntax works perfectly and is much more readable.

Now, here's the cool part: Because try is an optional now, you can benefit from the other features of optionals that you learned about in Chapter 6. For example, you can take advantage of the nil-coalescing operator ?? to assign a default value to division in case an error occurs. In Listing 12.20, division will be 0 due to the ?? 0 suffix. When the try? statement returns nil, the alternative fallback value of 0 is assigned instead.

```
enum MathError: Error {
    case divByZero
}

func divide(_ a: Double, _ b: Double) throws -> Double {
    guard b != 0 else { throw MathError.divByZero }
    return a / b
}

var division = (try? divide(10, 0)) ?? 0
print(division) // 0
```

Listing 12.20 Nil-Coalescing Operator with Try

Should You Use try?

We advise using try? only sparingly, because in many cases you will want to catch the error and deal with it. But there may be cases where the failure is not critical and you want a graceful fallback without a chunk of unnecessary code blocks. If that's the case, then try? offers concise syntax.

You can even implicitly unwrap a try statement: try!. This will tell the compiler that you are absolutely sure that the called function won't fail and take full responsibility if the app crashes due to an unforeseen error thrown.

```
enum MathError: Error {
    case divByZero
}

func divide(_ a: Double, _ b: Double) throws -> Double {
    guard b != 0 else { throw MathError.divByZero }
    return a / b
}
```

```
var division = try! divide(10, 2)
print(division) // 5
```

Listing 12.21 Implicitly Unwrapping Try Statement

Beware of the Danger of try!

Like any other implicit unwrap practice, try! statements are dangerous. They should be used sparingly and only when you are 100% certain that the function will not throw an error. For safety, you should avoid using try! in production code—especially with user inputs. Unit tests could be an appropriate spot for that syntax, where you can use it to reduce the boilerplate code.

Before moving to runtime checks, there is one more topic worth addressing: running cleanup code despite errors.

12.4 Cleanup with Defer

In Swift, defer is a powerful keyword targeting cleanup code. With defer, you can ensure that a cleanup code block runs no matter how a function exists—even if it throws an error. Which is, obviously, the topic at hand.

What do we mean by *cleanup code*? Table 12.2 features some examples to answer that question.

Function's Operation	Cleanup Code
File handling	Closing the file
Database transactions	Closing the database connection
Network session	Closing the socket
Temporary files	Deletion of the files
Locking resources	Unlocking resources

Table 12.2 Some Cleanup Code Examples

Focusing on the network session example, imagine that your function opens a network connection to exchange data. Logically, you need to close the connection at the end of the function, right? It doesn't matter if the function exists successfully or if it exists by throwing an error. In either case, the connection must be closed because the function's task has concluded. That's where defer could be used to ensure that the connection is closed eventually.

Temporary files present another case. Imagine that your function creates temporary files on the disk to perform its task. When the function is completed, those files would need to be deleted because they are no longer needed. Whether the function ends naturally or throws an error, the deletion requirement is the same. That's where defer would come in handy, ensuring you delete those files in the end, no matter how the function ended.

Now that the purpose of defer is clear, let's look at a hands-on example. We'll begin with a regular function without any errors involved and advance from there. Check the structure of readFile in Listing 12.22, noting the defer code block. Obviously, this function doesn't open and close any real file here; we're merely simulating the operation.

```
func readFile() {
    print("Opening file")

    defer {
        print("Closing file")
    }

    print("Reading file contents...")
}


readFile()
```

Listing 12.22 Deferred Code Without Error

The readFile function has three parts, which appear in the code in this order:

- Opening the file
- Running the defer code block
- Reading the file

But in what order are these parts executed? Can you guess? You can compare your guess with the actual output shown in Figure 12.10.

```
Opening file
Reading file contents...
Closing file
```

Figure 12.10 Execution Order of readFile

Note that the defer { ... } block was executed last, despite its place right in the middle of the readFile function. That's not a bug, but the exact purpose of defer; it will always be executed as the last code block of the function.

This mechanism becomes particularly useful if the function might throw an error. Listing 12.23 features an enhanced version of the same example, where readFile deliberately throws an error while reading the file to show the effect of defer.

```
enum FileError: Error {
    case readFailed
}

func readFile() throws {
    print("Opening file")

    defer {
        print("Closing file")
    }

    print("Reading file contents...")
    throw FileError.readFailed
}

do { try readFile() }
catch {  print("Error: \(error)") }
```

Listing 12.23 Deferred Code with Error

Once again, you are invited to guess the execution order! When you do, you can compare it with the actual output shown in Figure 12.11.

```
Opening file
Reading file contents...
Closing file
Error: readFailed
```

Figure 12.11 Execution Order of readFile with Error

Although `readFile` has thrown an error this time, the code in the `defer { ... }` block was still executed before `throw`, ensuring the completion of the function's cleanup.

Neat, right? Using `defer`, you can ensure that your cleanup code will always execute.

There is a catch, though: You can't throw from inside a `defer` block. The code sample in Listing 12.24 won't compile due to the `throw` statement in the `defer` block.

```
enum FileError: Error {
    case readFailed
    case closeFailed
}

func readFile() throws {
    print("Opening file")

    defer {
```

```
        print("Closing file")
        throw FileError.closeFailed // Won't compile
    }

    print("Reading file contents...")
    throw FileError.readFailed
}
```

Listing 12.24 You Can't Throw from Inside Defer Blocks

This catch makes sense, right? After all, defer is designed for a guaranteed cleanup, not for introducing new failure points. What you want to do is to conclude the function, not mask the original error.

If the defer code must call a risky function and you can tolerate an error at that point, making use of try? could be a concise solution. Listing 12.25 demonstrates such an example; note the try? closeFile() statement.

```
enum FileError: Error {
    case readFailed
    case closeFailed
}

func closeFile() throws {}

func readFile() throws {
    print("Opening file")

    defer {
        print("Closing file")
        try? closeFile()
    }

    print("Reading file contents...")
    throw FileError.readFailed
}
```

Listing 12.25 Optional Try in defer Block

In this example, even if closeFile() fails, that error will be silently ignored. Although the file will remain open at that point, if that's a tolerable situation, then you're good.

Before concluding this chapter, we'll address the topic of runtime checks, which are typically used to spot unrecoverable programmer mistakes: Programmers are human too, and human errors happen!

12.5   Runtime Checks

Defensive programming is a pillar of stable apps. Common errors, like connection failures, invalid user inputs, or a missing file can (and should) be caught and handled in whatever way possible. The app might solve some errors automatically, while others might need user interference.

However, not all errors are recoverable. There might be conditions such as logic bugs or violations of assumptions that should never happen in the program. For instance, you would assume a months array to have 12 members, but perhaps somewhere in the code, you encounter a case in which it has only 10 members: November and December are missing! This is probably a coding error that can't really be "handled"; instead, you have to fix the bug in your code and maybe release an update.

For such critical and unrecoverable program mistakes, Swift offers a set of runtime checks like assert, precondition, and fatalError. Using such checks, you can catch programming mistakes early and "crash safely" when assumptions are broken.

> **Safe Crash**
>
> A *safe crash* may sound like an oxymoron, but it's not. Would you prefer your app to crash before or after it erases the entire database due to a programming error? The crash happening before that point would be a safe crash. It's the lesser evil of two unpleasant choices.

In the following sections, you'll discover different types of runtime checks and see coding examples.

12.5.1   Debug Runtime Checks

Let's begin with runtime checks that only run during development and debugging, not in production. This means that such checks will execute while you're in Xcode, coding, building, and testing your app. However, once you release your app (such as to the App Store), target devices won't execute those checks.

In fact, when you compile your app for the final release, the compiler will strip out all debug runtime checks, such as assert calls, which we'll discuss in this section. They won't even make it to the released app.

Such lighter-weight checks are suitable to catch less critical (but still important) logic flaws during development, without impacting the user experience in the final product. By disabling them in production, you gain a performance advantage: Catch them during development and fix them, but don't let the final app suffer the runtime cost.

Assert

The most basic runtime check keyword is `assert`, which checks a condition and crashes in debug mode if the condition is `false`. Its behavior is similar to that of the `guard` keyword, with the obvious difference of crashing the app.

Listing 12.26 demonstrates the syntax of `assert`. Here you set the condition age >= 0 first and follow that with the crashing error message for when the condition is not met.

```
let age = 3
assert(age >= 0, "Age cannot be negative")
```

Listing 12.26 No Error on Assert

In this case, age is not negative, so the code will run normally without any interruptions. Listing 12.27, on the other hand, is a different story: Here, age is set as negative, which will cause the `assert` check to cause a crash.

```
let age = -3
assert(age >= 0, "Age cannot be negative")
```

Listing 12.27 Error on Assert

The playground output of that error is shown in Figure 12.12. Note that no matter where the assert crash occurs, the app will stop immediately.

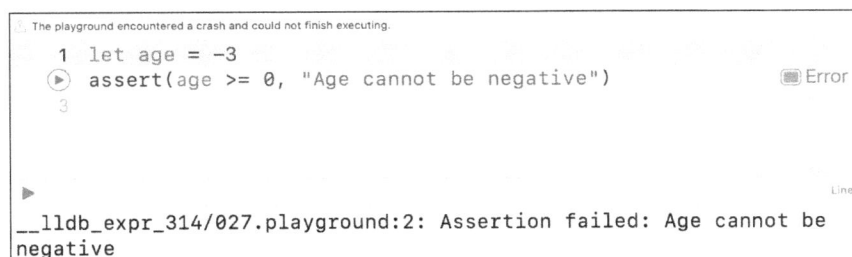

Figure 12.12 Crash Caused by Assert

As a further exercise, Listing 12.28 demonstrates some `assert` checks in a function. `subtractMonths` aims to find the difference between the given months, but it begins with some runtime checks to ensure the internal consistency of the app.

```
func subtractMonths(_ m1: Int, _ m2: Int) -> Int {
    assert(m1 >= 1 && m1 <= 12, "Invalid first month")
    assert(m2 >= 1 && m2 <= 12, "Invalid second month")
    assert(m2 >= m1, "Invalid month range")
    return m2 - m1
}
```

Listing 12.28 Assert Checks in a Function

Note that if those month values are given by the user, it would be wiser to throw an error instead, ultimately informing the user about their invalid entries. But if the source of those values is internal, then having an invalid value means that the app's internals have a malfunction, which the user can't do anything about. That's a case that can be approached with runtime checks.

Assertion Failure

A variant of assert is `assertionFailure`, which will crash the app unconditionally. It is typically used when the code reaches a point that should never be reached. Listing 12.29 demonstrates an example in which HTTP return codes are evaluated.

```
func evalHttpCode(_ code: String) {
    switch code {
    case "":
        assertionFailure("Empty HTTP code")
    case "200":
        print("HTTP Success")
    default:
        print("HTTP Failure: \(code)")
    }
}


evalHttpCode("200")
evalHttpCode("")
```

Listing 12.29 Demonstration of assertionFailure

If the HTTP return code is empty, this is most likely a programming mistake causing an empty value to be passed instead of the actual code. In that case, `assertionFailure("Empty HTTP code")` is executed, crashing the app in debug mode. In a released app, this command won't have an effect.

Here's a cool trick: You can apply an assertion for the development phase while still providing a fallback code for the final app. Check the demonstration in Listing 12.30.

```
func evalHttpCode(_ code: String) {
    switch code {
    case "":
        assertionFailure("Empty HTTP code")
        print("HTTP Failure: Empty")
    case "200":
        print("HTTP Success")
    default:
        print("HTTP Failure: \(code)")
    }
```

```
}

evalHttpCode("200")
evalHttpCode("")
```

Listing 12.30 Usage of assertionFailure with Fallback Code

In this example, assertionFailure will execute during development, crashing the app and demanding your undivided attention. However, it will be skipped in the final app and the fallback code, print("HTTP Failure: Empty"), will be executed instead.

This technique is useful to handle an unexpected situation gracefully—like installing ashtrays in an airplane bathroom. Although it's forbidden to smoke during flights, you'll have a fallback solution for the unlikely case in which someone will attempt to do so anyway.

In this technique, you can even combine the runtime check with an error! The code in Listing 12.31 implements this approach. This app will crash immediately during development, demanding the immediate attention of the programmer, but in production, it will gracefully throw an error, allowing the error-handling mechanism to take place.

```
enum HttpCodeError: Error {
    case emptyCode
}

func evalHttpCode(_ code: String) throws {
    switch code {
    case "":
        assertionFailure("Empty HTTP code")
        throw HttpCodeError.emptyCode
    case "200":
        print("HTTP Success")
    default:
        print("HTTP Failure: \(code)")
    }
}

try? evalHttpCode("200")
try? evalHttpCode("")
```

Listing 12.31 Usage of assertionFailure with Fallback Error

So far, we've covered runtime checks for the development phase. Now let's examine another species of runtime checks.

12.5.2   Release Runtime Checks

In this section, you'll learn about runtime checks that run during development as well as in the final app. Because these checks will crash the app on the user's device, they are highly unpleasant and should be used sparingly—only in cases in which the app crash is the lesser evil.

A security vulnerability is a good example: If you detect a nonadmin user executing an admin task, you might want to crash the app to limit the effects of the compromise.

Another example could be a finance app performing automatic transactions. If you encounter a negative loan interest value, you might want to crash the app to prevent further invalid financial transactions, applying some degree of damage control.

In this section, we'll walk through the syntax you would use in such cases.

Precondition

In Swift, precondition is the sibling of assert. They have nearly the same syntax, but precondition will crash the final app too. The code in Listing 12.32 contains an example.

```
func applyCredit(amount: Double, rate: Double) {
    precondition(amount >= 0, "Amount must be non-negative")
    precondition(rate >= 0, "Rate must be non-negative")
    print("Credit applied")
}


applyCredit(amount: 100, rate: 0.01)
applyCredit(amount: -100, rate: 0.01)    // Error
```

Listing 12.32 Syntax of Precondition

Precondition Failure

Similarly, preconditionFailure is the sibling of assertFailure. They have nearly the same syntax, but preconditionFailure will crash the final app too. The code in Listing 12.33 contains an example.

```
func deleteData(userRole: String) {
    switch userRole {
    case "admin":
        print("All data was deleted")
    default:
        preconditionFailure("Non admin shouldn't be calling this function")
    }
}
```

Listing 12.33 Syntax of preconditionFailure

Fatal Error

In Swift, `fatalError` is a runtime check that's very similar to `preconditionFailure`. Both will crash the app unconditionally in debug mode or in the final release. Their distinction is subtle and primarily lies in their semantic intent. Check out `fatalError` in action in Listing 12.34.

```
func someNewFeature() {
    fatalError("Not yet implemented!")
}
```

Listing 12.34 Syntax of fatalError

`preconditionFailure` is typically used to signal a precondition violation, as its name states. Meanwhile, `fatalError` is typically used to signal an unrecoverable error due to which the program simply can't execute safely or meaningfully any longer.

If the critical case occurred due to an unexpected condition, therefore, you can use `preconditionFailure`. If the critical case occurred independent from a condition, hitting a path that should never be executed, then you can use `fatalError`.

12.6 Summary

In this chapter, you learned about errors in Swift, which is a major pillar of developing consistent applications. You have the option to throw built-in or custom errors, which are basically data types implementing the `Error` protocol. If you need to include custom descriptions, the `LocalizedError` protocol can be implemented instead.

The function throwing the error can implement a final cleanup code block using `defer { ... }`, which is useful for tasks like closing files or concluding connections. Those blocks will execute even if the function ends with an error.

In the client code, you typically catch errors in `do-try-catch` blocks, where you can implement your error-handling logic. It's also possible to ignore errors by using the `try?` syntax, marking the function call as optional.

Finally, you have the option to implement runtime checks, which will stop and crash the app in case of unrecoverable errors. Some runtime checks are executed during development only, and others are executed in the final app too.

Now that you're empowered with the knowledge to tackle errors, the next chapter will cover the topic of file handling in Swift, which is a common task that's eventually needed by most developers.

Chapter 13
File Handling

This chapter will cover the basics of reading and writing local files with Swift.

Files and folders are core elements of operating systems. *Files* contain persistent data stored on the disk, while *folders* are used to organize files. In many apps, you are going to need file-access features. You might need to create folders, read existing files, create and fill new files, and so on. Naturally, Swift empowers programmers with the necessary tools for such operations. In this chapter, you are going to discover these tools and learn how to conduct folder/file operations.

As a heads-up, you should be aware of the sandboxing features of Apple operating systems, which are fundamental security features. In iOS, your app will get a unique *sandbox space* where it can freely read/write files, but it can't directly access files outside of that directory. In macOS, apps submitted to the App Store must also be sandboxed in the same way. To access further resources, your apps need to be granted permission by the user or settings.

Because the macOS examples in this chapter won't be targeting the App Store, they will be able to access any file that your user account has permission to access. Therefore, you won't encounter such a limitation at this time, but keep sandboxing in mind as a future topic to explore. Many iOS/macOS apps also make use of iCloud Drive to store and share files, which might be another future topic to study.

With that addressed, let's focus on Swift's core file-handling features.

13.1 Text Files

Arguably, the simplest type of files you will encounter are text files. Such files contain human-readable content as a string, saved with an encoding scheme like UTF-8. Some common text file types are described in Table 13.1; at least some of them should be familiar.

Text File Type	Contents	Sample File Name
Unformatted text	Plain text; saved as-is	*shopping_list.txt*
JSON file	Text-based structured data in JSON format	*settings.json*
XML file	Text-based structured data in XML format	*server_response.xml*
HTML file	Source code of a web page	*index.html*
Swift code file	Source code for a macOS app	*game_shapes.swift*

Table 13.1 Some Text File Examples

In this section, you will learn how to read and write text files using Swift.

13.1.1   Reading Text Files

Let's start with the operation of *reading* text files, which means loading the contents of a text file on the disk into the memory of your Swift app. We will use a file called *hello.txt*, the content of which is shown in Listing 13.1.

```
Hello world!
This is a text file.
Nothing more.
```

Listing 13.1 Contents of hello.txt

The easiest way to load this file to the memory is to make use of the `String(contentsOf-File:)` expression, which loads the file contents into a string variable. Listing 13.2 demonstrates the syntax for this task.

```
import Foundation

let filePath = "/Users/kerem/Desktop/hello.txt"
let hello = try? String(contentsOfFile: filePath, encoding: .utf8)
print(hello ?? "")
```

Listing 13.2 Loading Text File into a String

Now, let's break it down line by line. Initially, you need to know the exact path where the file resides. This simple example contains a literal path, but in a real-world app, the actual path would be within the app sandbox or provided by the user. When you are running those examples on your machine, you might change the literal path to point "your" file.

The next line is where the magic happens. `String(contentsOfFile: filePath, encoding: .utf8)` will return a string with the contents of *hello.txt*, reading it with the provided encoding. Text encoding was explained in Chapter 2; you may revisit that topic if you need a memory refresher.

Note that this function may throw an error. In exceptional cases, like the absence of the file or insufficient user authorization, you might get an error instead of a string. Due to this possibility, opening files requires error handling, which was the subject of Chapter 12. In this simple example, we merely applied an optional try to keep the code simple.

But if everything is OK, then the contents of *hello.txt* should be loaded into the hello variable and printed out as shown in Figure 13.1.

```
1  import Foundation
2
3  let filePath = "/Users/kerem/Desktop/hello.txt"
4  let hello = try? String(contentsOfFile: filePath, encoding: .utf8)
5  print(hello ?? "")
```

```
Hello world!
This is a text file.
Nothing more.
```

Figure 13.1 Output of hello.txt

This method is extremely simple and useful for small- to medium-sized text files. Beyond a simple line of code, you don't need to manage any details manually. But the disadvantage is that this method loads the entire file to the memory at once in full. If you encounter very large files, they might cause your computer to run out of free memory.

For larger files, you have the option to use the FileHandle object. Using FileHandle, you can read (or write) data in chunks instead of loading the file all at once. It's a bit more complex than loading the file as a string, so it's best to use this method only when necessary.

Listing 13.3 demonstrates the usage of FileHandle by loading the same file in eight-byte chunks.

```
import Foundation

let filePath = "/Users/kerem/Desktop/hello.txt"

if let fileHandle = FileHandle(forReadingAtPath: filePath) {
    let chunkSize = 8    // Read 8 bytes on each chunk

    while true {
        let data = fileHandle.readData(ofLength: chunkSize)
        if data.count == 0 { break }  // Reached end of file
        print(String(data: data, encoding: .utf8) ?? "")
```

```
    }

    fileHandle.closeFile()
} else {
    print("Couldn't open file.")
}
```

Listing 13.3 Loading Text File Using FileHandle

The example starts with `FileHandle(forReadingAtPath: filePath)`, which creates a handle for the given file. Note that this expression does not load the file immediately to the memory; instead, it finds the file and prepares to read it.

Following that initial step, you enter an infinite `while` loop, in which chunks of the file are read via `fileHandle.readData(ofLength: chunkSize)`. If no data is returned, it means that the end of the file is reached, and so you exit the loop. Otherwise, the `print` statement prints out the loaded data.

Finally, you close the file using `fileHandle.closeFile()`. This is a necessary action; otherwise, the file remains open within `FileHandle`, unnecessarily occupying system resources.

> **Reminder: Cleanup with Defer**
>
> Remember cleaning up with `defer` from Chapter 12? In a function reading a file, closing the file is a typical operation for the `defer { ... }` code block.

The output of this code sample is shown in Figure 13.2, where the output of each chunk was printed on a separate line.

```
Hello wo
rld!
Thi
s is a t
ext file
.
Nothin
g more.
```

Figure 13.2 File Broken into Chunks with FileHandle

It's a common requirement to load a text file completely and split it into an array, line by line. That's a very easy task, as demonstrated in Listing 13.4.

```
import Foundation

let filePath = "/Users/kerem/Desktop/hello.txt"
```

```
let hello = try? String(contentsOfFile: filePath, encoding: .utf8)
let helloLines = hello?.split(separator: "\n") ?? []
print(helloLines)
```

Listing 13.4 Splitting Text File into an Array

Once the file is loaded as a string variable, all you need to do is use `split` to split it into an array at the line break character. The output of this example is shown in Figure 13.3. Easy, right?

```
["Hello world!", "This is a text file.", "Nothing more."]
```

Figure 13.3 Text File Split by Line

Now that we have the common methods to read text files covered, let's advance to the topic of writing text files.

13.1.2   Writing Text Files

Swift offers various ways to write text data to files, and they're very similar to the ways to read text from files. The first and simplest method is to use the built-in `write` function of a string, which is demonstrated in Listing 13.5. As you can see, it's as easy as loading a text file into a string. This code sample might use better error handling in a real app, but let's keep things simple to focus on the subject at hand.

```
import Foundation

let txt = """
Hello world!
This is a text file.
Nothing more.
"""

let filePath = "/Users/kerem/Desktop/hello.txt"
try? txt.write(toFile: filePath, atomically: true, encoding: .utf8)
```

Listing 13.5 Writing Text to File via String Object

Here, `txt.write` will create a new file in the target path or replace an existing file. Replacing an existing file might be undesirable in some cases and may require extra care. You'll learn how to check for the existence of a file in Section 13.4, which will come in handy. The contents of the created file will appear as shown in Figure 13.4.

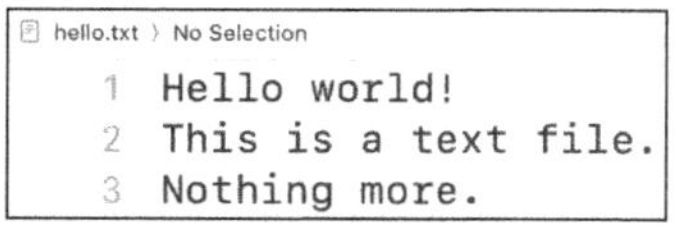

hello.txt ⟩ No Selection
```
1  Hello world!
2  This is a text file.
3  Nothing more.
```

Figure 13.4 File Created by String Object

The write function has a parameter called atomically, which is worth investigating in detail:

- When atomically is set as true, the function will first write the file to a temporary location and move to the actual destination if the write is successful. This ensures file consistency in unexpected cases, like a crash or full disk. This is an all-or-nothing approach.

- When atomically is set as false, the function will attempt to write to the actual destination directly. If anything goes wrong, the file might end up partially written or corrupted.

In nearly all cases, atomically should be set as true, unless you have a very good reason—like performance-sensitive code that writes huge files.

The method we just explored is used to write a string to a file all at once. The alternative method is to write data to a file chunk by chunk, using the familiar FileHandle option. This technique is demonstrated in Listing 13.6, with simplified error handling.

```
import Foundation

let texts = ["Hello world! ",
             "This is a text file. ",
             "Nothing more."]

let filePath = "/Users/kerem/Desktop/hello.txt"

if let fileHandle = FileHandle(forWritingAtPath: filePath) {
    for text in texts {
        fileHandle.write(text.data(using: .utf8)!)
    }

    fileHandle.closeFile()
} else {
    print("Couldn't open file.")
}
```

Listing 13.6 Writing Text to File via FileHandle

During the initialization of the FileHandle object, you need to apply a different mode than for reading a text file. FileHandle(forWritingAtPath: filePath) tells the computer that the file is being opened to write data (not read it). The follow-up is straightforward: Each string at hand is written to the file, and the file is eventually closed.

The resulting file is shown in Figure 13.5. Although each string is concatenated horizontally, you could have made each string a new line by writing the line break character after each string. You are more than welcome to try this as a small exercise!

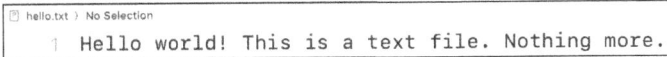

```
hello.txt ⟩ No Selection
  1   Hello world! This is a text file. Nothing more.
```

Figure 13.5 File Created by FileHandle

Another option is to *add* text to an existing file instead of replacing the file's contents completely. As the starting point of this demonstration, Figure 13.6 shows a text file with existing content.

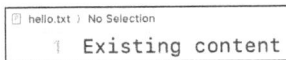

```
hello.txt ⟩ No Selection
  1   Existing content
```

Figure 13.6 Original Text File Content

To append new content to that file, let's make use of FileHandle once again. Check the demonstration in Listing 13.7, and see if you can spot the critical differences.

```
import Foundation

let texts = ["New content! ",
             "Here we go! "]

let filePath = "/Users/kerem/Desktop/hello.txt"

if let fileHandle = FileHandle(forUpdatingAtPath: filePath) {
    try fileHandle.seekToEnd()

    for text in texts {
        fileHandle.write(text.data(using: .utf8)!)
    }

    fileHandle.closeFile()
} else {
    print("Couldn't open file.")
}
```

Listing 13.7 Appending New Content to Text File

The first difference is in the object initialization step. `FileHandle(forUpdatingAtPath: filePath)` indicates that the file is being opened for updating. Following that statement, the computer will not replace the file. Instead, the file will be opened for modifications.

In the following line, `fileHandle.seekToEnd()` will move the cursor to the end of the file, where you want to add new content. Otherwise, you would be replacing different parts of the file—which might be a valid requirement in some cases, but we are aiming to append text to the file here.

The rest of the code is no different than the previous example: You write the content at hand and close the file. The result with the appended content is shown in Figure 13.7.

```
 hello.txt ⟩ No Selection
 1  Existing content
 2  New content! Here we go!
```

Figure 13.7 Text File with Appended Content

That example concludes our content on text files. Now that you're armed with the knowledge to read and write text files, we can advance to the same operations in binary files.

13.2   Binary Files

Binary files are files that store data in a non-human-readable format, as a sequence of bytes. In general, the purpose of binary files is to store nontext data. Unlike text files, which store characters with an encoding like UTF-8, binary files store data exactly as it exists in memory. The data is expected to be interpreted by the app or the operating system.

Some common binary file examples are shown in Table 13.2.

Binary File Type	Contents	Sample File Name
Image	Pixel color codes of an image	*portrait.png*
Office document	Formatted text and media	*homework.docx*
Audio	Digital recording of a sound	*my_song.mp3*
Video	Digital sequence of images and audio	*holiday.m4a*
Program	Compiled code, ready for execution	*tasks.app*

Table 13.2 Some Binary File Examples

As we did for text files, we will explain how to read and write binary files using Swift in this section.

Serialization in Swift

If you're looking for an easy way to serialize and deserialize your Swift objects as binary files, that topic will be covered in the next section. In this section, you'll learn how to read and write raw binary files, with control over bytes of data.

13.2.1   Reading Binary Files

Like you do for text files, you have multiple options to read binary files in Swift. For the upcoming examples, we'll use the *rheinwerk.png* binary file, which contains the logo of Rheinwerk Publishing. You may use any other binary file in your exercises.

The most basic and straightforward way to read a binary file is to read it all at once into a Data object. Data is the go-to type for working with binary content in Swift. Listing 13.8 features a demonstration of that approach.

```
import Foundation

let fileURL = URL(fileURLWithPath: "/Users/kerem/Desktop/rheinwerk.png")

do {
    let binaryData = try Data(contentsOf: fileURL)
    print("Read \(binaryData.count) bytes") // Read 7771 bytes
} catch {
    print("Failed to read binary file: \(error)")
}
```

Listing 13.8 Reading Binary File into a Data Object

First, you need to create a URL object pointing to your target file, which is called fileURL in this example. The next step is straightforward: Data(contentsOf: fileURL) will load the file and return its contents, and that's it! This initializer will naturally throw an error in an exceptional case, such as a missing file or insufficient authorization.

An alternative approach, as with text files, is to use the FileHandle object to read binary data in chunks. Because you're already familiar with FileHandle, we can jump directly to the example in Listing 13.9 as the syntax is mostly the same.

```
import Foundation

let fileURL = URL(fileURLWithPath: "/Users/kerem/Desktop/rheinwerk.png")

if let fileHandle = try? FileHandle(forReadingFrom: fileURL) {
    let chunkSize = 1024 // Read 1024 bytes on each chunk

    while true {
```

```
        let data = fileHandle.readData(ofLength: chunkSize)
        if data.isEmpty { break } // Reached end of file
        print("Chunk size: \(data.count) bytes")
    }

    try? fileHandle.close()
} else {
    print("Couldn't open file.")
}
```

Listing 13.9 Loading Binary File Using FileHandle

You attempt to open the file here with the `FileHandle(forReadingFrom: fileURL)` expression. Once the file is open, you read data in chunks of 1024 bytes until the end of the file is reached. Finally, the file is closed—as it should be.

The output of this example is shown in Figure 13.8. The sample file was 7,771 bytes long; if you add up the sizes of the chunks, the total is exactly 7,771. This means that the app ran successfully, leaving no byte behind!

```
Chunk size: 1024 bytes
Chunk size: 1024 bytes
Chunk size: 1024 bytes
Chunk size: 1024 bytes
Chunk size: 1024 bytes
Chunk size: 1024 bytes
Chunk size: 1024 bytes
Chunk size: 603 bytes
```

Figure 13.8 Output of FileHandle-Supported File Load

Binary Data Interpretation

Unlike text content, which can be viewed and printed out directly, binary content is a raw sequence of ones and zeros. This content is often not human-readable; instead, an app needs to process it. For instance, an image viewer may read a binary image file, interpret it as a sequence of color codes, and "paint" it to the monitor by assigning colors to pixels.

Apple frameworks, as well as third-party frameworks, contain many libraries to process and consume binary files. But because this book is about focusing on core Swift features, that part is best left as a topic for your own future exploration. We won't step into the details of binary file interpretation at this time.

That said, there is no reason not to go over the basics, as we'll do ahead!

If you're doing the hard work of interpreting binary content yourself, you'll certainly need to extract snippets of data from the file and assign them to Swift types.

Listing 13.10 features a basic demonstration, which extracts four-byte snippets from the binary file and interprets them as Int32 values. Although a PNG file is not a real sequence of Int32 values, we'll assume it to be and merely use the file as a mock data source.

```
import Foundation

let fileURL = URL(fileURLWithPath: "/Users/kerem/Desktop/rheinwerk.png")

do {
    let data = try Data(contentsOf: fileURL)
    var cursor = data.startIndex

    while cursor + 4 <= data.endIndex {
        let intBytes = data[cursor..<cursor+4]
        let value = intBytes.withUnsafeBytes { $0.load(as: Int32.self) }
        print("Int32: \(value)")
        cursor += 4
    }
} catch {
    print(error)
}
```

Listing 13.10 Interpreting Binary Data as Int32 Values

Because Int32 is four bytes long, you read four bytes of data on each iteration and print out the extracted integer value. Partial output of the example is shown in Figure 13.9.

```
Int32: 1196314761
Int32: 169478669
Int32: 218103808
Int32: 1380206665
Int32: 738263040
Int32: 738263040
Int32: 520
Int32: 421524992
Int32: 503316514
```

Figure 13.9 Int32-Based Output of Loaded Binary File

In reality, different parts of a binary file might contain integers, Booleans, characters, floats, and so on. The code reading the file must be aware of the format at hand and extract data accordingly. We will stop at that point and continue to writing binary files.

13.2.2   Writing Binary Files

As you may have guessed already, there are multiple approaches to create binary files with Swift. But most of the time, they will share similar steps:

- Creating a Data object
- Loading binary data into the Data object
- Writing the Data contents to a file

For starters, Listing 13.11 shows an example in which raw byte values in hex format are consolidated into a Data object and saved to a binary file with the simple data.write() function.

```
import Foundation

let fileURL = URL(fileURLWithPath: "/Users/kerem/Desktop/output.abc")

let bytes: [UInt8] = [0xDE, 0xAD, 0xBE, 0xEF]
let data = Data(bytes)

do {
    try data.write(to: fileURL)
    print("File written!")
} catch {
    print("\(error)")
}
```

Listing 13.11 Writing Raw Binary Values to a File

The target file is called *output.abc* in this instance, but the file name and extension are not important for now. To view the contents of *output.abc*, you can simply use the hexdump terminal command, as demonstrated in Figure 13.10. You can also use a hex viewer if you have one; there are many options in the App Store or available as VS Code plugins.

```
kerem3:Desktop kerem$ hexdump -C output.abc
00000000  de ad be ef
00000004
```

Figure 13.10 Hex Dump of Raw Binary Content

As you can see, the hex values shown by hexdump are exactly the same as the values in the sample code! Voilà!

Another approach is to write an array of numbers to a binary file, instead of raw hex values. The code in Listing 13.12 demonstrates that approach, with less defensive programming in place for the sake of simplicity.

```
import Foundation

let numbers: [Int32] = [42, 1337, 2025]
var data = Data()
```

```
for number in numbers {
    var value = number
    withUnsafeBytes(of: value) { buffer in data.append(contentsOf: buffer) }
}

let fileURL = URL(fileURLWithPath: "/Users/kerem/Desktop/output.abc")
try? data.write(to: fileURL)
```

Listing 13.12 Writing Numbers to a Binary File

The hexdump output of this operation is shown in Figure 13.11. Although those hex values may seem a bit puzzling at first sight, they are actually binary representations of integer numbers.

```
kerem3:Desktop kerem$ hexdump -C output.abc
00000000  2a 00 00 00 39 05 00 00  e9 07 00 00
0000000c
```

Figure 13.11 Hex Dump of Number Content

To understand the contents of the file, you can refer to the interpretation in Table 13.3. All the numbers were neatly packed into the binary file, you see? Here's a helpful hint: Many commercial hex viewers will help you with such decoding tasks, lest you get cross-eyed looking at terminal screens. You may want to use such offerings if you'll be doing a lot of binary work.

Number	Hex (32-Bit)	Little Endian Byte Order
42	0x0000002A	2A 00 00 00
1337	0x00000539	39 05 00 00
2025	0x000007E9	E9 07 00 00

Table 13.3 Interpretation of Int32 Binary Content

In the previous two examples, we wrote the entire content of the Data object to the file all at once, creating or overwriting the file in the process. But you can also write data in chunks, as we did many times in this chapter. Undoubtably, FileHandle will come to your aid once again! Listing 13.13 demonstrates an example in which the binary content of repeated 0xAB values (10K times) is written to a file in 1 KB chunks.

```
import Foundation

let fileURL = URL(fileURLWithPath: "/Users/kerem/Desktop/output.abc")
let largeData = Data(repeating: 0xAB, count: 10_000) // 10 KB
let chunkSize = 1024  // 1 KB chunks
```

```
do {
    FileManager.default.createFile(atPath: fileURL.path, contents: nil)
    let fileHandle = try FileHandle(forWritingTo: fileURL)
    var offset = 0

    while offset < largeData.count {
        let end = min(offset + chunkSize, largeData.count)
        let chunk = largeData.subdata(in: offset..<end)
        fileHandle.write(chunk)
        offset = end
    }

    try fileHandle.close()
} catch {
    print(error)
}
```

Listing 13.13 Writing Binary File in Chunks

Although the syntax has changed as needed, the skeleton structure didn't deviate much from the previous similar examples. You start by opening a file, extract and write chunks of data, and finally close the file. The output file is expected to be 10 KB because that's how many bytes you put into it; however, you dumped this content into the file in small chunks of 1 KB each.

As in previous sections, FileHandle can be used here to append data to an existing binary file too. Listing 13.14 acts as a follow-up example, appending new bytes to the file in the previous example and thus increasing its size to 11 KB.

```
import Foundation

let fileURL = URL(fileURLWithPath: "/Users/kerem/Desktop/output.abc")
let newData = Data(repeating: 0xAB, count: 1000) // 1 KB
let chunkSize = 1024   // 1 KB chunks

do {
    let fileHandle = try FileHandle(forWritingTo: fileURL)
    fileHandle.seekToEndOfFile()
    fileHandle.write(newData)
    try fileHandle.close()
} catch {
    print(error)
}
```

Listing 13.14 Appending Data to a Binary File

And that example concludes our content on binary files—for now! Next, we'll move forward to a section with arguably wider practical uses, in which you will learn about working with common file formats in Swift.

13.3   Working with Common Formats

The ecosystem of the software industry runs on many globally accepted common file formats. Some file types, like HTML and CSS, are centered on web pages. Others, like CSV and XLSX, revolve around tabular data. Naturally, there are many more widely accepted formats in use.

Using the techniques you learned before, you can technically read, interpret, and write any text or binary file format using manually written Swift code. But in practice, this is a redundant task that often comes with unforeseen difficulties because file format specifications might have many tiny details to consider.

To overcome this challenge, a typical programmer would import a tested and proven third-party library instead of reinventing the wheel. For example, you can import the popular open-source CoreXLSX library to your project and start reading Excel files immediately, or use CodableCSV, an open-source library targeting CSV files. After unlocking Swift modules in Chapter 15, you will be able to import and consume such libraries easily.

Having that said, you don't always need to resort to external libraries. The Swift language comes with built-in support for some common formats already. In this section, you will discover some of those formats and we'll offer demonstrations of how to read/write them.

13.3.1   JSON

Let's start with JSON, a widely used format to store and transfer lightweight data. JSON stands for *JavaScript Object Notation* and it is a text-based file format to represent structured data. To get you warmed up, Listing 13.15 shows a simple JSON file containing the details of a customer order.

```
{
    "orderId": 12345,
    "orderDate": "2025-06-15",
    "totalAmount": 79.97,
    "customer": {
        "id": 1001,
        "name": "John Doe"
    },
    "items": [
```

```
    {
        "productId": "P001",
        "productName": "Wireless Mouse",
        "quantity": 2,
        "unitPrice": 25.0
    },
    {
        "productId": "P002",
        "productName": "USB-C Cable",
        "quantity": 3,
        "unitPrice": 9.99
    }
  ]
}
```

Listing 13.15 Simple JSON File Representing an Order

As you can see, a JSON file supports three basic building blocks:

- Key-value pairs, as in `orderId`
- Flat structures containing key-value pairs, as in `customer`
- Arrays of structures, as in `items`

JSON content is widely used for backend-to-frontend data exchange in web applications, using RESTful APIs. However, you can store nonsensitive app data on the disk in JSON files too. After all, a JSON file will be a regular text file.

Now that you're familiar with the format, let's look at how to encode a Swift type as JSON or decode JSON to a Swift type in the following sections.

Encoding as JSON

Encoding a data type, such as a struct, as JSON is very easy and straightforward in Swift. There is a small prerequisite though: The struct (or class, or whatnot) must have implemented the `Encodable` protocol. Plus, all data types in the struct must also be compliant with the `Encodable` protocol. Otherwise, Swift wouldn't know how to encode the data at hand.

That's not a tall order at all. Many built-in Swift types are `Encodable` by default, so no manual work is needed. If you have custom data types in your struct, though, they must be `Encodable`-compliant.

The demonstration in Listing 13.16 shows a three-step approach to create a JSON file for an order with Swift, which we'll break down into digestible pieces. Note that the order's structure is totally made up; it is not an industry standard by any means.

```swift
import Foundation

// Structs
struct OrderItem: Encodable {
    let product: String
    let quantity: Int
    let price: Double
}

struct Order: Encodable {
    let id: Int
    let totalAmount: Double
    let customer: String
    let items: [OrderItem]
}

// Sample data
let order = Order(
    id: 12345,
    totalAmount: 79.97,
    customer: "John Doe",
    items: [
        OrderItem(product: "Wireless Mouse",
                  quantity: 2,
                  price: 25.0),
        OrderItem(product: "USB-C Cable",
                  quantity: 3,
                  price: 9.99)
    ])

// Encode as JSON and save to disk
let encoder = JSONEncoder()
encoder.outputFormatting = [.prettyPrinted, .sortedKeys]

do {
    let jsonData = try encoder.encode(order)
    let jsonURL = URL(fileURLWithPath: "/Users/kerem/Desktop/order.json")
    try jsonData.write(to: jsonURL)
} catch {
    print(error)
}
```

Listing 13.16 Creation of JSON File with Swift

The first part has the declarations for the structs. OrderItem represents an item in the order, while Order represents the order itself, containing some flat properties and an OrderItem array. That's common knowledge at this point, except for the fact that you add the Encodable suffix to both structs.

The Encodable protocol ensures that the struct can encode itself as an external representation, such as JSON. It expects the data type to implement a function called encode. When you declare a struct (or class) that contains only properties conforming to Encodable, Swift automatically provides a synthesized implementation of the encode function for you. In other words, you don't need to manually implement the encode function unless you need custom encoding behavior—for example, for renaming keys or skipping certain properties.

In the second part of the example, you create an order object containing sample data. There's nothing fancy here.

The third part is where the magic happens. Using the built-in JSONEncoder class, you can convert order to JSON format with a single line of code! The JSON representation of order was stored in jsonData. In the follow-up, the contents of jsonData were saved as a text file to the disk, as shown in Listing 13.17.

```
{
    "orderId": 12345,
    "orderDate": "2025-06-15",
    "totalAmount": 79.97,
    "customer": {
        "id": 1001,
        "name": "John Doe"
    },
    "items": [
        {
            "productId": "P001",
            "productName": "Wireless Mouse",
            "quantity": 2,
            "unitPrice": 25.0
        },
        {
            "productId": "P002",
            "productName": "USB-C Cable",
            "quantity": 3,
            "unitPrice": 9.99
        }
    ]
}
```

Listing 13.17 Contents of order.json

That's very easy and straightforward, right? Imagine the effort you would have to put in to create this file manually—and we didn't even address delicate details like JSON escape characters yet. It's good to know that JSONEncoder takes care of such things, leaving you with more headroom to focus on your app.

JSONEncoder Beyond Files

Although you used JSONEncoder to save JSON content to a file here, you can access the raw JSON string too! The syntax is as easy as jsonString = String(data: jsonData, encoding: .utf8). You're encouraged to try it yourself!

One line worth highlighting is encoder.outputFormatting = [.prettyPrinted, .sorted-Keys]. Here, you tell JSONEncoder to pretty-print keys in camelCase and sort keys by their names. Depending on the requirement at hand, you can add or exclude such options for formatting.

Decoding from JSON

Now that you've learned how to encode a JSON file, let's walk through how to decode one! To learn about JSON decoding, let's reuse the most recent example, with a twist. This time, you'll load the *order.json* file, parse it, and turn it into an Order instance, as shown in Listing 13.18.

```
import Foundation

// Structs
struct OrderItem: Decodable {
    let product: String
    let quantity: Int
    let price: Double
}

struct Order: Decodable {
    let id: Int
    let totalAmount: Double
    let customer: String
    let items: [OrderItem]
}

// Decode JSON from file
let jsonURL = URL(fileURLWithPath: "/Users/kerem/Desktop/order.json")
let decoder = JSONDecoder()

do {
    let jsonData = try Data(contentsOf: jsonURL)
```

```
    let order = try decoder.decode(Order.self, from: jsonData)
    print(order.items[0].product) // Wireless Mouse
} catch {
    print(error)
}
```

Listing 13.18 Loading and Parsing a JSON File

In the first part, you naturally have the definitions of the target structs, as before. The only change to notice is that you have to implement the Decodable protocol this time as the JSON data will be decoded into those structs. Like Encodable, you can take advantage of synthesized implementation with Decodable too.

Codable Protocol

For duplex JSON operations, where you need both Encodable and Decodable compliance, you can simply implement Codable, which contains both protocols.

In the second part, you take advantage of the JSONDecoder object, which kindly loads the *order.json* file and decodes that content into an order object. You see: It's as convenient as JSONEncoder!

JSONDecoder: Beyond Files

Naturally, you can use JSONDecoder to decode any JSON string, no matter what the source is. In this example, you filled jsonData from a file, but JSONDecoder doesn't care how jsonData was filled. Even if it was filled by a RESTful API call, you could decode it all the same!

When working with JSONDecoder, you need to be mindful about optional versus required keys. The code sample in Listing 13.19 will produce an error because Product has two required keys, name and price, while the JSON content has a name value only (no price).

```
import Foundation

struct Product: Decodable {
    let name: String  // Required key
    let price: Double // Required key
}

let json = """
{ "name": "Laptop" }
""".data(using: .utf8)!
```

```
do {
    let product = try JSONDecoder().decode(Product.self, from: json)
    print(product)
} catch {
    print(error) // keyNotFound error
}
```

Listing 13.19 JSON Error Due to Missing Key

If some properties, such as `price`, are optional in your JSON format, then they need to be optional in the corresponding data type too. Listing 13.20 demonstrates an enhanced version of the example, in which `Product.price` is optional, thus preventing key errors due to missing values in JSON.

```
import Foundation

struct Product: Decodable {
    let name: String    // Required key
    let price: Double?  // Optional key
}

let json = """
{ "name": "Laptop" }
""".data(using: .utf8)!

do {
    let product = try JSONDecoder().decode(Product.self, from: json)
    print(product) // Product(name: "Laptop", price: nil)
} catch {
    print(error)
}
```

Listing 13.20 Product Struct with Optional Price

That concludes our content on JSON, so let's move forward to other common data types.

13.3.2   Property List

Now we'll address the *property list*, which is a data format used extensively across Apple platforms. Although there is no technical limitation, you shouldn't expect to see widespread usage on other platforms and in other operating systems.

What Is a Property List?

A property list is a structured data format designed to store small amounts of configuration or state data. Such a file is expected to have the *.plist* extension and can be stored either as text (XML) or binary. Naturally, text format would be human-readable but larger in size, while binary format would be machine-readable only but smaller in size.

To illustrate this idea, Listing 13.21 shows an XML-based property list file. Note that the content is the same as in the earlier JSON example, which might help to highlight the format differences.

```xml
<?xml version="1.0" encoding="UTF-8"?>
<!DOCTYPE plist PUBLIC "-//Apple//DTD PLIST 1.0//EN" "http://www.apple.com/
DTDs/PropertyList-1.0.dtd">
<plist version="1.0">
<dict>
    <key>orderId</key>
    <integer>12345</integer>
    <key>orderDate</key>
    <date>2025-06-15T00:00:00Z</date>
    <key>totalAmount</key>
    <real>79.97</real>
    <key>customer</key>
    <dict>
        <key>id</key>
        <integer>1001</integer>
        <key>name</key>
        <string>John Doe</string>
    </dict>
    <key>items</key>
    <array>
        <dict>
            <key>productId</key>
            <string>P001</string>
            <key>productName</key>
            <string>Wireless Mouse</string>
            <key>quantity</key>
            <integer>2</integer>
            <key>unitPrice</key>
            <real>25.0</real>
        </dict>
        <dict>
            <key>productId</key>
            <string>P002</string>
            <key>productName</key>
            <string>USB-C Cable</string>
            <key>quantity</key>
            <integer>3</integer>
```

```
            <key>unitPrice</key>
            <real>9.99</real>
         </dict>
      </array>
</dict>
</plist>
```

Listing 13.21 Example Property List in XML Format

Although Swift saves you from worrying about building or parsing property lists with manual code, it doesn't hurt to know that the file stores a set of key-value pairs internally, as shown in the example. Like in JSON files, a property list supports three basic building blocks:

- Key-variable pairs, as in `orderId`
- Key-dictionary pairs, as in `customer`
- Key-array pairs, as in `items`

Editing a Property List

When saved as XML, you can edit a property list with any text editor you like. However, Xcode offers the benefit of editing the file in a user-friendly format as shown in Figure 13.12, preventing you from damaging the file due to manual syntax errors.

Key	Type	Value
021.plist ⟩ No Selection		
Key	Type	Value
⌄ Root	Dictionary	(5 items)
orderId	Number	12.345
orderDate	Date	2025-06-15T00:00:00Z
totalAmount	Number	79,97
⌄ customer	Dictionary	(2 items)
id	Number	1.001
name	String	John Doe
⌄ items	Array	(2 items)
⌄ Item 0	Dictionary	(4 items)
productId	String	P001
productName	String	Wireless Mouse
quantity	Number	2
unitPrice	Number	25
⟩ Item 1	Dictionary	(4 items)

Figure 13.12 Property List Editing in Xcode

Even if the property list is stored as a binary file, Xcode will still let you edit its contents as such. To give you a better idea, Figure 13.13 shows the hex dump of the same property list saved as a binary file. You couldn't possibly edit this file easily with manual work; it's more realistic to edit it in Xcode.

```
kerem3:Desktop kerem$ hexdump -C 021.plist
00000000  62 70 6c 69 73 74 30 30  d5 01 02 03 04 05 06 07  |bplist00........|
00000010  08 09 0e 57 6f 72 64 65  72 49 64 59 6f 72 64 65  |...WorderIdYorde|
00000020  72 44 61 74 65 5b 74 6f  74 61 6c 41 6d 6f 75 6e  |rDate[totalAmoun|
00000030  74 58 63 75 73 74 6f 6d  65 72 55 69 74 65 6d 73  |tXcustomerUitems|
00000040  11 30 39 33 41 c6 ff 22  40 00 00 00 23 40 53 fe  |.093A.."@...#@S.|
00000050  14 7a e1 47 ae d2 0a 0b  0c 0d 52 69 64 54 6e 61  |.z.G......RidTna|
00000060  6d 65 11 03 e9 58 4a 6f  68 6e 20 44 6f 65 a2 0f  |me...XJohn Doe..|
00000070  18 d4 10 11 12 13 14 15  16 17 5b 70 72 6f 64 75  |..........[produ|
00000080  63 74 4e 61 6d 65 58 71  75 61 6e 74 69 74 79 59  |ctNameXquantityY|
00000090  75 6e 69 74 50 72 69 63  65 59 70 72 6f 64 75 63  |unitPriceYproduc|
000000a0  74 49 64 5e 57 69 72 65  6c 65 73 73 20 4d 6f 75  |tId^Wireless Mou|
000000b0  73 65 10 02 23 40 39 00  00 00 00 00 00 54 50 30  |se..#@9......TP0|
000000c0  30 31 d4 19 1a 12 13 1b  1c 1d 1e 5b 70 72 6f 64  |01.........[prod|
000000d0  75 63 74 4e 61 6d 65 58  71 75 61 6e 74 69 74 79  |uctNameXquantity|
000000e0  5b 55 53 42 2d 43 20 43  61 62 6c 65 10 03 23 40  |[USB-C Cable..#@|
000000f0  23 fa e1 47 ae 14 7b 54  50 30 30 32 08 13 1b 25  |#..G..{TP002...%|
00000100  31 3a 40 43 4c 55 5a 5d  62 65 6e 71 7a 86 8f 99  |1:@CLUZ]benqz...|
00000110  a3 b2 b4 bd c2 cb d7 e0  ec ee f7 00 00 00 00 00  |................|
00000120  00 01 01 00 00 00 00 00  00 00 1f 00 00 00 00 00  |................|
00000130  00 00 00 00 00 00 00 00  00 00 fc                 |...........|
0000013b
```

Figure 13.13 Hex Dump of Binary Property List File

Size Comparison

Comparing the size of both files, the property list in text format is about 1 KB, while the binary file is merely 315 bytes. The size difference is merely the first difference; some other significant differences are described in Table 13.4.

Factor	Text File	Binary File
Readability	Human-readable	Machine-readable
Git-versioning	Natural	Difficult
Platform dependency	Low; anyone can parse XML	High; difficult to use outside Apple ecosystem
Editing	Possible with any text editor	Difficult outside Xcode
File size	High	Low
Parsing speed	Low	High

Table 13.4 Comparison of Text and Binary Property Lists

As general guidance, if you prioritize the highest performance and smallest file size on Apple platforms, then you can go binary; most of Apple's own system and application preferences reflect this. If you prioritize the simplicity and convenience of human-readable text files with natural Git support, though, you can go text.

JSON or Property List?

In practice, JSON is more popular for network communication, while property lists are often used to store local data on Apple platforms. But there is no rule stopping you from breaking the norm and using JSON for local storage or sending property lists over the network. Just be mindful of the industry standards and ask yourself if you have a good reason to deviate from the defaults.

On a multiplatform project, for instance, you might use JSON for local storage because it is well supported on any development environment and in any operating system. That might allow you to go forward with a single file format shared across platforms.

Now that you know the basics of property lists, let's discuss how to save and load them.

Saving a Property List

If you have a static list of properties that act like constants in your app, then you can simply prepare the file manually on Xcode and be done with it. All your app will need to do is to load it. Any change in future versions will also be handled via your manual edits.

But more often than not, you will also need to change the data in property lists. If you are storing user preferences, for example, then you'll have to save the new values when the user changes app settings.

For such cases, you should learn how to store property lists dynamically using Swift code. It's a welcome convenience that the method isn't too different from that for building a JSON file, as demonstrated in Listing 13.22.

```
import Foundation

struct UserSettings: Codable {
    let username: String
    let prefersDarkMode: Bool
    let fontSize: Double
}

let settings = UserSettings(
    username: "keremk",
    prefersDarkMode: true,
    fontSize: 14.5
)

let encoder = PropertyListEncoder()
encoder.outputFormat = .xml // or .binary

let fileURL = URL(fileURLWithPath: "/Users/kerem/Desktop/user-settings.plist")

do {
    let plistData = try encoder.encode(settings)
    try plistData.write(to: fileURL)
```

```
} catch {
    print(error)
}
```

Listing 13.22 Dynamically Creating a Property List File

In the first part, you declare a struct called `UserSettings` and create a corresponding object called `settings`; that's the exact same approach as in JSON encoding. Note that `UserSettings` needs to conform to the `Encodable` protocol, just as in JSON. In this instance, you can simply use the `Codable` protocol to cover both `Encodable` and `Decodable`.

In the second part, you create a `PropertyListEncoder` instance, which is responsible for encoding data types as a property list. This built-in class does all the heavy lifting for you. As the output format, you can pick `xml` or `binary`.

Finally, you encode your `settings` and write to a local file, which is very straightforward. The saved user settings will look as shown in Figure 13.14 when opened with Xcode.

Key	Type		Value
⊞ user-settings.plist ⟩ No Selection			
∨ Root	Dictionary	↕	(3 items)
fontSize	Number	↕	14,5
prefersDark...	Boolean	↕	YES
username	String	↕	keremk

Figure 13.14 Saved User Settings Shown in Xcode

Loading a Property List

Now, let's load that *user-settings.plist* file into the app! Naturally, after the app is restarted, you will have to load user settings again to make your app act accordingly. Listing 13.23 showcases how to do so.

```
import Foundation

struct UserSettings: Codable {
    let username: String
    let prefersDarkMode: Bool
    let fontSize: Double
}

let fileURL = URL(fileURLWithPath: "/Users/kerem/Desktop/user-settings.plist")

do {
    let plistData = try Data(contentsOf: fileURL)
    let decoder = PropertyListDecoder()
```

```
    let settings = try decoder.decode(UserSettings.self, from: plistData)

    print("\(settings.username)") // keremk
} catch {
    print(error)
}
```

Listing 13.23 Loading Property List File

First, you start by loading the contents of the file into `plistData`, which is a `Data` instance. That loads the file contents from the disk to the memory.

Next, you make use of the `PropertyListDecoder` class, which obviously decodes the values in memory and transforms them to a common Swift data type. In the case, the contents of the file are loaded into `settings`. After that point, you are free to use those settings across the app!

The subject of required versus optional keys, which was explained in the JSON section of this chapter, is also valid for property lists. If the property list looks like Figure 13.15, where `prefersDarkMode` is missing, then the last example in Listing 13.23 will fail because `UserSettings.prefersDarkMode` is not an optional.

13

Key	Type	Value
⊞ user-settings.plist ⟩ No Selection		
⌄ Root	Dictionary ⌄	(2 items)
fontSize	Number ⌄	14,5
username	String ⌄	keremk

Figure 13.15 Property List Missing prefersDarkMode

If the presence of `prefersDarkMode` in the property list is optional, then your corresponding struct should reflect that, as shown in Listing 13.24.

```
struct UserSettings: Codable {
    let username: String
    let prefersDarkMode: Bool?
    let fontSize: Double
}
```

Listing 13.24 UserSettings with prefersDarkMode as Optional

You've now learned about two common file formats that can be used to save data to the disk—and, naturally, load data from the disk as well. In many cases, using those built-in libraries is enough to store generic data.

Many other file formats can be encoded/decoded using Swift code. Some might be supported by Swift natively, while some might require a third-party library (the subject of

Chapter 15) or manual work to be unlocked. When the requirement arises, you will find a way to meet it for sure.

Now, let's leave the subject of file formats behind and move forward to file system operations.

13.4   File System Operations

Reading and writing files is an important aspect of file handling, which we covered in the previous sections. An equally important aspect is to conduct file system operations. In your applications, you might need to examine folders, move files, and perform other such tasks. In this section, we'll focus on such operations and discover how Swift empowers programmers to handle them.

13.4.1   Examining Folders

Swift has a built-in class called FileManager, which is the central stopping point for folder examination—and much more. In this section, we'll tackle some typical requirements and see how FileManager can be used to solve them.

Reading Folder Contents

Listing 13.25 demonstrates how to list all files and folders in a given folder. For this purpose, you access the contentsOfDirectory function of FileManager.

```
import Foundation

let folderURL = URL(fileURLWithPath: "/Users/kerem/Desktop")

do {
    let contents = try FileManager.default.contentsOfDirectory(atPath:
folderURL.path)
    print("Contents of folder:")
    for item in contents {
        print("- \(item)")
    }
} catch {
    print(error)
}
```

Listing 13.25 Accessing All Files and Folders in a Folder

That's very self-explanatory, right? contentsOfDirectory will return, as you'd expect, the contents of the directory! Who would have known? A sample output of this code is shown in Figure 13.16.

```
Contents of folder:
- .DS_Store
- .localized
- nataly.pdf
- Invoice
```

Figure 13.16 Printed Contents of Folder

You can use this feature any time you need a full list of folder contents and can even use some recursion to build file system maps if that's what you want to do.

Listing 13.26 features a slightly altered version of the example. This time, you fetch all items in the directory once again but follow up by applying a filter that will leave PDF files only. Using that template, you can quickly find files with a certain extension.

```
import Foundation

let folderURL = URL(fileURLWithPath: "/Users/kerem/Desktop")

let allItems = try FileManager.default.contentsOfDirectory(
    at: folderURL,
    includingPropertiesForKeys: nil)

let pdfFiles = allItems.filter { $0.pathExtension == "pdf" }

for file in pdfFiles {
    print("\(file.lastPathComponent)")  // nataly.pdf
}
```

Listing 13.26 Folder Contents Filtered by Extension

Differentiating Folders and Files

In a file system, you may encounter two basic components: either a folder or a file. When you get a path value from a configuration file or user input, you might need to understand if you have a folder path or a file path. The code template in Listing 13.27 has you covered for that purpose.

```
import Foundation

let folderURL = URL(fileURLWithPath: "/Users/kerem/Desktop")
var isDirectory: ObjCBool = false

let exists = FileManager.default.fileExists(
    atPath: folderURL.path,
    isDirectory: &isDirectory)

if exists {
    print(isDirectory.boolValue ? "This is a folder." : "This is a file.")
```

```
} else {
    print("Path does not exist.")
}
```

Listing 13.27 How to Distinguish a Folder from a File

Folder Properties

Finally, the code template in Listing 13.28 will help you access the properties of a folder, such as its size and last modification date. FileManager's attributesOfItem function will help you access such metadata easily.

```
import Foundation

let folderURL = URL(fileURLWithPath: "/Users/kerem/Desktop")

do {
    let attributes = try FileManager.default.attributesOfItem(atPath:
folderURL.path)
    if let size = attributes[.size] as? NSNumber {
        print("Size: \(size.intValue) bytes") // Size: 192 bytes
    }
    if let modified = attributes[.modificationDate] as? Date {
        print("Last modified: \(modified)") // Last modified: 2025-06-22
17:03:16
    }
} catch {
    print(error)
}
```

Listing 13.28 How to Access Folder Properties

Using Xcode's autocomplete feature, you can access many further attributes, which are partly shown in Figure 13.17.

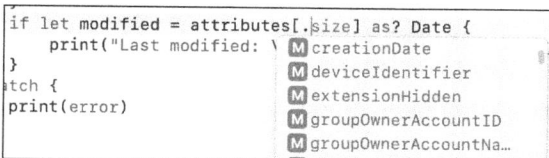

Figure 13.17 Some Further Folder Attributes

Those examples should give you a good idea of what's possible in terms of folder examination. Many operations you can perform via Finder on a Mac can be done using Swift too.

Next, let's discuss the possibilities of folder manipulation.

13.4.2   Manipulating Folders

For common folder-manipulation operations, such as creation, movement, renaming, and deletion, you can make use of the `FileManager` class once again. To see each operation in action, in this section, we'll go through them in their natural order.

Creating a Folder

Let's start with folder creation. Using the code template in Listing 13.29, you can create folders—so long as the user has the necessary authorizations for the given path. The main operation is obviously executed by `FileManager.default.createDirectory`.

```
import Foundation

let folderURL = URL(fileURLWithPath: "/Users/kerem/Desktop/NewFolder")

do {
    try FileManager.default.createDirectory(
        at: folderURL,
        withIntermediateDirectories: true)

} catch {
    print(error)
}
```

Listing 13.29 Creation of Folder

Although the code snippet is self-explanatory, the `withIntermediateDirectories` parameter deserves special attention.

When set to `true`, the function will create any missing intermediate folders before creating the final folder. For example, if the target path is */Users/kerem/Desktop/a/b/c*, then the function will create the folders *a* and *b* before creating *c* if they're missing.

If you set this parameter to `false`, then the function will throw an error if such intermediate folders are not present. This offers you two different strategies for different purposes.

An optional but welcome parameter for `createDirectory` is `attributes`, which is a dictionary expressing various folder attributes. The enhanced code in Listing 13.30 demonstrates how to create a folder with a specific creation date, modification date, and limited POSIX permissions.

```
import Foundation

let folderURL = URL(fileURLWithPath: "/Users/kerem/Desktop/NewFolder")

do {
```

```
    try FileManager.default.createDirectory(
        at: folderURL,
        withIntermediateDirectories: true,
        attributes: [.creationDate: Date(),
                     .modificationDate: Date(),
                     .posixPermissions: 0o700] )

} catch {
    print(error)
}
```

Listing 13.30 Creating Folder with Attributes

You can refer to Xcode's autocomplete feature for further attributes you can set for a folder. There are some really fun things in there!

POSIX Permissions

POSIX permissions are a fundamental system for controlling file and folder access on Unix-like operating systems. Because macOS is also based on Unix, permissions can be set that way. Using POSIX, you can define user- or group-level permissions for reading, writing, or executing.

File system permissions are subject to operating system administration knowledge and beyond the scope of this book. But once you know the necessary POSIX permission for a folder, you can apply it using Swift easily, as demonstrated in Listing 13.30.

Duplicating a Folder

To duplicate a folder with all its contents, such as files and subfolders, you need to invoke a simple FileManager function: copyItem will take care of that entire task. The demonstration in Listing 13.31 will duplicate NewFolder in NewFolderCopy.

```
import Foundation

let srcFolder = URL(fileURLWithPath: "/Users/kerem/Desktop/NewFolder")
let dstFolder = URL(fileURLWithPath: "/Users/kerem/Desktop/NewFolderCopy")

do {
    try FileManager.default.copyItem(at: srcFolder, to: dstFolder)
} catch {
    print(error)
}
```

Listing 13.31 Folder Duplication

Note that `copyItem` will throw an error if the target folder already exists. To overwrite an existing destination folder, you might need to delete the destination folder first; you'll learn how to do that shortly. For now, you can enjoy the result, as shown in Figure 13.18, where you can see how the folder was completely duplicated, including *my_file.pdf* within it.

```
KereM3:Desktop kerem$ ls -a New*
NewFolder:
.                       ..              my_file.pdf

NewFolderCopy:
.                       ..              my_file.pdf
```

Figure 13.18 Duplicated Folder, Including Contents

Moving a Folder

Moving a folder to another destination is as easy as duplicating a folder; a single function call is enough for the task. This time, you will call the intuitive `moveItem` function of the `FileManager` class. Listing 13.32 shows a demonstration of that.

```
import Foundation

let originalURL = URL(fileURLWithPath: "/Users/kerem/Desktop/NewFolder")
let destinationURL = URL(fileURLWithPath: "/Users/kerem/Desktop/ArchivedFold-
er")

do {
    try FileManager.default.moveItem(at: originalURL, to: destinationURL)
} catch {
    print(error)
}
```

Listing 13.32 Moving Folder to New Destination

In this case, because both the `NewFolder` source and the `ArchivedFolder` destination reside in `Desktop`, this operation looked like renaming a folder. But if the destination was somewhere else, you would see the entire contents moved there, assuming there is no conflict, such as missing permissions or a file being locked for editing.

Deleting a Folder

To delete a folder permanently, you can work with `FileManager` once again, invoking its `removeItem` function this time. The intuitive code sample in Listing 13.33 shows how to do it.

```
import Foundation

let targetURL = URL(fileURLWithPath: "/Users/kerem/Desktop/ArchivedFolder")
```

```
do {
    try FileManager.default.removeItem(at: targetURL)
} catch {
    print(error)
}
```

Listing 13.33 Deleting Folder Permanently

There is a catch, though: removeItem won't place the deleted items into the Trash for later recovery; instead, it irreversibly deletes the given folder and all subitems. Therefore, this function should be used with the utmost care.

For a lighter alternative, you can move the folder to Trash by invoking the trashItem function instead, preserving the option of recovery. Listing 13.34 demonstrates how; note that resultingItemURL will return the new path of the folder within Trash.

```
import Foundation

let targetURL = URL(fileURLWithPath: "/Users/kerem/Desktop/ArchivedFolder")
var resultingURL: NSURL?

do {
    try FileManager.default.trashItem(at: targetURL, resultingItemURL:
&resultingURL)
    print(resultingURL?.path)
} catch {
    print(error)
}
```

Listing 13.34 Putting Folder into Trash

That concludes our content on folder-level operations. Now, let's continue with file-level operations, which won't deviate too much from the recent examples for folders.

13.4.3   Examining Files

For file examination, Swift's FileManager will once again save the day, acting as the central point of information. In this section, we'll go over some typical tasks and how they work in Swift.

Checking File Existence

When you were working on the differentiation of folders and files, you had to use the fileExists function of FileManager, remember? As the function's name implies intuitively, it can be used to check whether a file exists in the given path or not. To keep this section self-contained, we will revisit this function in Listing 13.35.

```
import Foundation

let filePath = "/Users/kerem/Desktop/nataly.pdf"
var isDirectory: ObjCBool = false

if FileManager.default.fileExists(atPath: filePath, isDirectory: &isDirectory)
{
    if isDirectory.boolValue {
        print("This is a folder.")
    } else {
        print("This is a file.")
    }
} else {
    print("Path does not exist.")
}
```

Listing 13.35 Checking Existence of a File

As shown in the code sample, `fileExists` offers two features:

- It returns a Boolean value indicating if something exists in the given path or not.
- It determines if that something is a folder or file via the `isDirectory` parameter.

If `fileExists` returns `true` and `isDirectory` is `false`, then you can tell that a file exists in that path.

File Properties

Accessing file properties has nearly the same syntax as accessing folder properties. In both cases, you need to call the `attributesOfItem` function of the `FileManager` class. Listing 13.36 showcases a demonstration, in which various file properties are accessed and printed out.

```
import Foundation

let fileURL = URL(fileURLWithPath: "/Users/kerem/Desktop/nataly.pdf")

do {
    let attributes = try FileManager.default.attributesOfItem(atPath:
fileURL.path)

    if let size = attributes[.size] as? NSNumber {
        print("Size: \(size.intValue) bytes") // Size: 159076 bytes
```

549

```
    }
    if let created = attributes[.creationDate] as? Date {
        print("Created: \(created)") // Created: 2025-06-07 06:12:32
    }
    if let modified = attributes[.modificationDate] as? Date {
        print("Modified: \(modified)") // Modified: 2025-06-07 06:12:32
    }
    if let perms = attributes[.posixPermissions] as? Int {
        print(String(format: "Permissions: %04o", perms)) // Permissions: 0644
    }
} catch {
    print(error)
}
```

Listing 13.36 Accessing Various File Properties

For further available properties, you can check the autocomplete feature of Xcode.

Now, let's conclude this chapter with a look at file-manipulation techniques.

13.4.4   Manipulating Files

As you may have guessed already, file manipulation will also revolve around FileManager and will share a lot of common ground with folder manipulation. In this section, we'll go over examples of typical file-manipulation operations in Swift.

Creating a File

In the early sections of this chapter, you learned how to create files with text or binary content. That should be the typical approach for file creation in most cases. That said, FileManager also provides a neat function called createFile that you can use, of course, to create a file. Swift is a beautiful language in that sense; the built-in libraries are mostly intuitive to any English speaker.

Listing 13.37 demonstrates the steps to create a text file using FileManager.

```
import Foundation

let filePath = "/Users/kerem/Desktop/notes.txt"

let created = FileManager.default.createFile(
    atPath: filePath,
    contents: "Hello World!".data(using: .utf8))

print(created ? "File created." : "Failed to create file.")
```

Listing 13.37 Creation of Text File Using FileManager

The contents of *notes.txt* will be as you expect, as shown in Figure 13.19.

```
notes.txt ) No Selection
  1   Hello World!
  2
```

Figure 13.19 Contents of notes.txt

Likewise, Listing 13.38 demonstrates the steps to create a binary file using `FileManager`. Note that the code template doesn't change much; you just pass binary values to `contents` instead of text values this time.

```
import Foundation

let filePath = "/Users/kerem/Desktop/notes.bin"

let rawBinaryBytes: [UInt8] = [0x48, 0x65, 0x6C, 0x6C, 0x6F] // Hello in hex
let binaryContents = Data(rawBinaryBytes)

let created = FileManager.default.createFile(
    atPath: filePath,
    contents: binaryContents
)

print(created ? "File created." : "Failed to create file.")
```

Listing 13.38 Creation of Binary File Using FileManager

The binary content of *notes.bin* is shown in Figure 13.20. There are no surprises here either; the file merely contains the array of binary values you provided.

```
KereM3:Desktop kerem$ hexdump -C notes.bin
00000000  48 65 6c 6c 6f                                    |Hello|
00000005
```

Figure 13.20 Contents of notes.bin

While creating a file, you may also provide `attributes`, such as the creation date or permissions—just as you can during folder creation.

Duplicating a File

To duplicate an existing file, you can use `copyItem` of the `FileManager` class, just as you did for folders. The demonstration in Listing 13.39 will duplicate *notes.txt* to *notes_copy.txt* within the same folder.

```
import Foundation

let srcFile = URL(fileURLWithPath: "/Users/kerem/Desktop/notes.txt")
```

```
let dstFile = URL(fileURLWithPath: "/Users/kerem/Desktop/notes_copy.txt")

do {
    try FileManager.default.copyItem(at: srcFile, to: dstFile)
} catch {
    print(error)
}
```

Listing 13.39 File Duplication in Swift

The terminal output in Figure 13.21 shows that the duplication was a success.

```
KereM3:Desktop kerem$ ls notes*
notes.txt        notes_copy.txt
```

Figure 13.21 Original File and Its Clone

You may have to take into consideration that copyItem will throw an error if the target file already exists. If you want to overwrite the target file, you'll have to delete the target file before conducting the duplication. Listing 13.40 demonstrates how to duplicate a file with the option to overwrite.

```
import Foundation

let srcFile = URL(fileURLWithPath: "/Users/kerem/Desktop/notes.txt")
let dstFile = URL(fileURLWithPath: "/Users/kerem/Desktop/notes_copy.txt")

do {
    if FileManager.default.fileExists(atPath: dstFile.path) {
        try FileManager.default.removeItem(at: dstFile)
    }

    try FileManager.default.copyItem(at: srcFile, to: dstFile)
} catch {
    print(error)
}
```

Listing 13.40 File Duplication with Overwrite

Moving a File

As is the case with folders, file moving and renaming operations are performed using the moveItem function of the FileManager class. If the destination is a different folder, then the file will be moved; otherwise, the file will simply be renamed.

The example in Listing 13.41 will behave like renaming a file because both paths are in the same Desktop folder. Obviously, *notes.txt* will be moved to *archived_notes.txt*.

```
import Foundation

let originalURL = URL(fileURLWithPath: "/Users/kerem/Desktop/notes.txt")
let renamedURL = URL(fileURLWithPath: "/Users/kerem/Desktop/archived_
notes.txt")

do {
    try FileManager.default.moveItem(at: originalURL, to: renamedURL)
} catch {
    print(error)
}
```

Listing 13.41 Code to Move or Rename File

Deleting a File

To delete a file permanently, without the option of recovery from the trash, `FileManager.removeItem` is the function to use—as demonstrated in Listing 13.42. Note that this function will throw an error if the file to be deleted isn't present.

```
import Foundation

let fileURL = URL(fileURLWithPath: "/Users/kerem/Desktop/archived_notes.txt")

do {
    try FileManager.default.removeItem(at: fileURL)
} catch {
    print(error)
}
```

Listing 13.42 Deleting File Permanently

Deleting Sensitive Data

Note that `FileManager.removeItem` doesn't wipe file contents from the disk immediately. Instead, the file's directory entry is removed, so it disappears from Finder and becomes unavailable for regular user/API access.

However, the actual bytes on the disk may not be overwritten immediately. Until they are reused by new files, the original data may still reside in the physical storage. Such data might still be read with disk recovery tools.

If you need to delete sensitive data securely, here are some things you can do:

- Encryption is the easiest path. macOS has options for disk encryption, and computers carrying sensitive files ought to be setup to use them. Even if the computer is seized, the encrypted data will be meaningless. Decryption is not impossible, naturally, but it is very difficult.

- Before deleting the file, you may overwrite its contents with meaningless data of the same size multiple times. That will make the residual bytes on the disk useless.
- Some drives have secure erase commands to help wipe the data physically. You might trigger such commands if they're available.

For safer deletion with the option of recovery, you can use `FileManager.trashItem`. This function will place the file in the trash instead of destroying it. Check Listing 13.43 for a demonstration, remembering that the function will raise an error if the file doesn't exist.

```
import Foundation

let fileURL = URL(fileURLWithPath: "/Users/kerem/Desktop/notes_copy.txt")
var trashedURL: NSURL?

do {
    try FileManager.default.trashItem(at: fileURL, resultingItemURL: &trashedURL)
} catch {
    print(error)
}
```

Listing 13.43 Moving File to Trash

13.5   Summary

In this chapter, you have learned how to deal with files and folders using Swift. It's possible to read and write either text or binary files. Content can be read or written all at once or in manageable chunks. For common file formats, such as JSON or a property list, you can make use of libraries and let them do the heavy lifting of encoding/decoding data.

For file system tasks, such as creating folders, duplicating files, or accessing properties, Swift's `FileManager` class acts as a central point. Using the intuitive functions in that class, any operation can be handled easily.

The next chapter will address a completely different topic. By learning about concurrency in Swift, you will be able to develop high-performance multithreaded apps instead of single-thread structures with potential runtime bottlenecks.

Chapter 14

Concurrency

Swift provides powerful tools for writing asynchronous and concurrent code. This chapter introduces those tools and highlights best practices for building responsive, safe, and efficient apps.

This chapter is all about concurrency. You will learn how to write concurrent code in Swift and conduct multiple operations at once, without having them wait for each other in a single queue. Swift provides powerful tools for this. Once you learn the syntax, you won't need any manual thread management in most cases; you can comfortably let Swift do the heavy lifting and orchestration.

14.1 What Is Concurrency?

In simple programs, like most of our examples so far, it's enough to have a single flow at hand. The program would start, perform its tasks sequentially, and exit when everything is done, as sketched in Figure 14.1.

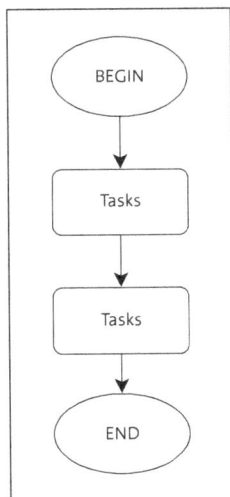

Figure 14.1 Single-Thread Program Flow

However, a comprehensive app might need to do multiple things at once. Imagine that you are developing the next big music-streaming platform. Your desktop app will surely need to stream audio, download album art, synchronize playlists, and respond to user input simultaneously—all at the same time—by multitasking. Such a flow is sketched in Figure 14.2. Note that subtasks are running simultaneously, in their own little realm. That's concurrency in a nutshell.

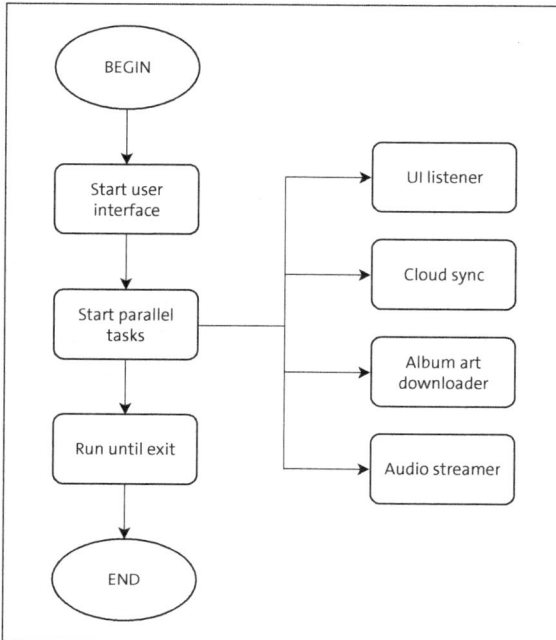

Figure 14.2 Multithreaded Program Flow

To give an official definition, *concurrency* means performing multiple tasks simultaneously and in harmony, letting them make progress together. It doesn't necessarily mean that tasks run at the exact same time, like in an Olympic race; it only means that each task has its own little lane in which to perform its tasks.

Naturally, realms of concurrent tasks are not totally isolated. They are governed by the main app flow, and they may communicate or share resources—such as variables or objects.

Concurrency may seem like a luxury in simple apps, but as the program's complexity grows, it starts to become a necessity. Table 14.1 shows sample scenarios in which concurrency shines.

Scenario	Without Concurrency	With Concurrency
Refreshing social media content	App blocked until content is downloaded	App keeps running, conducting the download in a separate thread

Table 14.1 Some Scenarios in Which Concurrency Shines

Scenario	Without Concurrency	With Concurrency
Backup service of a file manager	Once backup starts, app must wait until it is completed	Backup runs in a separate thread and can be paused or resumed
Batch image conversion	Images are converted one by one, in a row	Images can be converted simultaneously, reducing finish time

Table 14.1 Some Scenarios in Which Concurrency Shines (Cont.)

Now that you understand the concept of concurrency, let's talk about the basics of async functions and advance from there.

14.2 Async Functions

In Swift, async functions are the core building blocks or concurrency. An *async function* is a function that can pause in the middle of its execution to wait for something, without blocking the rest of the app. That *something* could be, for example, one of the following:

- A network response
- A large file read from the disk
- A timer ticking toward a deadline
- A database query

In such cases, you might want to execute the function in a separate thread lest it block the app until the operation is finished.

Multiple async functions may also be triggered in parallel; one async function might be waiting for a network response while another async function might be waiting for database query results, for example. In such a design, the main thread of the app won't be blocked by such operations and stays responsive for the user.

In this section, you will learn how to write and consume async functions.

14.2.1 Writing Async Functions

Let's start simply with the syntax of async functions, before jumping to real wait times. Listing 14.1 contains a very simple async function, which merely returns the literal `"Hello"`.

14

```
func sayHello() async -> String {
    return "Hello!"
}
```

Listing 14.1 Simple Async Function

As you have probably noticed already, the only different part is the `async` suffix in the function's signature. That's what it takes to mark a function as an async function. Simple enough, right?

As a more realistic example, Listing 14.2 features a fake weather forecast function. If this function was real, it would call a weather API, wait for the result, and return the forecast information. For the sake of simplicity, the function here is merely a simulation; it will wait for two seconds and return a mock forecast result.

```
func fetchWeatherForecast(for city: String) async -> String {
    try? await Task.sleep(nanoseconds: 2_000_000_000)
    return "Forecast for \(city): Sunny, 25°C"
}
```

Listing 14.2 Fake Async Weather Forecast Function

Although it's fake, this is a perfect example of where an async function would shine. If this code was part of a dashboard, you wouldn't want to lock the entire dashboard until the weather is determined, right? Instead, you would want to call this function in a separate thread and update the weather information whenever such information is retrieved.

Note that most function features are available for async functions too. For example, an async function may throw an error, just like a regular function. Such a function signature is shown in Listing 14.3.

```
func fetchWeatherForecast(for city: String) async throws -> String {
    return ""
}
```

Listing 14.3 Async Function Which Throws

Struct or class functions, as well as computed properties, can operate in async mode too! Listing 14.4 demonstrates a `RoomController` struct with async features, corresponding to an imaginary device controller.

```
struct RoomController {
    var currentTemparature: Double {
        get async throws { return 20 }
    }
```

```
    func wakeUp() async throws { }
    func sleep() async throws { }
}
```

Listing 14.4 Struct with Async Features

In this example, note the following:

- `currentTemparature` is an async computed property. You don't know how long it will take to measure the temperature, and you don't want to block the main app while it happens.
- `wakeUp` and `sleep` are async functions. You don't know how long those operations will take, and you don't want to block the main app while they're happening.

You get the logic of things, right? Async functions run in their own lane without blocking the main lane, letting the main app perform other tasks.

Marking a function as async is only part of the story. Now, we'll explain how to call such functions. It will be easy, but a bit different than what you're used to.

14.2.2 Calling Async Functions

First, a memory refresher for error handling: Remember that you had to add the `try` prefix in front of a function call if there was a possibility the function might throw an error. Likewise, calling an async function requires a prefix of its own: `await`. In this section, we'll go over some examples for a sound understanding of that seemingly simple keyword.

Async Calls from the Main Thread

The basic syntax of calling an async function from the main thread is demonstrated in Listing 14.5, which reuses the weather forecast function from earlier.

```
func fetchWeatherForecast(for city: String) async -> String {
    try? await Task.sleep(nanoseconds: 2_000_000_000)
    return "Forecast for \(city): Sunny, 25°C"
}

Task {
    let forecast = await fetchWeatherForecast(for: "London")
    print(forecast)
}

print("Doing other stuff...")
print("Doing even more stuff...")
```

Listing 14.5 Calling Async Function with Await

The fetchWeatherForecast async function artificially waits for two seconds to simulate a RESTful API call and returns a mock weather result, as before. But let's focus on the main flow now.

The main flow starts with a Task { ... } code block. You'll learn more about tasks in Section 14.3, but basically, this signals Swift to start a new concurrent thread and run the code inside the Task { ... } block in that new thread—without disrupting or pausing the main thread.

As the concurrent thread is started, the sample code will do two things simultaneously:

- In the main thread, the "Doing stuff" texts will be printed.
- In the concurrent thread, which was started as a Task { ... }, the following will happen:
  - fetchWeatherForecast will be called and that thread will wait until a response comes.
  - The weather forecast will be printed.

A visual representation of this flow is shown in Listing 14.5. Note how gracefully concurrent flows are executed in parallel: All you needed to do was to start a new Task { ... } with an await call. That's the power of Swift: You can really achieve a lot with minimal coding.

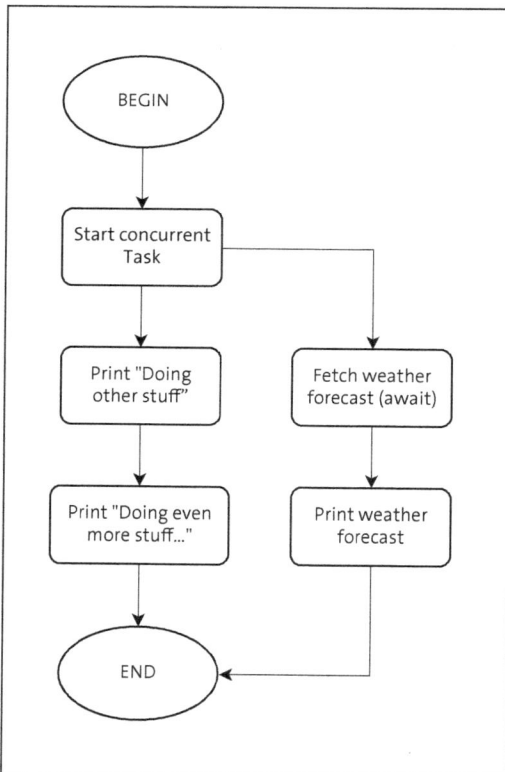

Figure 14.3 Flowchart of Concurrent Weather Call

The output of this sample is shown in Figure 14.4, which should be as expected. The main flow runs faster because there is an artificial delay of two seconds in the concurrent thread. After a delay of two seconds, the forecast is printed out, finishing the code in the Task { ... } block.

```
Doing other stuff...
Doing even more stuff...
Forecast for London: Sunny, 25°C
```

Figure 14.4 Output of Concurrent Weather Call

Now, what's so special about the await keyword? To call an async function, you must add the await prefix as we did in await fetchWeatherForecast(). This line suspends the execution of that Task { ... } until fetchWeatherForecast completes. But unlike a blocking call, Swift doesn't freeze the whole app: It allows the main flow or other concurrent tasks to continue to run.

In layman's terms, await means, "Pause the Task { ... } here, let other things happen, and resume that Task { ... } when the result is ready."

Async Calls from an Async Function

You learned that the main thread can't call async functions directly; that would block the app's flow and contradict the core purpose of concurrency. Instead, the main thread should start a Task { ... } block and make the async call from there, ensuring that it executes in its own lane.

However, an async function may call another async function directly, without necessarily starting a new Task { ... }. Let's explore why, via the sample in Listing 14.6.

```
func fetchWeatherForecast(for city: String) async -> String {
    try? await Task.sleep(nanoseconds: 2_000_000_000)
    return "Forecast for \(city): Sunny, 25°C"
}

func printForecast() async {
    print("Fetching forecast...")
    let forecast = await fetchWeatherForecast(for: "Boston")
    print(forecast)
}

Task { await printForecast() }
print("App started")
```

Listing 14.6 Async Function Call from Another Async Function

In this example, fetchWeatherForecast is the core async function you already know. On top of that, printForecast is another async function, which calls fetchWeatherForecast directly (without starting a Task { ... } block) and prints the result.

Note that printForecast is also async, which means it must be contained a concurrent thread anyway. Given that constraint, fetchWeatherForecast will also be contained in the same concurrent thread. Check Figure 14.5 for a visual representation of that logic.

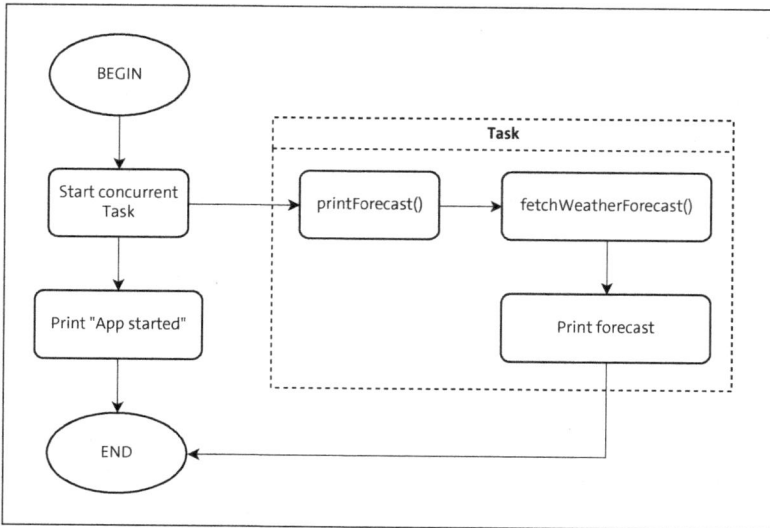

Figure 14.5 Async Call from Async Function

In this approach, printForecast is already in a Task { ... } block. Therefore, any async calls made in printForecast (or any other function) will be executed and set to await within the thread of that Task { ... } without disrupting other concurrent flows. That's why async functions may call other async functions directly; there will be a Task { ... } block to contain and govern them all anyway.

For the record, the output of Listing 14.6 is shown in Figure 14.6, which should be as expected.

```
App started
Fetching forecast...
Forecast for Boston: Sunny, 25°C
```

Figure 14.6 Async Within Async Output

So, you don't have to start a new task within printForecast to call fetchWeatherForecast. But you may, of course, if that's what you need. Check the demonstration in Listing 14.7.

```
func fetchWeatherForecast(for city: String) async -> String {
    try? await Task.sleep(nanoseconds: 2_000_000_000)
    return "Forecast for \(city): Sunny, 25°C"
```

```
}

func printForecast() async {
    print("Starting forecast request...")

    Task {
        let forecast = await fetchWeatherForecast(for: "Boston")
        print(forecast)
    }

    print("Forecast requested, waiting for results...")
}

Task { await printForecast() }
print("App started")
```

Listing 14.7 Starting New Task Within Async Function

In this case, the main thread starts a new task, which calls `await printForecast`. That function also starts a new task, which calls `await fetchWeatherForecast`. That way, you end up with three threads in total! A visual flow of this example is shown in Figure 14.7, which should clear up any potential confusion.

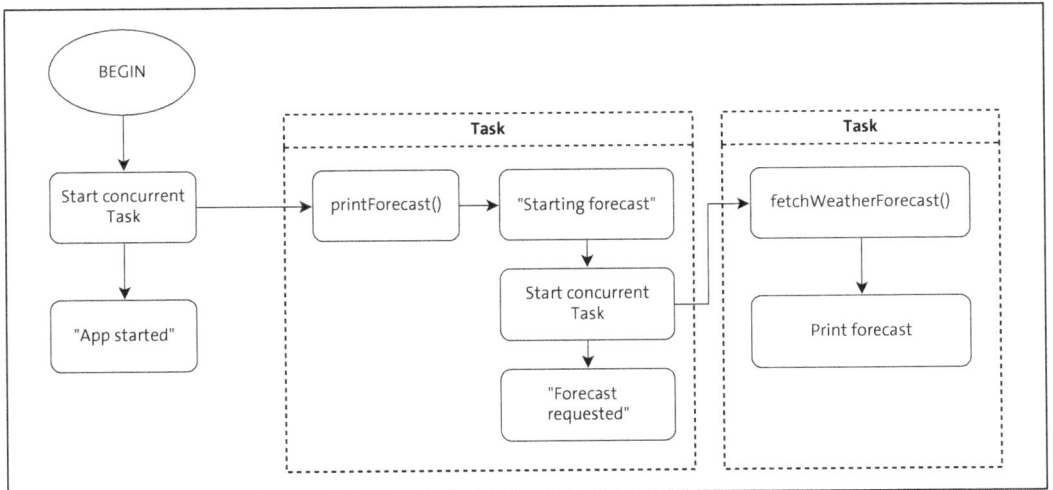

Figure 14.7 Async Function Starting New Task

The output of that example is shown in Figure 14.8. To match the app's thread flow with the output, follow these steps:

- The main flow triggers the task for `printForecast`; let's call it TASK1.
- As TASK1 is prepared, the main flow continues and prints the first line.

563

- Within TASK1's `printForecast`, the following occurs:
  - The second line is printed.
  - A new task is triggered for `fetchWeatherForecast`; let's call it TASK2.
  - As TASK2 is prepared, TASK1 continues and prints the third line.
- Within TASK2, the following occurs:
  - `fetchWeatherForecast` is called and set to `await`.
  - When the result comes, the fourth line is printed.

```
App started
Starting forecast request...
Forecast requested, waiting for results...
Forecast for Boston: Sunny, 25°C
```

Figure 14.8 Output of Async Function Starting New Task

Avoid Thread Complexity

As demonstrated in the last example, concurrency tends to build up complexity rather quickly. As your app grows and tasks/async calls interleave, the actual execution flow may become very difficult to predict and test. Occasional bugs surfacing due to unpredictable `await` times are a typical source of frustration for programmers because they like to play peekaboo with you. Race conditions, deadlocks, and resource bottlenecks are some other issues that may arise due to concurrency.

Note that those potential problems are not Swift-specific. They are fundamental, platform-independent challenges of concurrency for any programmer to consider. Although Swift does provide a robust framework to ease concurrent thread management and help build thread-safe apps, no platform can guarantee 100% safety. The programmer remains ultimately responsible for a solid architecture.

So, don't shy away from concurrency in Swift—but as with any technology, it is best to avoid overengineering and keep your concurrency designs as simple and straightforward as possible. You wouldn't want to find yourself drowning in a spaghetti bowl of concurrent threads.

Async Function Error Handling

Naturally, async functions may throw errors too—but luckily, handling those errors is no big `Task` (pun intended). The common error-handling mechanism of `do ... try ... catch` applies just as you would expect, so long as the async function is called in `Task` with `await`. Check Listing 14.8 for a clear demonstration.

```
func fetchWeatherForecast(for city: String) async throws -> String {
    try? await Task.sleep(nanoseconds: 2_000_000_000)
    return "Forecast for \(city): Sunny, 25°C"
}
```

```
Task {
    do {
        let forecast = try await fetchWeatherForecast(for: "London")
        print(forecast) // Forecast for London: Sunny, 25°C
    } catch {
        print(error)
    }
}
```

Listing 14.8 Handling Errors in Task

First, the async function call should be put into a Task { ... } block; that's the core requirement anyway. Within the task, you have a do ... try ... catch block as you would have for any other "thrower" function. That makes sense, right? There's nothing new here so far.

But you also know that an async function call requires the await prefix.

Combining both requirements, you end up calling the function with two prefixes: try await. The final form appears as let forecast = try await fetchWeatherForecast(for: "London").

It's not too complex, right? When you look at the structure, it merely combines two techniques into the same code: error handling and async call.

In the next section, you'll learn about async let, which starts multiple async functions concurrently within the same task.

14.2.3   Running Functions in Parallel

The examples so far have followed a simple approach, in which each task contained a single async function. For a second async function, we have created a second task. That architecture is sketched in Figure 14.9.

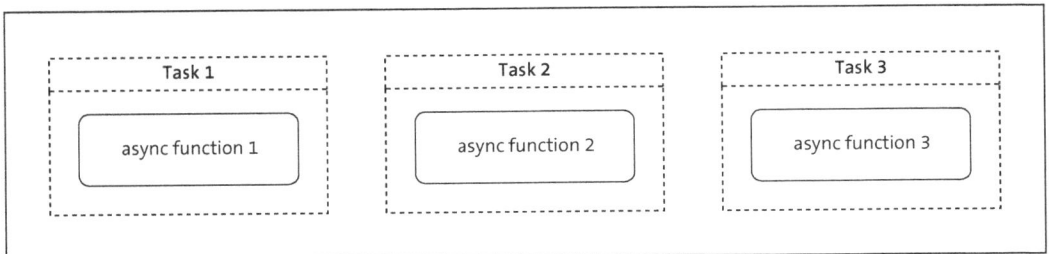

Figure 14.9 One Async Function per Task

Although this is one possible approach, it is not the only option. You also have the option to run multiple async functions concurrently, right within the same task. That architecture is sketched in Figure 14.10.

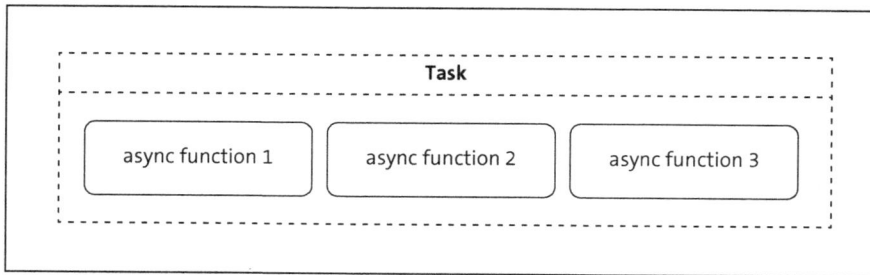

Figure 14.10 Multiple Async Functions in Same Task

Which Option Is Better?

Before learning the syntax to achieve this, you may already be wondering which option is better. We suggest postponing this question until closer to the end of the chapter; it may be too early to enforce a decision at this time. When you unlock the true power of tasks by learning about task groups and priorities, your ideal solution will probably shift. For now, let's focus on the syntax.

Now, how do you achieve the goal of starting multiple functions in parallel in the same task? Let's find the answer! Listing 14.9 contains two async functions, which are to be executed in the same task concurrently.

```
func fetchWeather() async -> String {
    try? await Task.sleep(nanoseconds: 2_000_000_000)
    return "Weather: Sunny"
}

func fetchNews() async -> String {
    try? await Task.sleep(nanoseconds: 1_000_000_000)
    return "News: Market is up"
}
```

Listing 14.9 Two Async Functions to Run Concurrently in Same Task

Those functions should be simple enough to follow through. fetchWeather simulates a server response time of two seconds and returns a mock weather forecast. Likewise, fetchNews simulates a server response time of one second and returns a mock news title.

Now, let's put them into the same task and run them concurrently. Focus on the Task { ... } block in Listing 14.10; that's where the magic happens.

```
func fetchWeather() async -> String {
    try? await Task.sleep(nanoseconds: 2_000_000_000)
    return "Weather: Sunny"
}
```

```
func fetchNews() async -> String {
    try? await Task.sleep(nanoseconds: 1_000_000_000)
    return "News: Market is up"
}

Task {
    async let weather = fetchWeather()
    async let news = fetchNews()

    print("Loading dashboard...")

    let w = await weather
    let n = await news

    print("Done!")
    print(w)
    print(n)
}
```

Listing 14.10 Running Two Async Functions Concurrently in Same Task

The task begins with two `async let` definitions:

- `async let weather = fetchWeather()` starts `fetchWeather` immediately and concurrently, setting `weather` as a handle for that concurrent execution.

- `async let news = fetchNews()` starts `fetchNews` immediately and concurrently, setting `news` as a handle for that concurrent execution.

In a nutshell, an `async let` statement tells Swift: "Start this async function right now, in its own lane, and I'll wait for the result later."

At that point, those two functions have already started and are active concurrently, as sketched in Figure 14.11. Unlike synchronous function calls, which block the execution of the main flow until completion, those functions will run in their own threads without blocking the main flow. Cool, right? Everything is in its own lane!

At this point, both functions are running concurrently. They will run in parallel without blocking the main flow of the task. But despite that advantage of concurrency, you will have to wait for the conclusion of those functions eventually. The following lines take care of that:

- `let w = await weather` will wait until `fetchWeather` is completed (if not already completed) and get the result into `w`.

- `let n = await news` will wait until `fetchNews` is completed (if not already completed) and get the result into `y`.

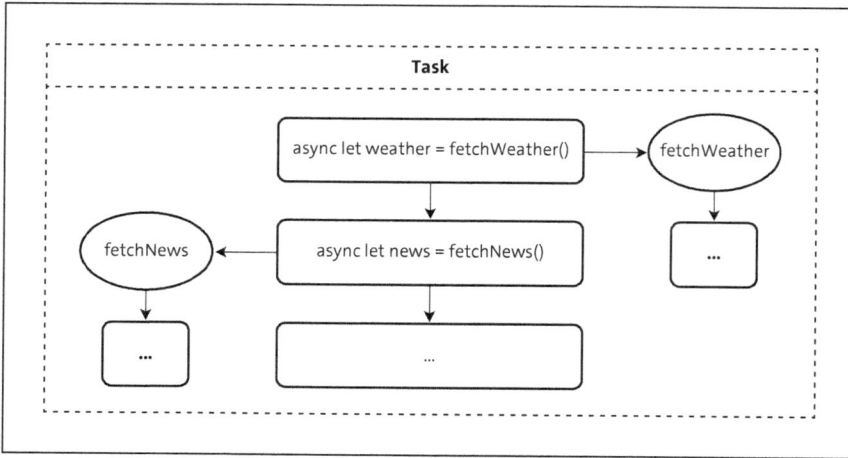

Figure 14.11 Starting Async Functions Immediately with async let

Note that `fetchWeather` and `fetchNews` were already started as soon as the initial `async let` statements were executed. The concluding `await` statements are merely landmarks to wait for the end of those functions in case they aren't completed yet. And those `await` statements will pause the current task only, not the entire app.

The entire flow is sketched in Figure 14.12, which gives the big picture of our little example.

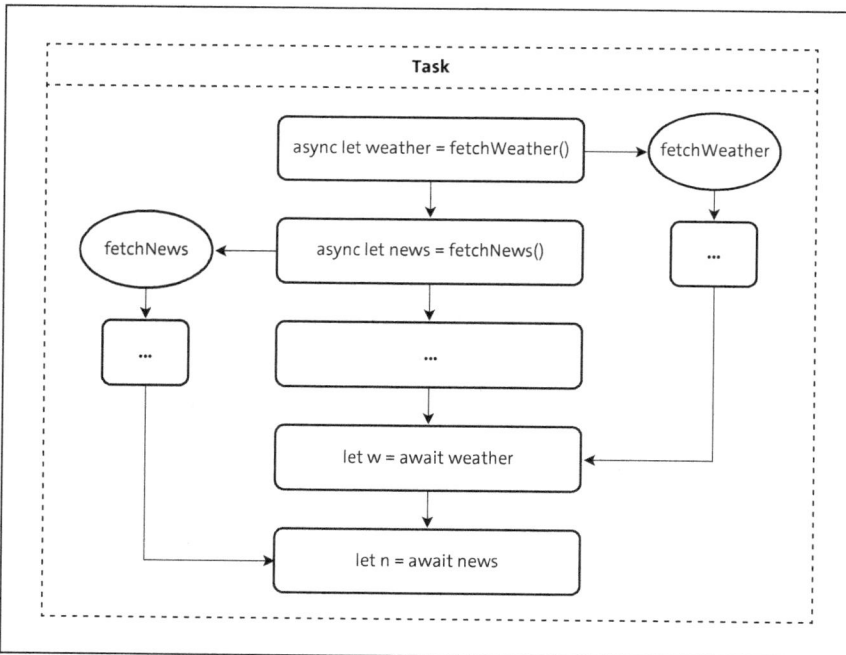

Figure 14.12 Concluding Concurrent Async Functions

For the record, the output of the example is shown in Figure 14.13.

```
Loading dashboard...
Done!
Weather: Sunny
News: Market is up
```

Figure 14.13 Output of Concurrent Weather and News Functions

At this point, you've learned about async functions running in tasks. Now, it's time to zoom into tasks themselves, which come with many options for control over concurrency.

14.3   Tasks

In our examples so far, we have already addressed tasks. We explained that you can think of Task { ... } code blocks as concurrent flow lanes in which async functions can run. And that is not wrong at all! But tasks offer much more than that. In this section, you will learn more about tasks and their useful features.

As an official definition, a *task* is a unit of asynchronous work. Tasks run in their own thread, independent from the main thread. Even if a task pauses or waits for something, it won't block the main flow or other tasks.

You can think of a task as a self-contained job that runs in the background, without blocking rest of the app. Tasks allow your code to do several things:

- Run multiple operations simultaneously, like downloading images while loading text
- Keep the main app responsive, running blocking operations like background jobs
- Perform structured, organized, parallel work

For an analogy, imagine the kitchen of a busy restaurant. Here, the chef would be your main thread. But a chef doesn't work alone, right? They have helpers! As the chef is preparing the meal, he may ask a helper to chop carrots (task 1) and another helper to toast the bread (task 2). If the chef had to do everything on their own, the customer would have to wait for a long time. Instead, three people work concurrently, shortening the meal preparation time.

Like a chef can control helpers, asking them to start, cancel, or prioritize work, tasks can be controlled in a similar way. It's possible to start or cancel tasks, as well as to prioritize them so that important things are completed first. It's even possible to group tasks together for easier orchestration in case you have to coordinate multiple tasks in tandem.

Now that you have a better idea of tasks, this section will get into the syntactical details of how to use them.

14

14.3.1   Creating Tasks

In Section 14.2, you learned how to create a basic Task { ... }, remember? In this section, we will revisit that basic method of task creation and then learn about alternative approaches.

Basic Task Creation

To keep this section self-contained, Listing 14.11 demonstrates the syntax for basic task creation.

```
Task {
    try? await Task.sleep(nanoseconds: 1_000_000_000)
    print("Task finished after 1 second")
}
```

Listing 14.11 Simple Task

This code sample will create a task and start running it immediately, like a chef's helper starting to chop carrots. The code block of the task is very simple: It waits for one second, simulating a runtime delay, then prints out a line. In the real world, the code within the task would do something more meaningful, of course—like downloading files or performing expensive calculations.

Now, that's one way of starting a task. Using this syntax, you can fire up a task and let it run in the background until it finishes without really caring about how things go and when the task concludes. It's the *fire-and-forget* type of task creation.

Task Handles

Sometimes, you need more control over a task instead of merely firing and forgetting it. You might need to wait for its results, cancel it, or inspect its state, something like that. In such cases, you can use a task handle.

Basically, a task handle is a variable that points to a fired-up task. You can use that handle to control the task—like a pet's leash! Listing 14.12 demonstrates the usage of task handles.

```
Task {
    print("Starting weather task")

    let weatherTask = Task {
        try await Task.sleep(nanoseconds: 3_000_000_000)
        print("Sunny, 32°C")
    }

    print("Weather task started, doing other work...")
```

```
    // Do other work here
    print("Other work finished, waiting for weather task...")
    try await weatherTask.value
    print("Weather task finished")
}
```

Listing 14.12 Task Handle Demonstration

In this example, you initially fire up a weather task, which simulates an API delay of three seconds and returns a mock weather value. But instead of firing and forgetting that task, you assign a handle to it using the `let weatherTask = Task` statement. From now on, `weatherTask` becomes a leash that can be used to control the task.

Once the task is started, and the `weatherTask` handle is pocketed, the code flows as necessary—doing other stuff concurrently while the weather task runs in its own lane. When the time comes to ensure that the weather task is completed, you have the `try await weatherTask.value` statement, which will wait until `weatherTask` is finished—if it's not already finished. Figure 14.14 contains a sketch of that flow for a better understanding.

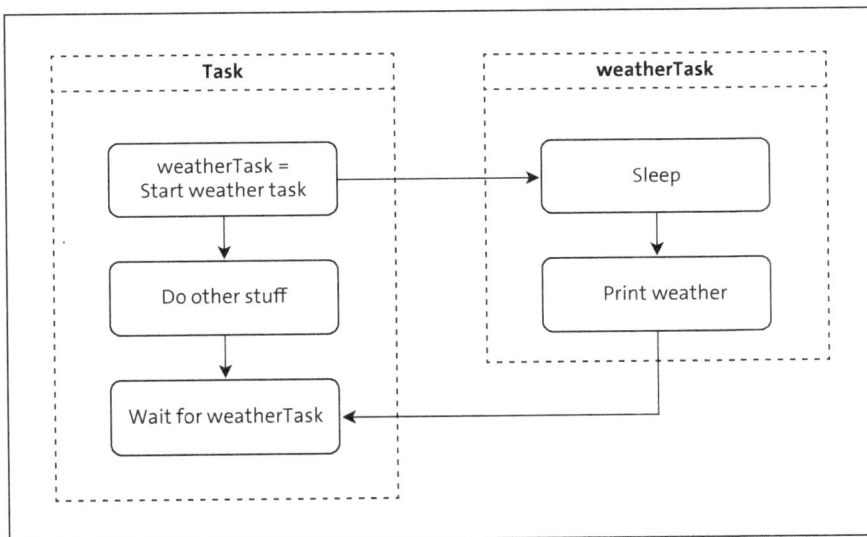

Figure 14.14 Task Handle Demonstration Flow

The output of this example is shown in Figure 14.15. Note that there will be a delay of three seconds before the weather forecast is printed out; you are encouraged to run the example and see it for yourself.

```
Starting weather task
Weather task started, doing other work...
Other work finished, waiting for weather task...
Sunny, 32°C
Weather task finished
```

Figure 14.15 Output of Weather Task Handle Example

Waiting for Results

As you can see, waiting for a task's conclusion is not too different than waiting for an async function's conclusion.

For an async function, you call the function and assign it to a *function handle*. Later, you use the await keyword to ensure or wait for the function's conclusion.

Likewise, for a task, you start the task and assign it to a *task handle*. Later, you use the await keyword to ensure or wait for the task's conclusion.

This example uses the task handle to wait for the task's conclusion. But there are more uses for task handles, such as task cancellation; you'll learn about these uses in the upcoming sections.

It's also possible for tasks to return values, and you can use task handles to fetch them. Listing 14.13 contains an enhanced version of the preceding example to demonstrate this functionality. In this case, weatherTask fetches the weather condition and temperature from async functions and returns them as a combined string.

```
func fetchCondition() async throws -> String {
    try await Task.sleep(nanoseconds: 3_000_000_000)
    return "Sunny"
}

func fetchTemparature() async throws -> Int {
    try await Task.sleep(nanoseconds: 1_000_000_000)
    return 32
}

Task {
    print("Starting weather task")

    let weatherTask = Task {
        let cond = try await fetchCondition()
        let temp = try await fetchTemparature()
        return "\(cond), \(temp)°C"
    }

    print("Weather task started, doing other work...")
```

```
    // Do other work here
    print("Other work finished, waiting for weather task...")
    let weather = try await weatherTask.value
    print(weather)
    print("Weather task finished")
}
```

Listing 14.13 Fetching Task Results via Handle

To get the result of the task, you use the `let weather = try await weatherTask.value` statement. With this statement, you instruct the computer to wait until `weatherTask` is completed (if it's not already completed) and get its result in the `weather` variable. Cool, right? You can refer to Figure 14.16 for a visual representation of that flow.

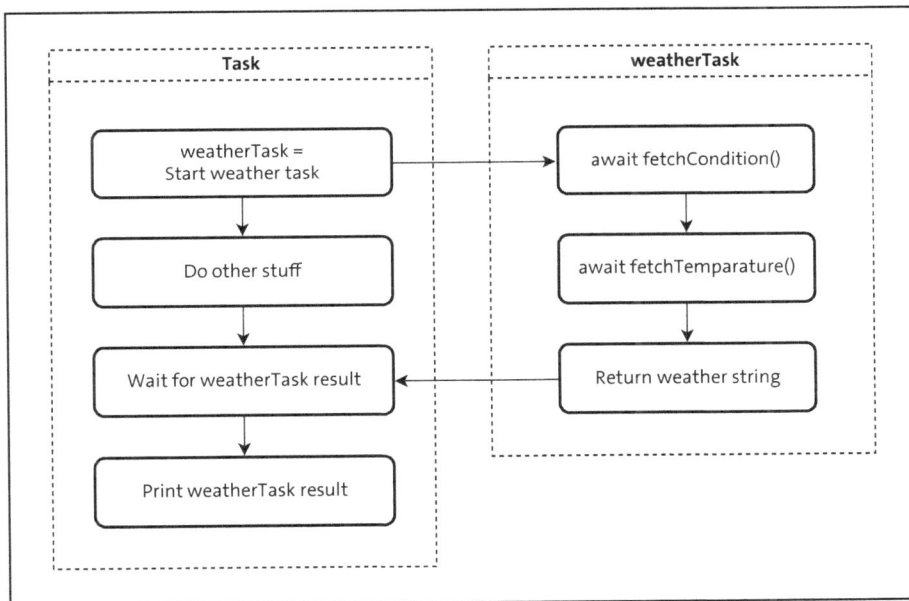

Figure 14.16 Flowchart of Task Result Example

The output of this example will be identical to the former one, as shown in Figure 14.17.

```
Starting weather task
Weather task started, doing other work...
Other work finished, waiting for weather task...
Sunny, 32°C
Weather task finished
```

Figure 14.17 Weather Task with Return Value Output

Task handles can also be consolidated as an array of tasks that need to be started up and orchestrated simultaneously. The example in Listing 14.14 will start by fetching the weather forecasts of four different cities simultaneously, wait for their conclusion, and then print out the forecast for each city.

```
func fetchForecast(for city: String) async -> String {
    try? await Task.sleep(nanoseconds: 1_000_000_000) // Simulate delay
    return "Forecast for \(city): 25°C and sunny"
}

var forecastTasks: [Task<String, Never>] = []

for city in ["London", "Paris", "Tokyo", "Sydney"] {
    let cityTask = Task { await fetchForecast(for: city) }
    forecastTasks.append(cityTask)
}

Task {
    for task in forecastTasks {
        let forecast = await task.value
        print(forecast)
    }
}
```

Listing 14.14 Fetching City Weather Forecasts Simultaneously

In this example, forecastTasks is defined as an array of task handles. In the for loop, you start a new weather task for each city, fetching the handle into cityTask. Then, cityTask is appended to forecastTasks.

Once all tasks are started, you loop through forecastTasks, waiting for the forecast of each city and printing it out. The output is shown in Figure 14.18.

```
Forecast for London: 25°C and sunny
Forecast for Paris: 25°C and sunny
Forecast for Tokyo: 25°C and sunny
Forecast for Sydney: 25°C and sunny
```

Figure 14.18 Output of Multicity Weather Forecast

Although putting task handles into an array and processing them that way is a viable option, there is a better built-in mechanism for this purpose—which is our next topic.

14.3.2   Task Groups

As its name implies, a *task group* is a container for tasks that groups relevant tasks together. Within a typical task group, you would start tasks together and then wait for them to finish, collecting their results as they come in. In this section, you will see how to use task groups and their different features.

Regular Task Groups

In the preceding example in Listing 14.14, you placed tasks into an array to process them as a batch. Task groups offer a similar built-in mechanism. Let's convert the same example to a task group solution, as shown in Listing 14.15.

```
func fetchForecast(for city: String) async -> String {
    try? await Task.sleep(nanoseconds: 1_000_000_000)
    return "Forecast for \(city): 25°C and sunny"
}


let cities = ["London", "Paris", "Tokyo", "Sydney"]

Task {
    await withTaskGroup(of: String.self) { group in
        // Add forecast task for each city
        for city in cities {
            group.addTask { await fetchForecast(for: city) }
        }

        // Collect and print results
        for await forecast in group {
            print(forecast)
        }
    }
}
```

Listing 14.15 Fetching City Weather Forecasts in Task Group

This example starts with the `await withTaskGroup` statement. This statement creates a new task group that will contain tasks running concurrently—and the group will end only when all tasks are completed.

The declaration of the task group is made with the `withTaskGroup(of: String.self)` signature. That's because the tasks will return strings (weather forecasts). If the tasks were performing an operation that doesn't return a value, then you would use `Void.self` instead.

Within the task group block, `group` acts as a handle for the task group itself. By calling `group.addTask`, you can add new tasks to the group and start them. In this case, you add the async `fetchForecast` function for various cities. Each task added to the group starts running right away; there's no need to manually start or schedule them.

Once all tasks are placed into the group, all you need to do is to wait until all tasks are completed. To ensure that, you use the `for await forecast in group { ... }` statement. That statement will await each task in the group, and as tasks are completed, the code within

14

the { ... } block is executed. Note that forecast is a made-up variable here, containing the result of the finished task.

The flowchart of this task group is illustrated in Figure 14.19. In this instance, the group has four concurrent tasks that are expected to run in parallel. The purpose of the task group will be fulfilled when all those tasks are finished or canceled.

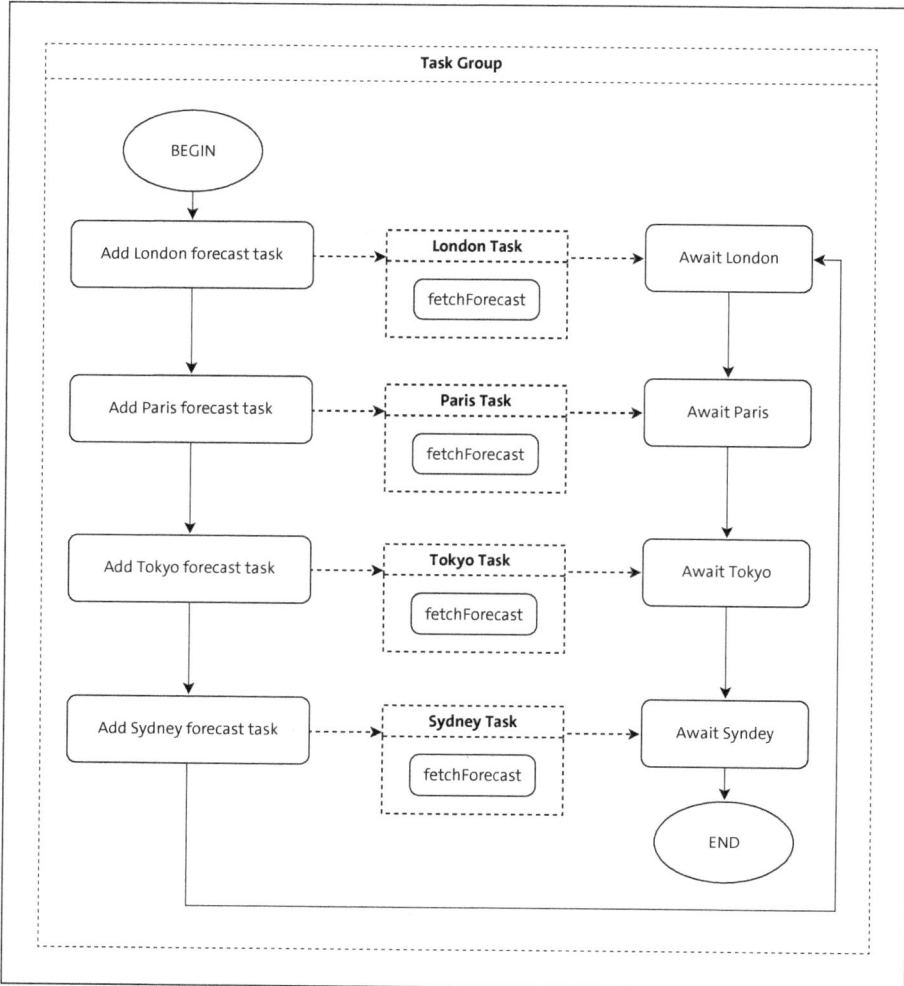

Figure 14.19 Flowchart of Task Group

The output of this task group example is shown in Figure 14.20.

```
Forecast for London: 25°C and sunny
Forecast for Paris: 25°C and sunny
Forecast for Sydney: 25°C and sunny
Forecast for Tokyo: 25°C and sunny
```

Figure 14.20 Output of Task Group

Throwing Task Groups

Another kind of task group is a *throwing task group*. This type is used to contain tasks that might throw errors. The syntax required to create a throwing task group is nearly the same, with only minor differences, as shown in Listing 14.16.

```
func fetchForecast(for city: String) async throws -> String {
    try? await Task.sleep(nanoseconds: 1_000_000_000)
    return "Forecast for \(city): 25°C and sunny"
}

let cities = ["London", "Paris", "Tokyo", "Sydney"]

Task {
    do {
        try await withThrowingTaskGroup(of: String.self) { group in
            // Add forecast task for each city
            for city in cities {
                group.addTask { try await fetchForecast(for: city) }
            }

            // Collect and print results
            for try await forecast in group {
                print(forecast)
            }
        }
    } catch {
        print(error)
    }
}
```

Listing 14.16 Usage of withThrowingTaskGroup

Beyond the naturally expected error handling via do … try … catch, the only significant difference is in the declaration of the task group. Instead of await withTaskGroup, you make the await withThrowingTaskGroup declaration, indicating that those tasks may throw errors.

If any task in a throwing task group throws an error, the remaining tasks are automatically canceled.

Further Task Group Operations

In the preceding examples, you have seen how to add tasks to a group and wait for their completion—and that's the basic task group behavior. However, task groups offer other operations too, as described in Table 14.2.

Task Group Operation	Purpose
addTask { ... }	Adds a new child task to the group
await next()	Waits for the next completed result of a child task
await waitForAll()	Waits for all child tasks of the group to complete
cancelAll()	Cancels all remaining child tasks of the group
isEmpty	Checks if the group has any remaining tasks
isCancelled	Checks if the group has been canceled

Table 14.2 Task Group Operations

Now, let's see examples of some of those operations. Listing 14.17 contains a demonstration of await next. Every time a child task is finished, this code block will be executed for the finished task.

```
Task {
    await withTaskGroup(of: String.self) { group in
        group.addTask {
            try? await Task.sleep(nanoseconds: 4_000_000_000)
            return "Task 1 finished"
        }

        group.addTask {
            try? await Task.sleep(nanoseconds: 2_000_000_000)
            return "Task 2 finished"
        }

        while let result = await group.next() {
            print(result)
        }
    }
}
```

Listing 14.17 Demonstration of Await Next

The output of that example is shown in Figure 14.21. Although Task 1 was started first, Task 2 finished earlier, triggering await next before Task 1. For each finished task, await next will be triggered once.

```
Task 2 finished
Task 1 finished
```

Figure 14.21 Output of await next Example

If you simply want to wait for all child tasks to finish without processing anything they return, then you can call await waitForAll, as demonstrated in Listing 14.18. Note that this is only useful when you don't care about the results; otherwise, you'd be better off using for await or await group.next().

```
Task {
    await withTaskGroup(of: String.self) { group in
        group.addTask {
            try? await Task.sleep(nanoseconds: 4_000_000_000)
            return "Task 1 finished"
        }

        group.addTask {
            try? await Task.sleep(nanoseconds: 2_000_000_000)
            return "Task 2 finished"
        }

        await group.waitForAll()
    }
}
```

Listing 14.18 Demonstration of await waitForAll

If a need to cancel all remaining tasks in the group, you can simply invoke group.cancelAll(), as demonstrated in Listing 14.19.

```
Task {
    await withTaskGroup(of: String.self) { group in
        group.addTask {
            try? await Task.sleep(nanoseconds: 4_000_000_000)
            return "Task 1 finished"
        }

        group.addTask {
            try? await Task.sleep(nanoseconds: 2_000_000_000)
            return "Task 2 finished"
        }

        // Following some condition;
        group.cancelAll()
    }
}
```

Listing 14.19 Canceling All Remaining Tasks in Group

Finally, the isEmpty and isCancelled group properties are demonstrated in Listing 14.20, which should be self-explanatory.

```
Task {
    await withTaskGroup(of: String.self) { group in
        group.addTask {
            try? await Task.sleep(nanoseconds: 4_000_000_000)
            return "Task 1 finished"
        }

        group.addTask {
            try? await Task.sleep(nanoseconds: 2_000_000_000)
            return "Task 2 finished"
        }

        print(group.isEmpty)        // false
        print(group.isCancelled)    // false

        group.cancelAll()
        print(group.isCancelled)    // true
    }
}
```

Listing 14.20 isEmpty and isCancelled Properties

When To Use Task Groups

Now that you've learned about task groups, you have multiple options at hand to run multiple concurrent code blocks. Table 14.3 provides a general comparison of those options.

Method	When to Use	Advantages	Disadvantages
Multiple functions with async let	When you need to perform a fixed number of simple independent operations	Simple, readable, automatic cancellation	Not suitable for dynamic number of tasks, limited control over progress or cancellation
Multiple tasks with task handles	When you have a fixed number of tasks and need more control	Fine-grained control for awaiting completion, cancellation, and priorities	Manual management for awaiting or cancellation, more boilerplate code

Table 14.3 Comparison of Multitask Concurrency Methods

Method	When to Use	Advantages	Disadvantages
Creating a task group	When you need to perform a dynamic, unknown number of operations that need to complete together	Dynamic number of tasks, ensuring all tasks finish, bulk cancellation of tasks	Less control over individual tasks, even more boiler-plate code

Table 14.3 Comparison of Multitask Concurrency Methods (Cont.)

Which Method?

For the simplest option with a fixed number of concurrent operations, you can choose async let. If you have a handful of related concurrent tasks to run in parallel and eventually finish together, then you can use a task group. Direct task creation can be used for scenarios where you need explicit control over a task's priority or cancellation.

Now that you've learned how to group tasks, let's continue with a look at prioritizing tasks.

14.3.3   Task Priorities

Concurrency enables you to run multiple tasks simultaneously. However, some of concurrent tasks may be more important than others and need to be executed with a higher priority, while others may be trivial and can be run whenever possible. You can't expect Swift to make that differentiation automatically, though; programmer guidance is needed. In this section, you'll learn how to set priorities for tasks.

Why Priorities Are Needed

The need for priority comes from the fact that system resources are finite. Each computer has limited CPU power, memory, I/O bandwidth, and so on. Even if you have a supercomputer, the operating system may assign limited resources to your app, leaving room for other processes.

If your concurrent tasks collectively exhaust the available resources, then Swift needs to choose between them. Inevitably, some tasks will be paused temporarily to leave room for others.

That's where priorities step in. In such a case, Swift will attempt to pause or slow down tasks with lower priority while running and completing tasks with higher priority. That makes sense, right? But this is not a guarantee that important tasks are finished as fast as you hope; it's just a priority list that Swift tries its best to abide by. The final decision might still be made by the operating system.

It's also important to understand that priorities are relative, not absolute. If you mark all tasks as important, then Swift will behave as if you marked them all as trivial because all the tasks are equal now. For priorities to make sense, there needs to be a hierarchy in which some are important and others are not.

How To Set Task Priorities

Setting a task's priority starts with picking the appropriate priority level first. Swift's priority levels are defined in the TaskPriority enumeration, as explained in Table 14.4.

Priority	When to Use
userInitiated	A task triggered by the user that needs a quick response. This is the default priority and is typically used for tasks that the user will observe, like UI updates, gesture responses, or app launch tasks.
utility	Ongoing tasks that are important for the app but don't require user attention. Downloads or data synchronization are typical examples.
background	Typical backgrounds tasks that are not time sensitive, like backups or database maintenance.
high	This is a similar priority level as userInitiated. It's used for truly critical real-time operations, like audio processing in a music app or game logic.
low	This is a similar priority level as background. It's used for tasks that can be deferred indefinitely without impacting the app, like cleaning up a local cache.

Table 14.4 Priority Levels for Tasks

Once you decide on the priority of a task, assigning it to that task is almost trivial. As demonstrated in Listing 14.21, all you need to do is to mention the priority while creating the task.

```
let task = Task(priority: .userInitiated) {
    // Do important stuff
}
```

Listing 14.21 Setting Priority of a Task

You see? Just like that, you can determine a task's priority. If the task was within a task group, the syntax to determine the priority would nearly be the same, as demonstrated in Listing 14.22. Note that different tasks in the task group may have different priorities.

```
Task {
    await withTaskGroup(of: String.self) { group in
        group.addTask(priority: .background) {
            // Do less important stuff
```

```
        return "Task 1 finished"
    }

    group.addTask(priority: .utility) {
        // Do utility staff
        return "Task 2 finished"
    }

    await group.waitForAll()
    }
}
```

Listing 14.22 Setting Priorities of Tasks in Task Group

If a parent task contains a child task, then the child task will inherit the parent's priority unless it's explicitly overridden. In Listing 14.23, the first child task will inherit the user-Initiated priority, while the second child task will have the low priority.

```
Task(priority: .userInitiated) {
    Task {
        // Inherits .userInitiated
    }

    Task(priority: .low) {
        // Overrides with .low
    }
}
```

Listing 14.23 Child Task Priorities

Because we're talking about task priorities, we should also discuss the Task.yield() function. Using that function, a task can voluntarily suspend itself during a long-running operation, giving other tasks a chance to proceed for a while before execution of the yielding task resumes.

To understand that process better, check out Listing 14.24, in which yield wasn't used at all. Two concurrent tasks are fired, and both will run as fast as possible.

```
import Foundation

Task {
    for i in 1...5 {
        print("Task 1.\(i)")
    }
}
```

```
Task {
    for i in 1...5 {
        print("Task 2.\(i)")
        try? await Task.sleep(nanoseconds: 1000)
    }
}
```

Listing 14.24 Two Concurrent Tasks Without Yield

The output of this example is shown in Figure 14.22. Note that the first task ran without any breaks, finishing completely before the second task even had the chance to debut.

```
Task 1.1
Task 1.2
Task 1.3
Task 1.4
Task 1.5
Task 2.1
Task 2.2
Task 2.3
Task 2.4
Task 2.5
```

Figure 14.22 Output of Two Concurrent Tasks Without Yield

Now let's use yield to make the first task yield, as shown in Listing 14.25. Note that the first task yields after each iteration, giving some breathing space for other concurrent tasks.

```
import Foundation

Task {
    for i in 1...5 {
        print("Task 1.\(i)")
        await Task.yield()
    }
}

Task {
    for i in 1...5 {
        print("Task 2.\(i)")
        try? await Task.sleep(nanoseconds: 1000)
    }
}
```

Listing 14.25 Two Concurrent Tasks with Yield

The output of this version is shown in Figure 14.23. Note that the task outputs are mixed this time, proving that yielding the first task gave some runtime space to the second task, allowing them to run fairly without one dominating the resources.

```
Task 1.1
Task 2.1
Task 1.2
Task 2.2
Task 1.3
Task 2.3
Task 1.4
Task 1.5
Task 2.4
Task 2.5
```

Figure 14.23 Output of Two Concurrent Tasks with Yield

Setting priorities for concurrent tasks is important: It gives Swift a chance to allocate resources accurately, favoring important tasks over trivial ones. But an equally important mechanism is cancellation of tasks: No matter how important a task is, the need to cancel it may arise. We'll inspect that topic in the next section.

14.3.4   Task Cancellation

Although task cancellation was briefly addressed with task groups, it deserves its own section, so here we go! In real-world apps, not all tasks need to finish. Sometimes, the requirement arises to cancel a working task, making it stop whatever it's doing and end its lifecycle gracefully. Some possible scenarios for task cancellation are shown in Table 14.5.

App	Action	Cancelable Task
Video editor	User closes their video project	Background render tasks
Image editor	User closes an image	Autosave task
Game	Main character dies	Animation and sound tasks
Cloud file browser	User deletes a cloud file	Sync task for that file
Music player	User skips to the next song	Streaming task for that song

Table 14.5 Some Reasons to Cancel Tasks

For such cases, Swift offers a built-in mechanism to safely cancel tasks. Unlike killing a thread, Swift tasks aren't forcefully stopped. Instead, tasks are politely asked to cancel themselves. When a task notices that it needs to cancel, it has the chance to conduct any necessary cleanup steps and stop gracefully. This is called *cooperative cancellation*: Tasks must agree to stop and may decide how/when to stop.

Therefore, a task might never listen to cancellation requests or might hear them only to ignore them. When coding the task, the programmer is responsible for adding the cancellation logic if necessary; otherwise, the task will be uncancelable.

Now, let's cancel some tasks! Listing 14.26 features an imaginary virus scan task, which is canceled after running for a while.

```
let virusScan = Task {
    print("Virus scan started")

    while true {
        if Task.isCancelled {
            print("Cancelling scan...")
            print("- Cleaning up temporary files")
            print("- Scan was cancelled")
            break
        }

        print("Scanning next file")
        try? await Task.sleep(nanoseconds: 500_000_000)
    }
}

Task {
    try? await Task.sleep(nanoseconds: 1_500_000_000)
    virusScan.cancel()
}
```

Listing 14.26 Canceling a Virus Scan Task

First, let's focus on the virusScan task. As the task proceeds through files, it checks for a cancellation request with if Task.isCancelled before scanning each file. That property returns true only if a cancellation was requested. By checking that property occasionally, the task gains the ability to stop in the middle of the process.

If the task's code didn't contain an if Task.isCancelled { ... } block, then the task would never cancel—even if a cancellation was requested. The task is responsible for listening for requests, and it has the freedom to decide whether to honor requests or ignore them. But in this case, the task complies with the cancellation request, cleaning up temporary files and leaving the scan loop.

The client task merely waits for a short while and then asks the virusScan task to cancel, simulating a user canceling the scan via the UI.

The output of this example is shown in Figure 14.24, which looks as expected. After virusScan task starts and scans a couple of files, it notices the cancellation request, taking the necessary cleanup steps and exiting the scan loop.

```
Virus scan started
Scanning next file
Scanning next file
Scanning next file
Cancelling scan...
- Cleaning up temporary files
- Scan was cancelled
```

Figure 14.24 Output of Virus Scan Cancellation

Checking `Task.isCancelled` is one option for task cancellation. An alternative is to call `Task.checkCancellation()`, which will deliberately throw an error if the task is canceled. That alternative might be a good option if you already have an error-handling mechanism, in which case task cancellation would be just another reason to end the task.

Listing 14.27 demonstrates the usage of `Task.checkCancellation()`. In this example, if any error occurs during the virus scan, including a task cancellation, the task will cleanup and exit.

```
let virusScan = Task {
    print("Virus scan started")

    while true {
        do {
            try Task.checkCancellation()  // throws if cancelled
            print("Scanning next file")
            try? await Task.sleep(nanoseconds: 500_000_000)
        } catch {
            print("Cancelling scan...")
            print("- Cleaning up temporary files")
            print("- Scan was cancelled")
            break
        }
    }
}

Task {
    try? await Task.sleep(nanoseconds: 1_500_000_000)
    virusScan.cancel()
}
```

Listing 14.27 Cancelling a Task with checkCancellation

At this point, you know how to start, group, and cancel concurrent tasks. Now, let's discuss a special genre of tasks, which focuses on streamed data.

14.4   Async Streams

In Swift, Sequence types offer a list of values, which you can step through one at a time. That's how for-in loops work with arrays, sets, dictionaries, and many other data types.

Adding concurrency to that leads to the AsyncStream data type. In Swift, *async streams* let you step through a stream of values over time, one item at a time, asynchronously. Instead of accessing values directly, the client needs to use await to receive values as they become available.

Some typical scenarios in which to use async streams are described in Table 14.6, which should give you a sound idea about their value.

Scenario	Benefit
Stock updates	Instead of repeatedly calling an API, you can open a single connection and receive a stream of price changes as they happen.
Chat messages	You can receive new messages in real time as they are sent to a conversation.
Car sensor data	You can receive continuous updates from a car sensor, such as speed, altitude, coordinates, and the like.
Fitness tracking	Heart rate, step count, and similar biometric data can be received from a wearable device continuously.
Large remote files	Instead of downloading a file all at once, you can provide partial file data as the download is running.

Table 14.6 Some Scenarios for Async Sequences

In the upcoming sections, you will learn how to start and cancel async streams and how to handle errors that might occur.

14.4.1   Starting an Async Stream

To understand the syntax of async streams, Listing 14.28 contains a simple example. Instead of a complex stream source, such as a sensor, we will work here with the numbers array for the sake of simplicity.

```
let numbers = [1, 2, 3]

let numberStream = AsyncStream<Int> { continuation in
    for number in numbers {
        continuation.yield(number)
    }
    continuation.finish()
}
```

```
Task {
    for await number in numberStream {
        print("Received:", number)
    }
}
```

Listing 14.28 Async Sequence Implementation Example

In this example, `numberStream` stands as the async stream. This object, like any `Async-Stream` instance, is expected to return streamed sequential values asynchronously. The `numberStream = AsyncStream<Int>` definition indicates that `numberStream` will return `Int` values.

In the following code block, `continuation.yield` is called every time a new value becomes available:

- If this was a network stream, you would call `continuation.yield` every time a new package is received.
- If this was a sensor, you would call `continuation.yield` for every new sensor value.
- And so on.

In this simple case, you merely loop through `numbers` and call `continuation.yield` as though values were being received from a real source. Finally, `continuation.finish` will finish the sequence when necessary.

In the task, `for await number in numberStream` is the accurate syntax to receive async sequence values. Because the numbers are returned asynchronously, you naturally need to use the `await` mechanism, as in any concurrent operation.

The output of this example is shown in Figure 14.25. This was a simulation returning pre-defined numbers, and the simple output reflects that.

```
Received: 1
Received: 2
Received: 3
```

Figure 14.25 Number-Based Async Sequence Output

As a second exercise, Listing 14.29 features an async stream simulation in which a sequence of new members is processed. As new people register for this imaginary social media platform, their IDs are returned by `newMemberStream`.

```
import Foundation

struct Member { let id: String }

let newMemberStream = AsyncStream<Member> { continuation in
    for _ in 1...5 {
```

```
            continuation.yield(Member(id: UUID().uuidString))
        }
        continuation.finish()
    }

Task {
    for await newMember in newMemberStream {
        print("New member ID:", newMember.id)
    }
}
```

Listing 14.29 Async Stream Following New Members

In a real-world app, newMemberStream would have to listen to the member database, but in this case, it merely returns five random members as a stream. As you can see, the skeleton structure of the code is the same; the only difference lies in returning a Member instead of an Int. The output of this example is shown in Figure 14.26.

```
New member ID:  E788094A-7D91-4ECD-9ACA-75B1E4DE8831
New member ID:  61001123-D572-48BF-925F-33A791430EF0
New member ID:  99501C88-75B0-44D4-BAD0-3DAF295BC592
New member ID:  EBD31B73-639D-44C5-9B01-8FD3579F9F9D
New member ID:  B0D2A8F2-1294-44C1-9F4A-F2F992F4C38E
```

Figure 14.26 Output of New Member Stream

14.4.2   Canceling an Async Stream

Now you know how to start an async stream and read values. But in some cases, async streams might run indefinitely because values keep coming in. If you are listening to a network port, sensor, or video camera, for example, you may do so indefinitely because the stream might never end on its own.

In such cases, you can cancel the async stream manually when it's no longer needed. Listing 14.30 contains an enhanced version of the preceding example, containing the cancellation now as well.

```
import Foundation

struct Member { let id: String }

let newMemberStream = AsyncStream<Member> { continuation in
    continuation.onTermination = { @Sendable reason in
        print("Stream terminated: \(reason)")
    }

    Task {
```

```
        while true {
            try? await Task.sleep(nanoseconds: 1_000_000_000)
            continuation.yield(Member(id: UUID().uuidString))
        }
    }
}

let streamTask = Task {
    for await newMember in newMemberStream {
        print("New member ID:", newMember.id)
    }
}

Task {
    try? await Task.sleep(nanoseconds: 2_500_000_000)
    streamTask.cancel()
}
```

Listing 14.30 Cancellation of Async Stream

In this case, newMemberStream became a new code block: continuation.onTermination { … }. This is the spot where you would do any cleanup tied to the cancellation, such as closing the source network port or disconnecting from a sensor. In this example, you simply print out a cancellation message.

You transform the body of newMemberStream here to an infinite loop, which returns mock members. An infinite loop is not a requirement for stream cancellation; a finite stream can be canceled just the same. But an infinite loop makes it more dramatic!

In the main flow, after letting the task run for a while, you cancel the stream with the streamTask.cancel() expression. That should be self-explanatory, right? In a real app, the stream would probably be canceled upon user request or due to a state change making it so that the stream is no longer needed.

The output of the async stream cancellation is shown in Figure 14.27, which has no surprises.

```
New member ID: BCEF8F26-DBB5-44E8-8121-6F74F91A8F88
New member ID: 61213F84-979D-4D01-B916-7FFBBA3F828B
Stream terminated: cancelled
```

Figure 14.27 Async Stream Cancellation Output

14.4.3   Handling Async Stream Errors

As you learned in Chapter 12, errors are a natural part of programming—and async streams are not immune to errors either. Network connections may drop, gadget batteries may run out, and many similar unforeseeable situations may occur. In the face of such fatal errors, you need to end the async stream gracefully, notifying the client about the error.

An async stream that might throw such an error must implement `AsyncThrowingStream`. This implementation tells Swift that the stream is risky and that any clients must use `try await` to access it.

Listing 14.31 showcases an enhanced version of the `newMemberStream` example, containing such an error-handling mechanism.

```
import Foundation

struct Member { let id: String }
enum StreamError: Error { case serverDisconnected }

let newMemberStream = AsyncThrowingStream<Member, Error> { continuation in
    Task {
        for i in 1...5 {
            if i == 3 {
                continuation.finish(throwing: StreamError.serverDisconnected)
                return
            }
            continuation.yield(Member(id: UUID().uuidString))
        }
        continuation.finish()
    }
}

Task {
    do {
        for try await newMember in newMemberStream {
            print("New member ID:", newMember.id)
        }
    } catch {
        print("Stream ended with error:", error)
    }
}
```

Listing 14.31 Async Stream Error-Handling Demonstration

In this version, `newMemberStream` simulates a server disconnection error on the third member. Upon the error, instead of calling the concluding `continuation.finish()` function in its vanilla form, it is called as `continuation.finish(throwing: StreamError.serverDisconnected)`. This variant triggers a graceful conclusion of the stream, ultimately throwing the provided error.

In the client code, the async stream had to be called in a do-catch block. Because AsyncThrowingStream might throw an error, appropriate error-handling blocks are needed. Except the required do-try-catch decoration, the client code is the same.

The output of this sample is shown in Figure 14.28 and doesn't contain any surprises.

```
New member ID: D8217064-6881-436E-853B-CB28AB023B8E
New member ID: 6FA6D53E-E06F-41D4-8154-A57AF0D8C4D7
Stream ended with error: serverDisconnected
```

Figure 14.28 Async Stream Error Output

Next, we'll look at what happens when multiple tasks access a shared state and how to prevent typical pitfalls.

14.5   Shared State Safety

Running multiple tasks concurrently is all fun and games until concurrent tasks need to access the same variables. What if multiple tasks need to read and modify the same piece of data? How can you ensure intertask value consistency in such cases?

That's what this section is going to be about. If two tasks try to read and write to the same data simultaneously, it leads to a *data race*, one of the most common problems in concurrent programming. Data races are also notoriously hard to spot and debug as they occur when concurrent tasks are active. You can't always review the problem in slow motion using debug tools.

Listing 14.32 demonstrates a simple case of data races, in which two concurrent tasks attempt to access counter simultaneously.

```
var counter = 0
Task { counter += 1 }
Task { counter += 1 }
```

Listing 14.32 Simple Data Race

Note that counter += 1 is actually a sequence of operations:

- The value of counter is fetched.
- 1 is added to that value.
- The result is written back to counter.

If you're hoping to see 2 as the final value of counter, you might be sorely disappointed. If both tasks run at the same time with an unfortunate overlap, then they might both read the same initial counter value (0) and write back the same result (1). Because you don't have exact control over when tasks begin and end, such data race situations might occur.

In traditional concurrent programming, you could use locks to prevent data races. In such an approach, when `counter` is being accessed by a task, other tasks requesting access would wait for their turn. Thus, simultaneous access would be prevented. But locks come with their own issues, such as *deadlocks* caused by conflicting locks on multiple variables.

As a modern language that implements lessons from the past, Swift offers powerful tools to eliminate data races. In this section, you will discover those tools and see how they help with safe data access in concurrency.

14.5.1   Task-Local Values

Task-local values are special types of shared variables, designed specifically for Swift's concurrency model. They aim to solve a common concurrency problem, typically for read-only global variables: data races caused by a shared global state. Consider a case like the one illustrated in Figure 14.29, where a global variable is accessed by the main thread and two tasks.

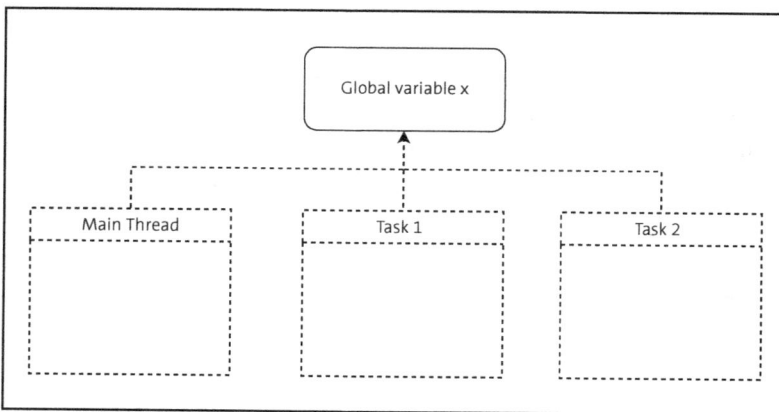

Figure 14.29 Simultaneous Access to Global Variable

You may have started Task 1 when global variable x was 5, assuming that the code in Task 1 would run accordingly. However, a sneaky line of code in the concurrent Task 2 may change x to 6, ruining the flow in Task 1.

Global variables are fragile like that; their power in single-thread apps may turn against you as a weakness in concurrency. If any thread changes the value of a global variable, then all other threads will also see and access that changed value.

So how can you have a global variable specific to Task 1 that can't be seen or changed by Task 2?

That's the exact purpose of task-local variables. They look like global variables, are defined like global variables (except for the `@TaskLocal` prefix), but their global nature is limited to a task. As shown in Figure 14.30, each task will get its own copy of global variable x, which can neither be accessed nor be modified by other tasks.

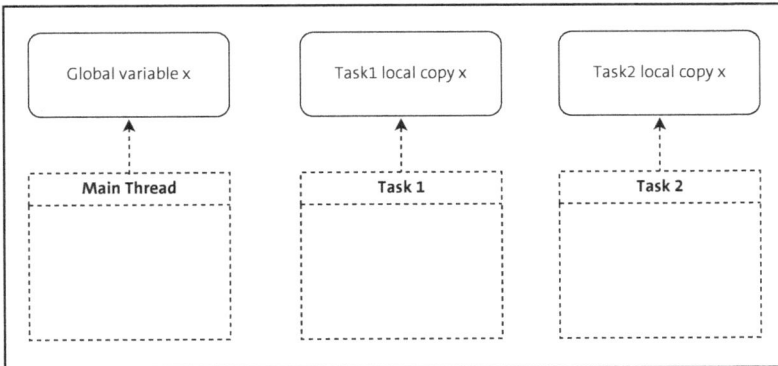

Figure 14.30 Task-Local Values

That allows you to start Task 1 with x = 5 and never worry about Task 2 changing that value. Simultaneously, you can start Task 2 with x = 6 and never worry about Task 1 changing that value. Each task gets its own global copy of x in its own context. That task, and all its child tasks, will be able to access x as an isolated global variable, with its value protected from other tasks.

Now, let's see task-local values in action in Listing 14.33.

```
enum Context {
    @TaskLocal static var userID: String?
}

Context.$userID.withValue("UserA") {
    Task {
        print("Task 1 - User ID: ", Context.userID ?? "none")
    }
}

Context.$userID.withValue("UserB") {
    Task {
        print("Task 2 - User ID: ", Context.userID ?? "none")
    }
}
```

Listing 14.33 Task-Local Variables Used by Tasks

The approach starts with a static global variable in an enumeration, struct, or class. This example uses the Context.userID variable. Note that the variable was marked with the @TaskLocal prefix, which is a requirement for task-local variables.

After preparing the value, you create an isolated scope for the first task, with the `Context.$userID.withValue("UserA")` statement. This statement tells Swift that for the task within that scope, the value of `Context.userID` should be assumed to be "UserA". Any piece of code running in that scope may access `Context.userID`, reading the "frozen" value of "UserA".

Likewise, you create an isolated scope for the second task too, with the `Context.$userID.withValue("UserB")` statement. This statement tells Swift that for the task within that second scope, the value of `Context.userID` should be assumed to be "UserB". Any piece of code running in that second scope may access `Context.userID`, reading the frozen value of "UserB".

A task in the first scope will never have to worry about another task changing the "UserA" value. Likewise, a task in the second scope will never have to worry about another task changing the "UserB" value. The output of these tasks is shown in Figure 14.31.

```
Task 1 - User ID:   UserA
Task 2 - User ID:   UserB
```

Figure 14.31 Different Task-Local Values

Naturally, you can put multiple tasks into a scope, too; they all would share the same task-local variable. This scenario is demonstrated in Listing 14.34, where tasks A1 and A2 will run with value "UserA", and tasks B1 and B2 will run with value "UserB".

```
enum Context {
    @TaskLocal static var userID: String?
}

Context.$userID.withValue("UserA") {
    Task { print("Task A1 - User ID: ", Context.userID ?? "none") }
    Task { print("Task A2 - User ID: ", Context.userID ?? "none") }
}

Context.$userID.withValue("UserB") {
    Task { print("Task B1 - User ID: ", Context.userID ?? "none") }
    Task { print("Task B2 - User ID: ", Context.userID ?? "none") }
}
```

Listing 14.34 Multiple Tasks in Same Task-Local Scope

You'll be pleased to know that async functions called from the task will also share the same task-local values. They will be considered to be running in the same task-local scope. Check the demonstration in Listing 14.35, where `printCurrentUserID` is called from two different tasks with two different task-local values.

```
enum Context {
    @TaskLocal static var userID: String?
}

func printCurrentUserID() async {
    print("Current User ID: ", Context.userID ?? "none")
}

Context.$userID.withValue("UserA") {
    Task { await printCurrentUserID() }
}

Context.$userID.withValue("UserB") {
    Task { await printCurrentUserID() }
}
```

Listing 14.35 Async Functions Accessing Task-Local Values

Note that printCurrentUserID has accessed Context.userID, just like a regular global variable. However, its value will be determined by the task-local value, as shown in Figure 14.32.

```
Current User ID:    UserA
Current User ID:    UserB
```

Figure 14.32 Async Function Task-Local Value Output

When the first task calls printCurrentUserID, the function will see Context.userID as "UserA". When the second task calls printCurrentUserID, the function will see Context.userID as "UserA".

You see? All you need to do is to provide the @TaskLocal prefix. Swift does all the heavy lifting to ensure that tasks get their own protected copies of the global variable.

14.5.2 Detached Tasks

In Swift, child tasks are created as *structured* by default. This means that they automatically inherit their parent task's features, such as priority, cancellation state, and @TaskLocal values.

But as we all know, some children choose to reject their parents' wishes and follow an independent path. That's an option for Swift's child tasks, too; they may reject inheriting their parents' features and be independent instead. Such tasks are called *detached tasks*.

Listing 14.36 shows an example, in which a regular task and a detached task run concurrently.

```
enum Context {
    @TaskLocal static var userID: String?
}

Task(priority: .userInitiated) {
    Context.$userID.withValue("User123") {

        // Regular child task (inherits everything)
        Task {
            print("Regular - userID:", Context.userID ?? "none")   // User123
            print("Regular - Priority:", Task.currentPriority)     // high
        }

        // Detached task (inherits nothing)
        Task.detached {
            print("Detached - userID:", Context.userID ?? "none") // none
            print("Detached - Priority:", Task.currentPriority)   // medium
        }
    }
}
```

Listing 14.36 Detached Task Sample

In this example, the first child task is a regular task. It inherits the task-local variable and priority of the parent task. The second child is a detached task, declared via the Task.detached { ... } syntax. As proven by the generated output, the detached task inherited neither the task-local variable nor the priority; it acts as an independent child.

In Swift, it is suggested to generally prefer regular tasks over detached tasks. Detached tasks should only be reserved for cases where you have a specific, compelling reason to break out of the norm. Some justifiable cases include the following:

- A long-running, standalone background work task
- Dodging a parent task's context
- Escaping actor isolation (a topic we will address in the next section)

And here are some cases in which you should avoid detached tasks:

- Operations tied to a UI/view lifecycle that should be canceled if the component is deallocated
- Situations in which automatic child cancellation is needed
- Short-lived tasks

14.5.3   Actors

You learned that task-local variables offer a safe option for read-only global variables for the task. But what if concurrent tasks need to read *and write* the same global variables? How can you ensure thread safety in that case, preventing data races?

Actors are a common answer for that.

In Swift, an *actor* can be considered a special kind of class with built-in thread safety. Unlike a class, an actor will protect its state/properties from concurrent access. Even if multiple tasks attempt to access an actor, the actor will automatically put them in a queue, ensuring that only one task's access is allowed at a time.

Database Locks

If you're familiar with database programming, then you can imagine an actor's lock as akin to a database lock. When multiple queries are executed simultaneously, the database lock will ensure that they are processed in a row. Such is the behavior of a Swift actor too.

Therefore, an actor may only be accessed *sequentially* by tasks, not *simultaneously*. And the good news: You don't need to do much coding to achieve this! Simply replacing the `class` keyword with the `actor` keyword is nearly enough to let Swift do all the heavy lifting for you.

As a reminder of data races and the platform for actors, Listing 14.37 contains a class with data race risks. If both tasks access `counter` simultaneously, then `counter.value` might be calculated as 1 instead of the expected value of 2.

```
class Counter {
    var value = 0
    func increment() { value += 1 }
}

let counter = Counter()

Task { counter.increment() }
Task { counter.increment() } // Data race!
```

Listing 14.37 Class with Data Race Risks

Now, inspect the improved code in Listing 14.38, which features a thread-safe actor.

```
actor Counter {
    var value = 0
    func increment() { value += 1 }
}
```

```
let counter = Counter()

Task { await counter.increment() }
Task { await counter.increment() } // No data race!
```

Listing 14.38 Actor Without Data Race Risks

Now that `Counter` is an actor (instead of a class), it will automatically ensure singular access among tasks, keeping its `value` safe from data races. But actors come with some natural side effects:

- Actors may only be called from tasks because they become asynchronous.
- Actor functions need to be called with the `await` prefix because the caller must wait for its turn.

This should also address why we don't automatically use actors instead of classes all the time: Actor instances are mostly suitable for tasks, not the main thread.

But what if you need to make a class thread-safe? No worries—we've got you covered! As demonstrated in Listing 14.39, you may wrap a complex data type, such as a struct or class, into an actor. If you already have a handful of such data types that need to be made thread-safe, you can use this approach.

```
struct CounterPoint {
    var value: Int = 0
    mutating func increment() { self.value += 1 }
}

actor Counter {
    private var point = CounterPoint()
    func increment() { self.point.increment() }
}

let counter = Counter()

Task { await counter.increment() }
Task { await counter.increment() }
```

Listing 14.39 Struct Wrapped in an Actor

When talking about actors and thread safety, `@MainActor` is a mechanism worth mentioning; it enforces execution on the main thread. As a typical example, in apps with a user interface, most UI frameworks are not thread-safe. You can update UI components from the main thread only. If you attempt to do so from a background task, your app may crash or behave unpredictably.

To tackle that problem, you may simply add the @MainActor prefix. In Listing 14.40, updateLabelText is marked as @MainActor.

```
@MainActor
func updateLabelText(_ text: String) {
    print("Label text updated")
}

updateLabelText("Hello from main thread")

Task {
    updateLabelText("Hello from task")
}
```

Listing 14.40 Main Actor Function Demonstration

In this example, even when updateLabelText is called from within the task, Swift will hop it back to the main thread before running the function.

Beyond functions, you can annotate an entire struct or class as @MainActor too, as shown in Listing 14.41.

```
@MainActor
class MyView {
    func updateLabelText(_ text: String) {
        print("Label text updated")
    }
}

let myView = MyView()

myView.updateLabelText("Hello from main thread")

Task {
    myView.updateLabelText("Hello from task")
}
```

Listing 14.41 Main Actor Class Demonstration

If you need to execute a small code snippet in a task on the main thread, this is possible via an await MainActor.run { … } code block. Check the demonstration in Listing 14.42.

```
Task {
    print("Start of task")

    await MainActor.run {
        print("Code executed in the main thread")
```

```
    }

    print("End of task")
}
```

Listing 14.42 Execution of Task Code Block in Main Thread

In this example, the second `print` statement will be pushed to the main thread for execution. That would be a safe place for UI updates.

Actors are the go-to way to access mutable objects from concurrent threads because they offer convenient safety features against data races. However, they don't come without some drawbacks, such as increased complexity due to their asynchronous nature and performance overhead due to thread queueing.

Actors justify the cost of such drawbacks for mutable objects, but in some cases, all you want is to simply pass values between tasks safely. In such cases, you can avoid the drawbacks of actors by using sendable types, which is our next topic.

14.5.4   Sendable Types

In Swift, a type is *sendable* if it can be passed safely between concurrent tasks. By *safely*, we mean that it is safe against data race conditions. When you mark a data type `Sendable`, you are declaring that the type is concurrency-safe.

To mark a data type as such, all you need to do is to apply the `Sendable` protocol, as demonstrated in Listing 14.43.

```
final class Point: Sendable {
    let x: Int
    let y: Int

    init(x: Int, y: Int) {
        self.x = x
        self.y = y
    }
}
```

Listing 14.43 Sendable Class

As you can see, all properties of the `Point` class are immutable. That alone makes instances of this class thread-safe, because no matter what concurrent tasks do, no data race can occur because `Point.x` and `Point.y` will act as constants. Still, you need to apply the `Sendable` protocol to use `Point` as a parameter in tasks: That's how secure Swift is!

Listing 14.44 contains a complete example, in which `Point` instances are used as parameters in concurrent detached tasks. As a personal exercise, you can remove the `Sendable` protocol from `Point` and see how Swift complains—even before compiling your app.

```
final class Point: Sendable {
    let x: Int
    let y: Int

    public init(x: Int, y: Int) {
        self.x = x
        self.y = y
    }
}

func drawPointToScreen(_ point: Point) async {
    print("Drawing (\(point.x), \(point.y))")
}

Task {
    let myPoint = Point(x: 5, y: 5)

    for _ in 1...3 {
        Task.detached { await drawPointToScreen(myPoint) }
    }
}
```

Listing 14.44 Using Point Instances as Parameters in Concurrent Detached Tasks

In the scope of Sendable objects, Swift is clever enough to act relaxed where possible. For example, the program in Listing 14.45 runs without compilation errors, despite not marking the Point struct as Sendable.

```
struct Point {
    let x: Int
    var y: Int
}

func drawPointToScreen(_ point: Point) async {
    print("Drawing (\(point.x), \(point.y))")
}

Task {
    var myPoint = Point(x: 5, y: 5)

    for _ in 1...3 {
        Task.detached { await drawPointToScreen(myPoint) }
    }
}
```

Listing 14.45 Sendable Struct Without Sendable Implementation

Surprised? You shouldn't be! Remember that structs are value-based data types, unlike classes, which are reference-based data types. In this example, a distinct copy of myPoint arrives at drawPointToScreen as an immutable argument for each concurrent task. Therefore, Swift sees no risk of data races and liberates you from applying the Sendable protocol to Point. But you can, if you want to!

If you deliberately want to anger Swift, you can try sending myPoint to a mutating async function, as shown in Listing 14.46. That will definitely cause a data race compilation error, which you can't bypass even if you add the Sendable suffix to Point. Swift is really sensitive about data races because Apple doesn't want unsafe apps on its devices.

```swift
struct Point: Sendable {
    let x: Int
    let y: Int
}

func drawPointToScreen(_ point: inout Point) async {
    print("Drawing (\(point.x), \(point.y))")
}

Task {
    var myPoint = Point(x: 5, y: 5)

    for _ in 1...3 {
        Task.detached { await drawPointToScreen(&myPoint) } // Error
    }
}
```

Listing 14.46 Deliberately Causing Data Race Compilation Error

There may be some cases in which you want to silence Swift's compiler checks for data races. In Listing 14.47, the Point class has a mutable property called drawnAtLeastOnce, which causes a compiler error for potential data races.

```swift
final class Point: Sendable {
    let x: Int
    let y: Int
    var drawnAtLeastOnce = false // Error

    public init(x: Int, y: Int) {
        self.x = x
        self.y = y
    }
}
```

```
func drawPointToScreen(_ point: Point) async {
    print("Drawing (\(point.x), \(point.y))")
    point.drawnAtLeastOnce = true
}

Task {
    let myPoint = Point(x: 5, y: 5)

    for _ in 1...3 {
        Task.detached { await drawPointToScreen(myPoint) }
    }
}
```

Listing 14.47 Mutable Class Without Concurrency Risks

Most of the time, you should take compiler errors into consideration and look for better architectures. But in this example, Point.drawnAtLeastOnce merely points out that Point was displayed on screen, by any task. If you don't have any critical mechanism depending on that variable, then you can assume that it is safe to let concurrent tasks modify this value simultaneously because it may be true only once. Even if it's overwritten multiple times, it will ultimately be true after being drawn; there will be no further complications.

In such cases, you may add the @unchecked annotation to calm the compiler down, as demonstrated in Listing 14.48.

```
final class Point: @unchecked Sendable {
    let x: Int
    let y: Int
    var drawnAtLeastOnce = false // No error

    public init(x: Int, y: Int) {
        self.x = x
        self.y = y
    }
}

func drawPointToScreen(_ point: Point) async {
    print("Drawing (\(point.x), \(point.y))")
    point.drawnAtLeastOnce = true
}

Task {
    let myPoint = Point(x: 5, y: 5)
```

```
    for _ in 1...3 {
        Task.detached { await drawPointToScreen(myPoint) }
    }
}
```

Listing 14.48 Application of Unchecked Annotation

Applying the @unchecked annotation is like telling the compiler, "Trust me, I know what I'm doing." Swift will then reluctantly let you pass Point instances around, but now it's completely your responsibility to prevent data races using your custom program logic. Be careful: You might be opening a door for data race errors with your own hands!

14.6 Summary

In this chapter, you learned how to apply concurrency in Swift, as well as how to avoid common pitfalls like data races.

Swift supports async functions, which are ideal for operations with unpredictable completion times, such as fetching data from the internet or reading files. Such functions are typically called from within tasks, which are additional threads with individual runtime flows.

An app may have multiple concurrent tasks running simultaneously, each performing a separate operation. You have the option to set higher priorities for more important tasks and to group relevant tasks under a common task group, allowing for easier bulk management.

In any programming language, a typical pitfall for concurrency is a data race situation, in which multiple threads attempt to access and modify the same value, leading to potential conflicts. Swift has multiple built-in measures against data race risks, such as task-local values, detached tasks, actors, and sendable types.

If you are new to programming, concurrency may seem overwhelming—but don't worry! You can start small and apply concurrency only where it adds real value. Keeping concurrency architectures simple is a good idea anyway. As you gain more experience and feel more confident, you may introduce more features of concurrency into your programs.

Concurrency is an inevitable necessity in most real-world apps, though, so it's best to get used to it.

In the next chapter, we will continue with another necessary topic in the real-world of Swift programming: modules.

Chapter 15
Modules in Swift

In many cases, Swift programmers develop reusable modules and use the same module in multiple projects. This chapter will teach you how to develop custom modules and import them into your projects.

Throughout this book, you've learned about different methods of code modularization. In Chapter 5, you learned about functions, which are reusable blocks of executable code. In Chapter 8 and Chapter 9, you learned about structs and classes, which feature properties and functions as reusable object templates.

Such methods are essential and useful within a codebase for one program. But in some cases, the need for modularization extends beyond a single program. For example, you may want to reuse the same utility class in multiple projects, or you might have a useful collection of functions that you want to isolate and share with other developers.

For such scenarios, where code chunks need to be reused beyond the scope of a single project, Swift offers a built-in feature called *modules*.

In this chapter, we'll explore what modules are and how to use them to organize Swift projects more effectively. You'll learn how to create custom modules, import existing ones, and handle dependencies using Swift Package Manager. By the end of the chapter, you'll understand how a modular approach can make your code more reusable and maintainable, even in beginner-level projects.

Without further ado, let's go!

15.1 Introduction to Modules

In this debut section, we'll warm you up to the idea of interproject modularity. You'll learn about the core logic of Swift modules, their benefits, and how they work within the Apple development ecosystem. Once those concepts become clear, we'll move to hands-on examples.

15.1.1 What Is a Module?

Swift modules are best understood with the support of their satellite concepts.

Let's start with libraries. *Library* is a generic abstract term, defining a collection of reusable code. A library, as sketched in Figure 15.1, may contain enumerations, functions, structs, classes, and similar components under its umbrella, waiting to be consumed.

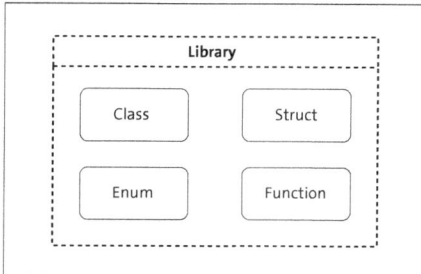

Figure 15.1 Sketch of Library

In your app, you may have developed a local library for image processing, for example. That library could contain Swift code that can resize, invert, desaturate, or blur images.

When such a library is made ready for use by multiple apps, it's called a *module*. You can think of a module as a labeled box containing a library that you can open and use, as sketched in Figure 15.2.

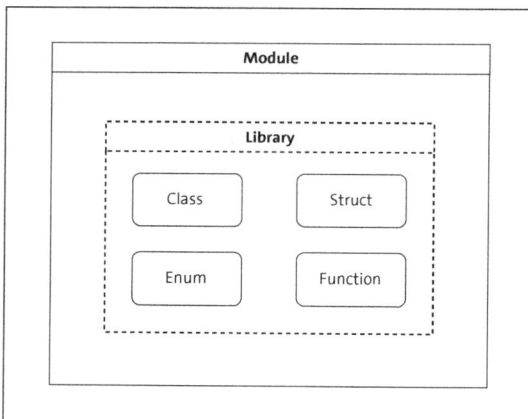

Figure 15.2 Sketch of Module

In Swift, modules are commonly delivered either as frameworks or as packages. Let's go over their core differences, starting with frameworks.

Fundamentally, a *framework* is a module compiled as a binary file. It contains reusable Swift components, for sure, but it may also contain extras like pictures, sounds, and data files, all bundled together nicely for the benefit of the client code. Check Figure 15.3 for a sketch.

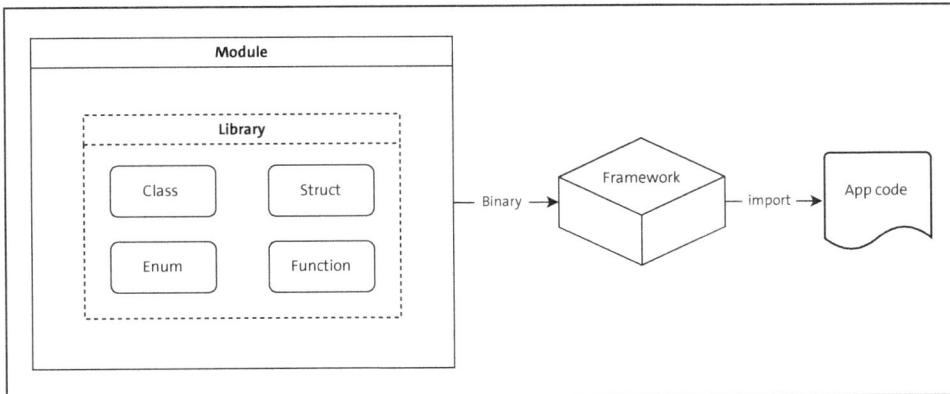

Figure 15.3 Sketch of Framework

A framework is the standard way to publish a compiled Swift module so that it can be consumed by different apps and developers, without necessarily making the source code available.

Throughout the code examples in previous chapters, we occasionally wrote the `import` `Foundation` expression. That expression opens the `Foundation` framework and makes its components available to your custom code in the editor. Once the module is imported, you can use all classes, structs, and so on included in that module in your custom program.

Apple has many frameworks presented to developers like you, such as SwiftUI, Cloud-Kit, AVFoundation, Metal, MapKit, ARKit, HealthKit, and others. In your custom apps, you can import such frameworks and start utilizing their contents, assuming that you know your way around those frameworks. As mentioned before, such frameworks may demand a dedicated time investment to master.

A *package*, on the other hand, mainly differs from a framework by being open source. Basically, a Swift package can be considered an organized collection of open-source code libraries, as sketched in Figure 15.4.

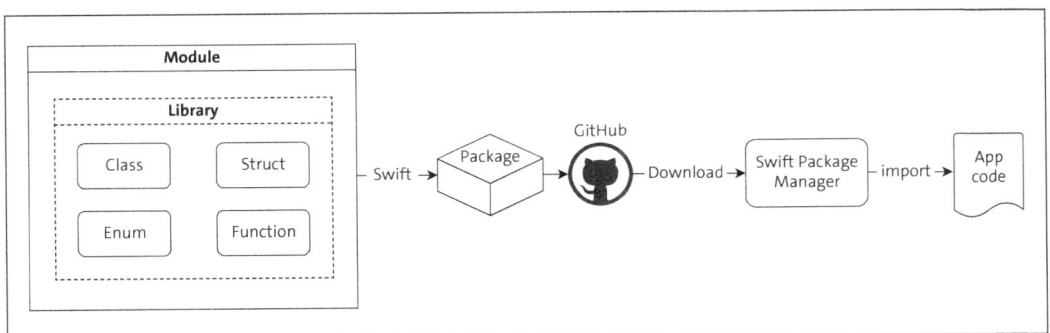

Figure 15.4 Sketch of Package

15

Note that the contents of a package are typically not compiled; packages contain raw, human-readable Swift code. Thus, they are preferred for open-source projects. Packages are typically handled via Swift Package Manager, which we will address in due course.

For your convenience, Table 15.1 contains a summary of the concepts addressed so far.

Term	Description	Example
Library	General term for reusable code	Collection of math helper functions
Module	Library ready to be consumed by apps	–
Framework	Compiled module with resources	Foundation
Package	Collection of open-source libraries, available via Swift Package Manager	CodableCSV

Table 15.1 Summary of Modularization Terms

Now that you're familiar with the terminology, let's discuss why modules are used in the first place.

15.1.2   Why Use Modules?

In Swift, there are some typical scenarios in which resorting to modules makes sense, such as the following:

- **Building a reusable module**
  If your code snippets can be reused in multiple apps, then you can pack them up as a module and import them into multiple projects, or even make them open source and share them with other developers.

- **Consuming a reusable module**
  Beyond Apple's built-in modules, there are countless third-party packages available online. If you find a package helpful for your app, then you can import it into your project instead of reinventing the wheel by coding everything from the scratch.

- **Splitting an app into layers**
  If you have a very large app, it makes sense to split it into distinct modules, such as UI, networking, persistence, AI, and so on. That way, your codebase becomes more manageable, and different teams may focus on different modules instead of running a one-man show.

If you run into such a scenario, building a module-based architecture will surely bring many benefits, which we will address in the following sections.

Reusability

One of the most significant advantages of modules is code reusability. Once you write a module and all tests run correctly, you can reuse that module as a safe and dependable library.

Imagine a custom date-formatting module. Once you develop and publish it, you don't have to recode that logic redundantly in your apps. Instead, you can simply import the module into each relevant app, as sketched in Figure 15.5.

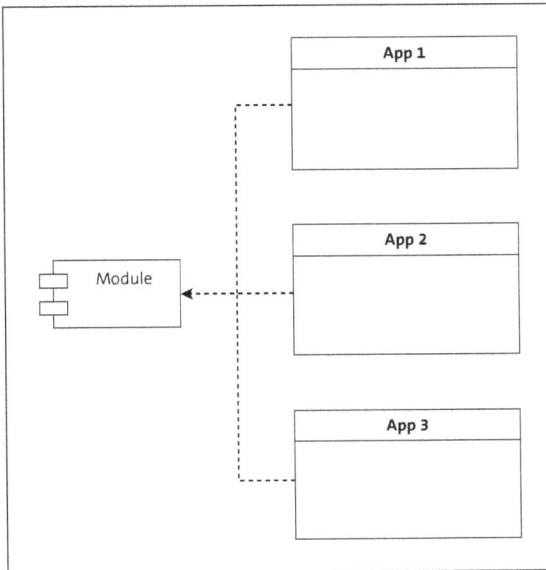

Figure 15.5 Interapp Reusability Sketch of Module

That approach pays off in two ways: You save both development time and test time because both were spent already during the development of the module.

If you make an improvement in the module, such as a new feature, a bug fix, or a performance improvement, then all relevant apps will be improved as well. Be careful, though: A central bug in a widely used library may also crash dependent apps.

Organization

As your codebase grows, the need for proper code organization grows too. At the very least, you should organize relevant code files into folders. Figure 15.6 shows the partial structure of a real iOS game, which is carefully organized.

Although folder-based organization was good enough for this single-developer indie game, a module-based approach would be more suitable in a larger project with multiple members.

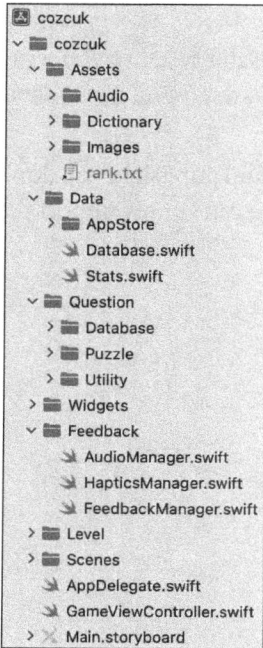

Figure 15.6 Code Organization of Real iOS Game

In this example, **Data**, **Question**, **Widgets**, **Feedback**, **Level**, and **Scenes** could be set up as different modules developed by different teams of people. The main app would merely act as a controller, merging and orchestrating those modules.

That way, you get the flexibility to develop, version, test, and maintain modules separately, allowing for a clean codebase. On top of that, if you abstract module access via protocols, you can easily plug in and take out different implementations as well as mock classes during tests.

That kind of organization and flexibility is certainly an upgrade over hundreds of code files looking like a bowl of spaghetti in a single project.

Encapsulation

Modules offer a built-in system for access control, which we will address in Section 15.4. In a nutshell, a module can hide its implementation details and expose only the parts necessary, similar to a class with public/private sections.

As with classes, module encapsulation prevents unexpected errors due to client misuse.

Compilation Speed

If you have a huge single-project app, then building and running the app during development can take a long time. In a project with dozens of daily rebuilds, the compilation time may add up, endangering your sensitive deadlines.

But when your app is split into modules, compilation durations may decrease dramatically, especially with frameworks:

- When you rebuild a module, only the module's code is compiled, not the entire app.
- When you rebuild your app, only the app's code is compiled if the modules were already compiled.

As you can see, programming decisions are not merely technical but involve other factors, like delivery time.

Dependency Management

In many cases, avoiding circular dependencies is a good idea. For example, your UI library may depend on your Data library. But if your Data library depends on the UI library too, that should be considered a *code smell*, which points to an immature architecture.

Concrete dependence is another typical pitfall. In most cases, your code components should be *loosely coupled*, meaning that you should be able to replace one component without affecting others—as you learned in Chapter 10 with protocols.

A single-project app doesn't necessarily enforce applying such best practices to your codebase. When you work with modules, though, they inevitably define explicit relationships between different parts of your app. That leads you to a cleaner architecture and prevents some typical design problems that may be overlooked in a big project with a huge pile of code blocks.

Hopefully, you've been convinced of the benefits of a module-based approach in Swift projects. Now that you're warmed up with those core concepts, it's time to finally get hands-on practice with module development!

15

Heads-Up: Screenshots

You're about to enter a section in which step-by-step instructions are provided with screenshots. By its nature, software evolves constantly, and the screens shared here might change slightly at some point. If that happens, you still can adapt the core knowledge presented in this chapter to new Xcode versions.

15.2   Working with Frameworks

In this section, you'll learn how to prepare and consume frameworks. We will go through step-by-step instructions to create a custom framework, then import this framework into an app project to make use of its contents.

15.2.1   Setting Up a New Framework

To create a new framework, you first need a new framework project. In Xcode, follow menu path **File • New • Project**. In the template window, select **Framework**, as shown in Figure 15.7, and click **Next**. Note that the top section contains the platforms you can target; we will focus on macOS for this demonstration. You can play around with different platforms later.

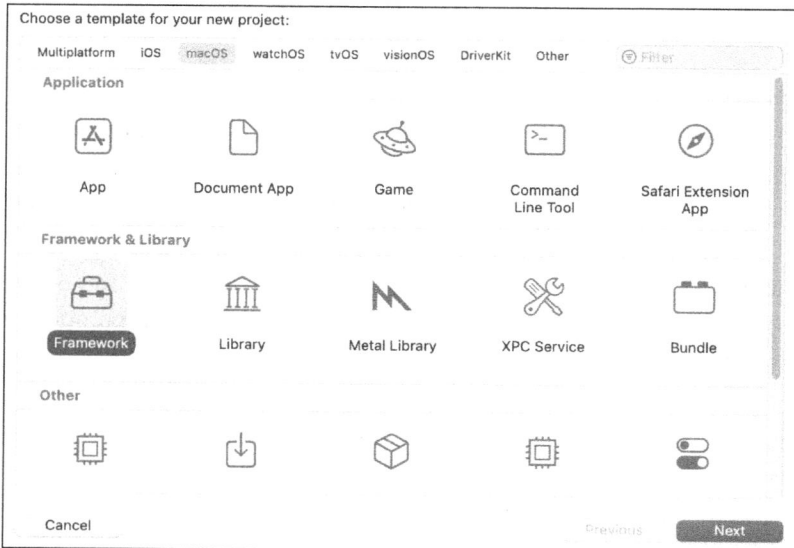

Figure 15.7 Selecting Framework Project Type

In the next window, enter a meaningful name for your product. In our case, we entered "KeremUtil", as shown in Figure 15.8. Click **Next** when you're done.

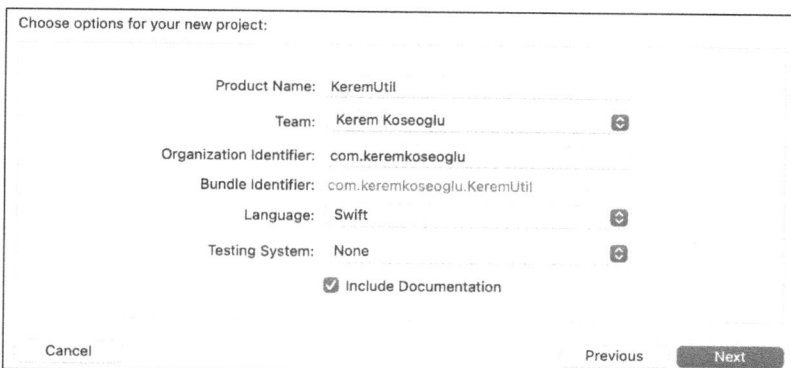

Figure 15.8 Entering Product Name

In the next window, Xcode will ask for a target folder in which to place the project, as shown in Figure 15.9. Pick an appropriate folder and click **Create**.

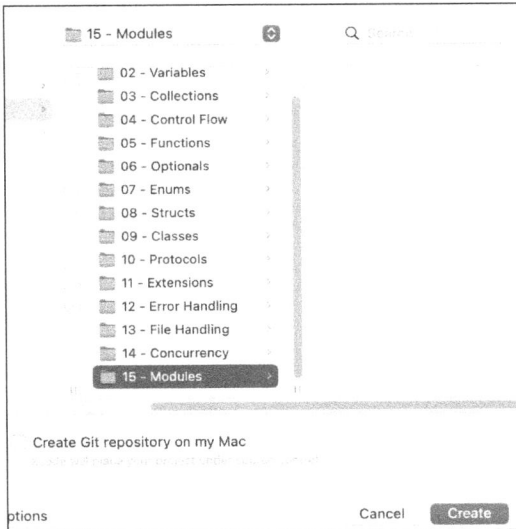

Figure 15.9 Selecting Folder to Create Your Project

And voilà! You have a blank slate for a framework project, as shown in Figure 15.10, waiting for you to code your libraries.

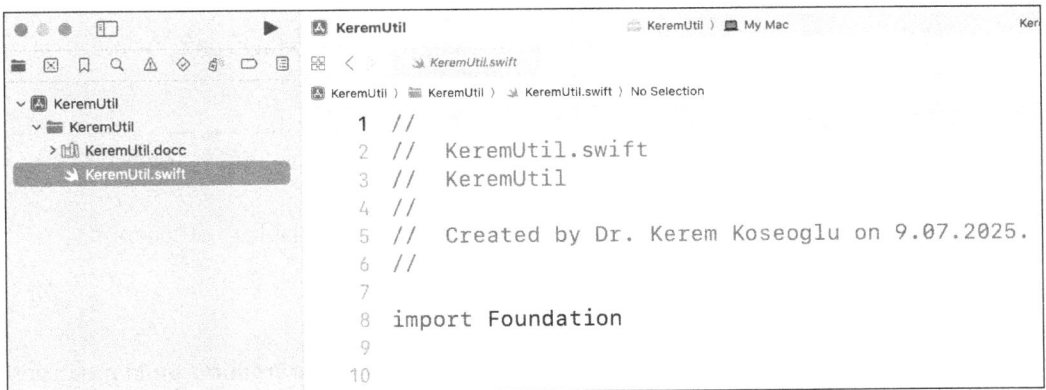

Figure 15.10 Empty Framework Project

Let's make the framework a little more useful by adding some code in the next section.

15.2.2 Adding Code to a Framework

Now that you have an open editor in Xcode, you can apply your Swift knowledge to develop a library! You can code classes, structs, data types, and any other form of reusable content here. Also feel free to open or create new *.swift* files for better code organization.

To keep the focus on Swift modules here, let's merely code a simple struct called Math-Helper at this time, as shown in Listing 15.1.

```swift
public struct MathHelper {
    public static func add(_ v1: Int, _ v2: Int) -> Int {
        return v1 + v2
    }

    public static func square(_ value: Int) -> Int {
        return value * value
    }
}
```

Listing 15.1 Simple Struct Coded in Your Library

At this point, your Xcode screen should look like Figure 15.11.

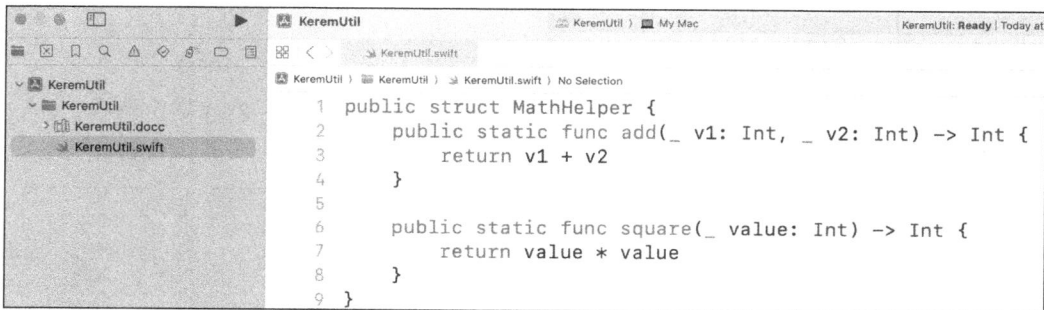

Figure 15.11 MathHelper in Xcode

Now you have to build this project to make MathHelper available to other apps.

15.2.3 Building a Framework

To compile the framework, all you need to do is to follow the **Product • Build** menu path in Xcode. Depending on the size of the library and the speed of your computer, it may take a while. If there are no errors, the build will complete successfully, writing the desired files to the disk.

To see those files, follow menu path **Product • Show Build Folder in Finder**. A Finder window will open, as shown in Figure 15.12, showing the *.framework* file that you'll import and consume soon.

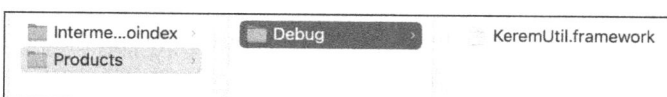

Figure 15.12 Compiled Framework File Shown in Finder

You may have noticed that the folder containing your framework file is called **Debug**. When you build a project in Xcode, it uses the active scheme configuration, which is **Debug** by default. The name of the target folder reflects that.

To generate a release-quality-optimized build, you need to explicitly use the **Release** configuration. Let's do that to see the difference.

In Xcode, follow menu path **Product · Scheme · Edit Scheme**. In the popup that opens, select **Run**, and change **Build Configuration** from **Debug** to **Release**, as shown in Figure 15.13.

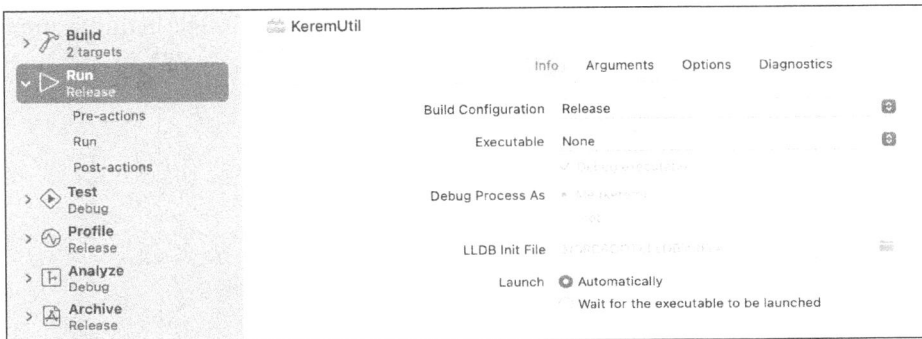

Figure 15.13 Changing Build Configuration to Release

If you rebuild the project this way, a **Release** folder will appear next to the **Debug** folder, containing binaries suitable for release.

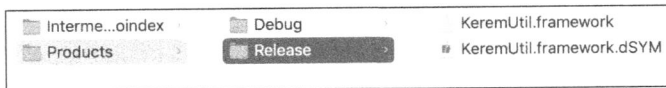

Figure 15.14 Compiled Framework File Suitable for Release

Both the debug and release build options produce a *.framework* file that contains the compiled module and can be imported into apps. But there are some notable differences, as highlighted in Table 15.2.

Feature	Debug	Release
Optimization	Minimal	Full
Can be debugged	Yes	No
Performance	Slow	Fast
Symbols	Full	Minified
File size	Large	Small

Table 15.2 Feature Comparison Between Debug and Release Builds

Therefore, debug builds are meant for local development and debugging. The output of a release build is what you want to distribute or share externally.

Whether debug or release, you have the compiled framework ready to go at this point. Now, let's walk through how to import it into a project.

15.2.4   Importing a Framework

First, you need a separate project to import the framework into. In Xcode, click the **File • New • Project** menu option; then, under the **macOS** tab, select **Command Line Tool**, as shown in Figure 15.15, and click **Next**. We're picking a simple project type in this example to keep the focus on modules.

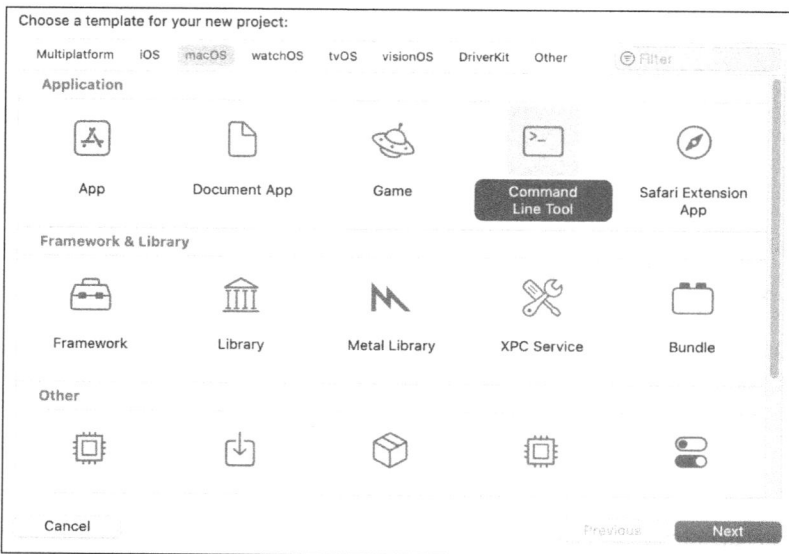

Figure 15.15 Selecting Command Line Tool Template

In the next window, provide a name for your product, as shown in Figure 15.16, and click **Next**. In this case, we picked a very obvious name; you can pick a cooler name if you like!

Finally, pick the destination folder of the project, as shown in Figure 15.17. We have picked the same parent folder, **KeremUtil** in our example, for convenience, but this is not a requirement.

At this point, you will have an empty consumer project, as shown in Figure 15.18, eagerly waiting for the framework from the previous section.

Figure 15.16 Picking Product Name

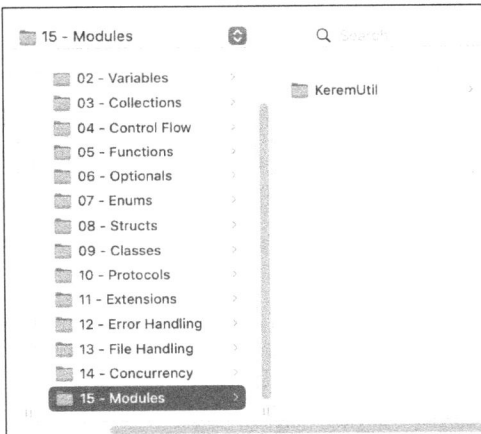

Figure 15.17 Destination Folder of Project

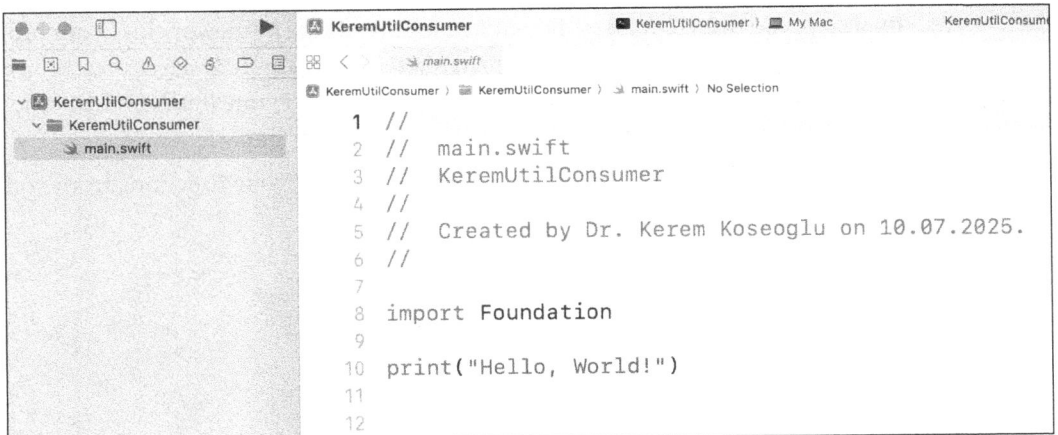

Figure 15.18 Empty Consumer Project

To import the framework into that project, simply drag and drop the framework file (**KeremUtil.framework** in our example) from Finder into your Xcode project navigator. A popup will open, as shown in Figure 15.19, giving you some options for the import. You can choose to either copy or move files to the destination; pick **Copy** if you are unsure. Close the popup by clicking **Finish**.

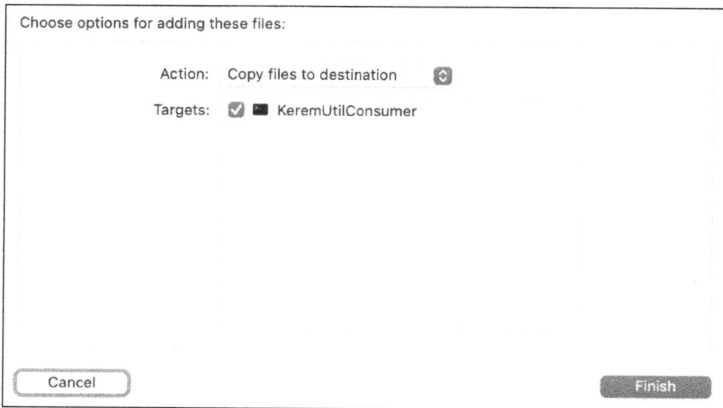

Figure 15.19 Picking an Action for Your Framework

If all goes well, the framework will appear in the project navigator, as shown in Figure 15.20.

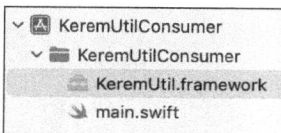

Figure 15.20 Framework in Project Navigator

As a final step, you need to make sure that Xcode embeds the framework into your project binaries, adopting it as part of the app. As shown in Figure 15.21, select your project in the navigator and browse to **General · Targets**. Under **Frameworks and Libraries**, ensure that the imported framework is marked as **Embed & Sign**.

That's all! Now you're ready to write some Swift code to invoke functions from your framework. Listing 15.2 demonstrates an example.

```
import KeremUtil
print(MathHelper.square(16))    // 256
print(MathHelper.add(10, 32))   // 42
```

Listing 15.2 Consuming Functions in Framework

Note that we started here with the `import KeremUtil` statement. This statement declares that the contents of `KeremUtil` should be made available in this context.

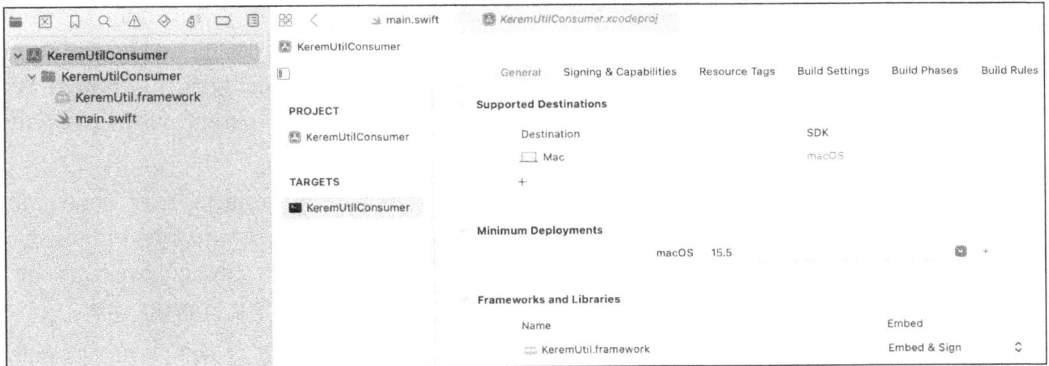

Figure 15.21 Embedding a Framework into Your Target

Consuming Standard Frameworks

In examples in earlier chapters, we occasionally used a similar statement: import Foundation. That has the exact same purpose! It declares that Apple's Foundation module should be made available.

Because Apple's frameworks are readily available in the operating system, you don't need to take the step of dragging and dropping files for them into your project; instead, you can directly import such modules. But the logic is the same: You are ultimately importing additional modules into your context.

Once your framework is imported and available, its public components can be accessed and consumed as if they were components within the client project! The print(Math-Helper.square(16)) expression here will work perfectly, as if MathHelper was developed within KeremUtilConsumer.

Cool, right? This step-by-step example highlights the path to develop, build, and import a framework to conduct modular development. Your real-word libraries may be more complex than this one, but the method you'll follow will be the same.

In the next section, we'll focus on Swift packages.

15.3   Working with Packages

In Swift, a *package* is a collection of raw, open-source code, organized and shared with Swift Package Manager. The biggest difference between a framework and a package is that frameworks are compiled, while packages contain the source code openly, waiting to be compiled in the client's project.

Packages are suitable for code sharing, such as through open-source projects on GitHub. Some popular Swift packages available on GitHub include Alamofire (for networking),

CodableCSV (for CSV parsing), and Kingfisher (for image loading). Naturally, there are many more. Once you learn how, you can import such third-party open-source libraries to your project and enjoy the productivity boost. Mind the licensing terms, though! You can also develop and publish your own Swift packages for the benefit of the programming community.

However, packages aren't just for open-source community projects. They are equally useful for internal modularization between teams in the same companies—or even in sole-developer projects! Although frameworks offer a similar approach, packages are a more transparent alternative for cases where open-source code is appropriate.

In either case, packages give clients the option to contribute as well. You can make improvements to a package and send your changes back to GitHub via a pull request. If your modifications are approved, the entire community will reap the rewards.

Now that the distinct role of packages has been highlighted, in this section, we'll work on packages via hands-on examples.

15.3.1   Setting Up a New Package

The creation of a package is very similar to the creation of a framework. In Xcode, select menu path **File · New · Package**. In the popup window, select **Library**, as shown in Figure 15.22, and click **Next**.

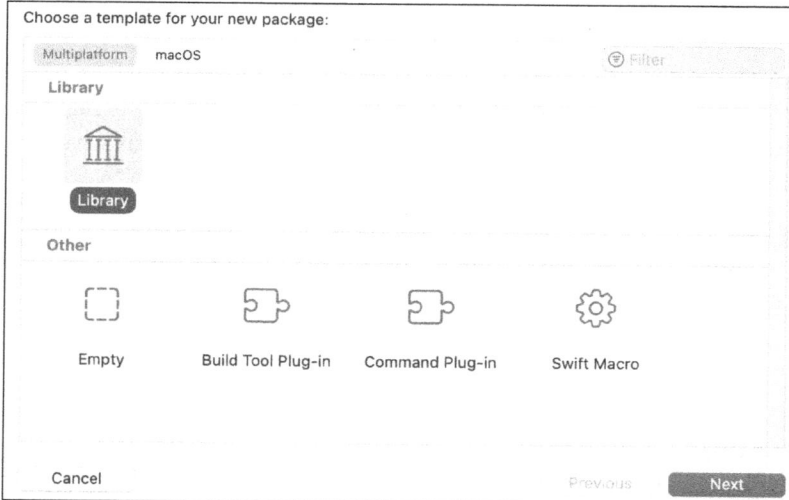

Figure 15.22 Selection of Appropriate Template

The next window might ask for a testing system. Select **None** for now, as shown in Figure 15.23, then click **Next**. You'll learn about Swift testing in Appendix A.

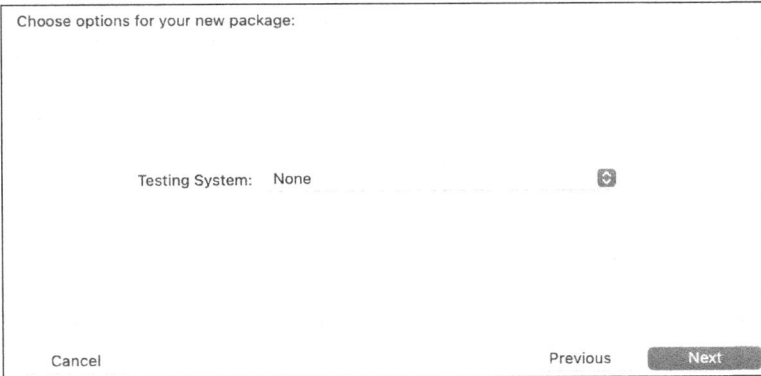

Figure 15.23 Selecting Testing System

In the next window, pick your target folder, give it a meaningful name, as shown in Figure 15.24, and click **Create**.

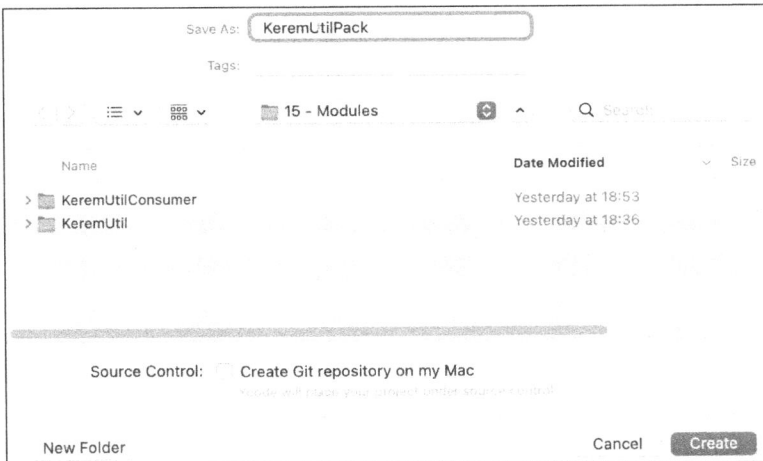

Figure 15.24 Picking Project Name and Target

If everything goes well, you should have an empty package, as shown in Figure 15.25. Congratulations!

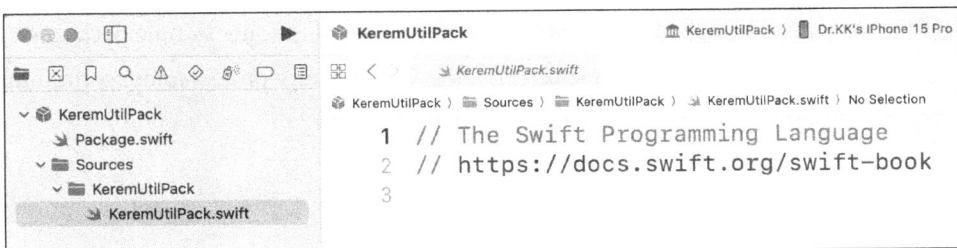

Figure 15.25 Empty Package

15.3.2   Adding Code to a Package

Now that you have an empty package, let's make it a little more useful by adding some code to it. Coding a package is not too different from coding a framework. You simply write reusable code using your Swift knowledge. To keep things simple, we'll add the same code from the framework example to KeremUtilPack.swift, as shown in Listing 15.3.

```
public struct MathPack {
    public static func add(_ v1: Int, _ v2: Int) -> Int {
        return v1 + v2
    }

    public static func square(_ value: Int) -> Int {
        return value * value
    }
}
```

Listing 15.3 Contents of KeremUtilPack.swift

The only difference is that we named the struct MathPack instead of MathHelper to emphasize the difference a little better. This way, if you import both of these into the same app as an exercise, you won't experience any conflicts and confusion.

The contents of a package can be as complex as you need, but we will limit this example to this simple struct and move forward to consuming the package.

15.3.3   Importing a Package

To import a package into a project, you naturally need a project first. Following the steps outlined in Section 15.2.4, create a new command line tool project. In our example, we'll call it **KeremUtilPackConsumer** for the sake of name consistency. Now, let's look at the methods for importing local versus remote packages.

Importing a Local Package

As an example for a local package, we'll use the **KeremUtilPack** package created in the previous step. In the project window, click the **File · Add Package Dependencies** menu option. A popup will appear, as shown in Figure 15.26, listing some sample packages.

In this window, click the **Add Local** button. Select your package project folder containing the **Package.swift** file, then click **Add Package** (see Figure 15.27).

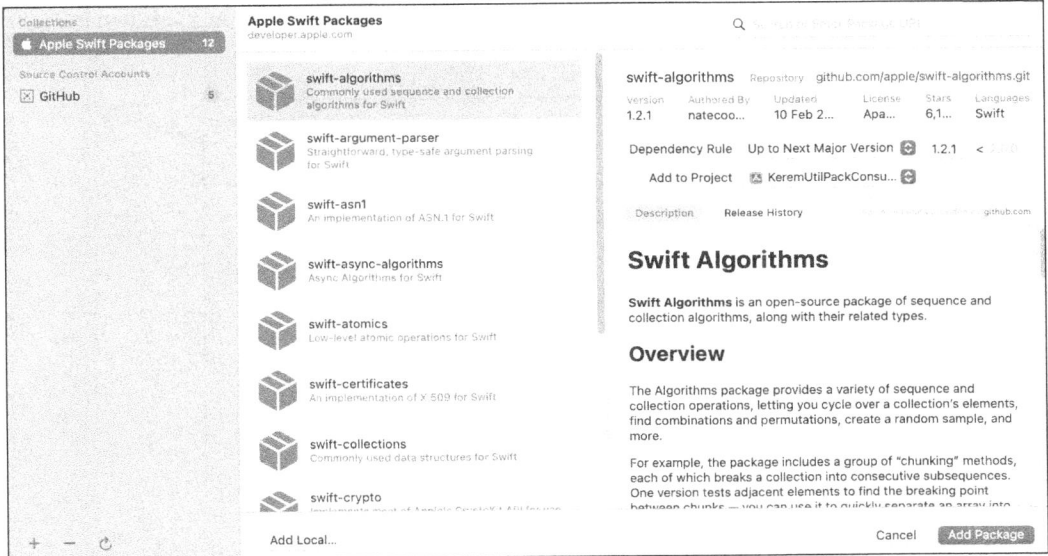

Figure 15.26 Package Import Popup

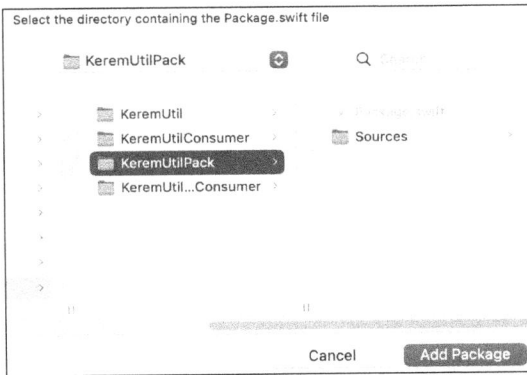

Figure 15.27 Selecting Local Package

In the next popup, Xcode will ask you to select the products you need in the package, as well as the target for each product. In our example, we have a single product and a single target, so we can just select the default values, as shown in Figure 15.28, and click **Add Package**.

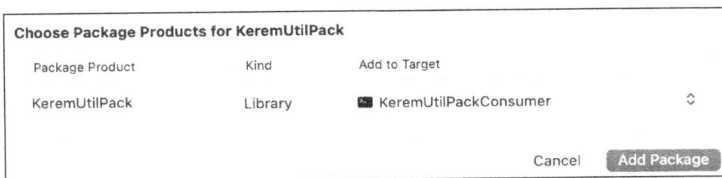

Figure 15.28 Selecting Package Products

Now the package should appear in the project tree as a dependency, as shown in Figure 15.29.

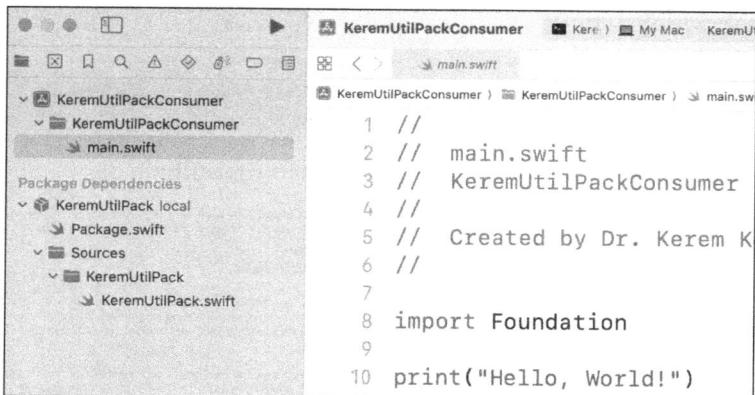

Figure 15.29 Pack Dependency in Project Tree

That's it! Now you can consume the contents of the package and write your own Swift code on top of it. Listing 15.4 demonstrates some simple code that imports the package module and calls functions from the package.

```
import KeremUtilPack
print(MathPack.square(16))    // 256
print(MathPack.add(10, 32))   // 42
```

Listing 15.4 Sample Swift Code Calling Package Functions

As you can see, once the package is imported, using it as a module is not too different than using a framework.

Importing a Remote Package

To import a remote package into the project, the steps to follow are very similar to those for a local project. As an example, we will use the popular CodableCSV library on GitHub, but you can proceed with any Swift package you want.

In the project window, click the **File • Add Package Dependencies** menu option. When the popup appears, paste the target URL into the search bar, as shown in Figure 15.30. When the repository is loaded, click **Add Package**.

Xcode will connect to that repository, downloading and organizing source code files as necessary. Once that is complete, you will be asked to choose package products as you would for local packages (see Figure 15.31). For this example, you can proceed with the defaults, so just click **Add Package**.

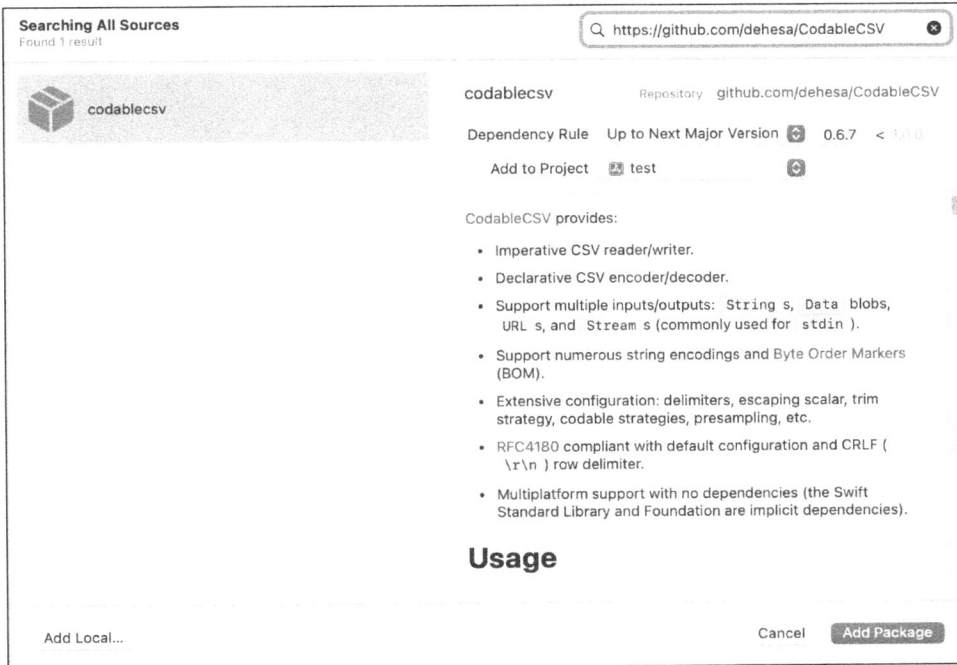

Figure 15.30 Entering URL of Remote Package

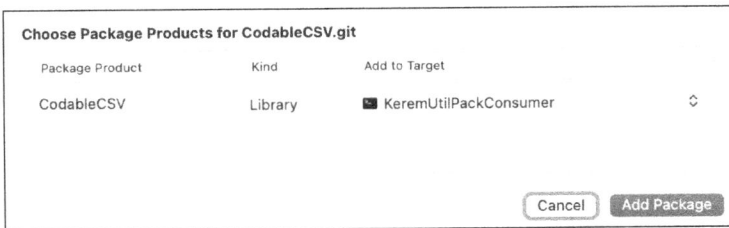

Figure 15.31 Selecting Remote Package Product

Now CodableCSV will be part of the project and shown in the project tree, as you can see in Figure 15.32.

From now on, you can benefit from the features of CodableCSV, as shown in Listing 15.5.

```
import KeremUtilPack
import CodableCSV

print(MathPack.square(16))    // 256
print(MathPack.add(10, 32))   // 42

let encoder = CSVEncoder()
```

Listing 15.5 Sample Code Using Local and Remote Packages

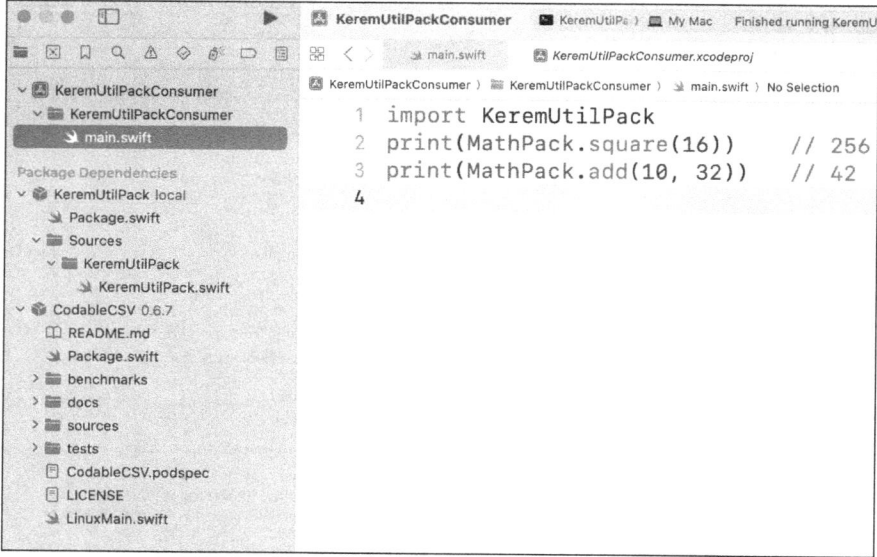

Figure 15.32 External Package, Listed as Part of Project

If a new version of CodableCSV is released in the future, you can easily update your project with that version! All you need to do is to right-click the package name and select **Update Package**, as shown in Figure 15.33. Xcode will check the remote repository for any updates and refresh your project with the new content if possible.

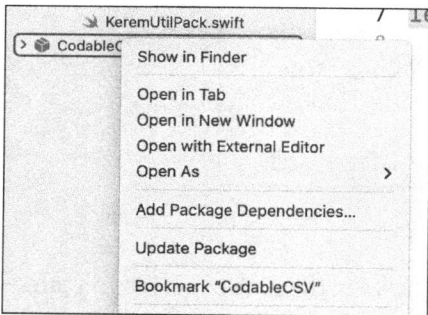

Figure 15.33 Updating Package

As you can see, Swift Package Manager is a great helper in terms of package management. Whether you are working with local packages or remote packages, Swift Package Manager helps with importing, organizing, and bundling external source code.

At this point, you know how to create, publish, and consume modules in the form of frameworks or packages. In the simple examples so far, the modules have been very permissive; all module content was made publicly available. But that's not what you'll want all the time; encapsulation is a requirement in many cases. In the next section, you'll learn about access control in modules.

15.4   Access Control

Access control is an integral part of Swift modules, which offers various degrees of code encapsulation. You already learned about the concept of encapsulation for structs and classes, where public parts were available to all clients while private parts were reserved for that component's internal usage only.

Swift modules offer various access levels, which you will learn about in order of the most open to the most restricted. The examples in the following sections are assumed to be part of a framework called AccessLevelDemo.

15.4.1   Open

This is the most permissive access level, without any limitations. Open components are available anywhere within or outside the module, such as other apps, packages and frameworks. They can also be subclassed and overridden outside of the module too. Listing 15.6 demonstrates the syntax for open components.

```
open class Animal {
    open func makeSound() {
        print("Some sound")
    }
}
```

Listing 15.6 Open Class

The client app or another module will be allowed to subclass Animal or override make-Sound.

Open components are mainly used for APIs, which are meant to be extended by others—for example, a module to create invoice PDFs. You may want to allow clients to inject their own content before the final PDF is published, which makes it suitable for the open access level.

15.4.2   Public

Public components are nearly as permissive as open. They are also available anywhere within or outside the module, such as in other apps, packages, and frameworks. However, they can't be subclassed or overridden outside the module, so they are closed for modifications. Listing 15.7 demonstrates the syntax for public components.

15

```
public struct MathHelper {
    public static func add(_ a: Int, _ b: Int) -> Int {
        return a + b
    }
}
```
Listing 15.7 Public Struct

The client app or another module can use `MathHelper.add`, but it can't extend `MathHelper` or override any behavior.

Public components are mainly used for APIs that are not meant to be extended by others—for example, a module communicating with a certain weather API. Because the protocol and format of the weather API is constant, there is no benefit to altering the module's behavior.

15.4.3   Internal

The internal access level is the default. If you don't explicitly declare an access level for a module component, Swift will assume it to be `internal`. Such components are accessible anywhere within the same module, but not visible to other modules. Listing 15.8 demonstrates the syntax for internal components.

```
struct InternalType {
    func internalMethod() {
        print("Internal use only")
    }
}
```
Listing 15.8 Internal Struct

The client app or another module can't access `InternalType` in any usual way.

The internal access level is perfect for app code that should be publicly available within the module, but hidden from the outside. For example, consider a `LogFormatter` class in an `AppLogger` module: You'd want the formatter to be widely available within `AppLogger`, but clients would access high-level read/write functions only.

15.4.4   File Private

This is a light variant of the private access level. It limits the usage of a component even within the module itself. File private components are only accessible within the same *.swift* source file. Other files within the same module can't access those components, and client code certainly can't either. Listing 15.9 demonstrates the syntax for file private components.

```
fileprivate func helperFunction() {
    print("Used only in this file")
}
```

Listing 15.9 File Private Function

Neither other files in the module nor the client app can access `helperFunction` in this case.

File private access offers a niche option for components, which makes them almost private, except to their immediate families within the same file. In a file containing human classes like `Employee`, `Manager`, and `Student`, a common file private `formatHumanName` function could come in handy. All classes would benefit from that utility function, while it's invisible to the outside world.

15.4.5   Private

Private is the most restrictive access level, with which you are already familiar from enumerations, structs, and classes. A private component is accessible only to its owner, such as the class it belongs to. Any other component can't access it. Listing 15.10 demonstrates the syntax for private components.

```
class SecretKeeper {
    private var secret = "hidden"

    private func revealSecret() {
        print(secret)
    }
}
```

Listing 15.10 Private Variable

In this example, `SecretKeeper.secret` is only available to components of `SecretKeeper`.

The private access level is suitable for the internal mechanisms of data types, exposure of which is meaningless or dangerous.

15.5   Summary

In this section, you learned about modules in Swift, which enable code sharing between different projects in a standardized way. Modules offer various advantages, such as reusability, organization, encapsulation, compilation speed, and dependency management.

Modules come in two main flavors: They can be published either as binary frameworks or as open-source packages.

Frameworks are coded and built within the source project and imported as a binary file into the target project.

Packages, on the other hand, are coded within the source project and typically published to a code platform like GitHub. To import the package to the target project, you typically use Swift Package Manager, which also offers convenient features such as automatically updating the package from the source.

Not all components within a module need to be made completely accessible. Instead, you can mark them for various access levels, like open, public, internal, file private, and private.

Now, we're nearly done! In the next chapter, we will offer a summary of the book and go over some suggestions for your future learning path.

Chapter 16
Conclusion

This chapter will summarize the content of the book, what you learned, and how to advance further in the programming journey you started with Swift.

Congratulations on reaching the end of the book! We certainly hope that your learning journey was at least as enjoyable as the writing journey was on our side. At this point, you are armed with the core knowledge of the Swift language. Here is a recap of what you learned about throughout this book:

- General Swift knowledge
- Variables and data collections
- Control flow statements
- Functions for code modularization
- Complex data types like enums, structs, classes, and protocols
- Extensions to enhance existing data types
- Error handling and defensive programming
- File handling to read and write persistent data
- Concurrency to develop multithreaded apps
- Modules for sharing libraries among projects

In any book, including this one, it is impossible to cover every single feature, component, nook, corner, and alley that a programming language has to offer. But nowadays, it's arguable if that's a necessity at all. Once you are adept with the general architecture and structure of a programming language, you can consult online resources or AI tools about the syntax of something you haven't done before.

Relying 100% on AI for code generation, which is also called *vibe coding*, isn't a recommended practice at this time. If you can't understand the generated code and how/if it fits into the language structure, you're taking a dangerous gamble with your app. Gamble enough times and you'll eventually lose.

However, if you know a programming language well already, AI can be a helpful tool to complete some missing pieces of the puzzle when necessary. In other words, if you understand the language's architecture, AI can help fill in the gaps. Having completed this book, you are now in a position to use AI as a code suggestion engine when necessary.

Building such a Swift framework was the core goal of this book in the first place. We aimed to educate readers like yourselves about the architecture and practical features of the Swift language, preparing you well for your future Apple-oriented programming adventures. The content of the book was curated to target typical requirements and challenges in app development. That's a more practical choice than being a cold syntax reference resource, right?

Now that you have a solid knowledge of the Swift language, you can continue your learning journey by focusing on Apple's standard frameworks and developing experimental apps for yourself. After all, a typical app is developed using both Swift knowledge *and* Apple's frameworks. Those frameworks can include, for example, the following:

- SwiftUI for user interfaces
- SpriteKit for 2D graphics
- Core Data for persistence
- URLSession for HTTP networking
- AVFoundation for audio and video processing
- Core ML for machine learning
- MapKit for embedded maps

There are many other frameworks, of course. You can pick and study Apple frameworks that serve your app's purpose. You can even bring in third-party modules or develop your own! The idea is that you can import and consume such frameworks easily and combine them with your app's business logic, coded in Swift.

In your programming journey, developing your theoretical knowledge is also important. You can learn from the best practices of programmers before you by studying principles like YAGNI, KISS, DRY, and SOLID. Awareness about code smells and antipatterns can protect you from typical programmer mistakes too. Knowledge about typical design patterns is also a strong asset if you are working with object-oriented architectures. Programming methods like test-driven development or pair programming are also worth knowing; you might find them useful in your own practices.

Before moving to Apple frameworks or programming theory, though, we encourage you to visit this book's appendixes. Appendix A is about unit testing in Swift, which is an important topic for building solid and dependable code bases. Appendix B is about debugging in Xcode, which is an essential skill for any programmer.

That's all for now! We fare thee well and hope to meet you in another book!

Appendix A
Unit Testing

Swift provides built-in mechanisms to debug and test code.
This appendix will cover those mechanisms.

Programming is not about writing code; it's about writing *correct* and *reliable* code. In that regard, one of the most helpful tools is *unit testing*, which is the art and science of writing test code to validate your app code.

Most modern programming ecosystems offer some sort of unit test toolkit. Java has JUnit; Python has Pytest; JavaScript has Jest; and so on. Likewise, Swift offers two distinct testing frameworks: XCTest (older) and Swift Testing (newer).

In this appendix, you will learn the basics of unit testing and how to conduct your own unit tests using either tool.

A.1 Introduction to Unit Testing

In software development, programmers aren't only expected to write and deliver code. They are expected to deliver code that works correctly. All programmers are human, and even the best of us will make mistakes from time to time; that's to be expected.

To catch bugs early and stop them from leaking into production apps, programs will be tested by humans eventually. But they can't test everything—even if they use the best test automation tools.

For instance, imagine an invoice number generation algorithm that uses numbers 0 to 9 and characters A to F for new sequences. It's nearly impossible for a human to sit down and generate billions of invoices to see if the app really jumps from 9 to A every single time.

Besides, testers usually have access to the outer interface of the app only. They can't access and test features deep within the bowels of the app, which reduces the coverage rates of external tests.

To overcome such limitations, the concept of unit testing emerged in the software industry. Unit tests allow programmers to automatically test their own code before delivering it to people further along the pipeline.

Let's start by discussing what a unit test is. Then we will address the programming paradigm called *test-driven development* (TDD) before moving into the testing features of Swift.

A.1.1   What Is a Unit Test?

In a unit test, you basically write *tester code* that lives alongside your *app code*. When you run your unit tests, they test and pressure your app code to ensure that your algorithms work as intended. This architecture is sketched in Figure A.1.

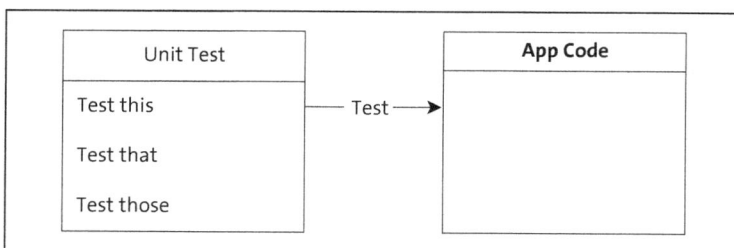

Figure A.1 Unit Test Alongside App Code

With an appropriate architecture, you can test even the deepest internal components of your apps, which might be nearly impossible for an external tester to test.

Unit tests prove to be useful especially after future modifications. After a modification, you can simply rerun your tests to ensure that the new version of your codebase works as well as before. If that's not the case, then either a new bug was injected into your app's code, or your tests need to be updated—a good catch either way!

Naturally, writing and maintaining unit tests take time and effort. Some programmers tend to avoid unit tests to ship fast, accepting the risks and potential technical debt. Some might write unit tests for everything possible, accepting the additional cost in exchange for having an extremely solid app. Some chose a middle route, writing unit tests for critical or externally untestable parts only.

Once you become experienced in unit testing, you can pick the best approach per project with the consensus of your team.

Test Coverage

In unit testing, *test coverage* is a term that pops up frequently. It is a measurement of how much of your code is executed when you run your tests. A high test-coverage rate suggests that most of your code is tested. A high test coverage doesn't guarantee that your code is bug-free, but low coverage usually signals untested, riskier code areas.

A.1.2   Test-Driven Development

Although not mandatory for unit tests, TDD ought to be mentioned in a discussion of unit tests. TDD is a methodology in which you write unit tests before you write the actual production code.

In this approach, you develop a fragment of your app (like a function) in three steps:

- **Red step**
  You write the unit test for a new empty function, declaring the expectation for that function at the very beginning. That test should naturally fail because the function's behavior is missing—thus proving that your test catches errors correctly.

- **Green step**
  You write just enough production code to make the unit test pass, implementing the minimum expectancy into the function.

- **Refactoring step**
  Once the unit test passes, you refactor your code to improve readability, maintainability, and so on. After refactoring, you can rerun your unit tests to ensure that you didn't ruin anything that worked before.

When applied as suggested, TDD brings various benefits to the table:

- Ensuring that the tests themselves run correctly
- Enforcement to write modular, testable pieces of code, thus improving your architecture
- Insurance that code fragments are inherently well-tested and robust
- Reduced debugging effort and external test bugs

However, TDD may also come with some compromises:

- Increased up-front time and effort, which might be risky for extremely tight deadlines
- Difficulties applying the approach to legacy systems
- Risk of overtesting trivial implementation details instead of behavior
- Possibility that tests may become stale and eventually obsolete if not maintained as well as the app

Despite the potential compromises, TDD stands as a solid approach for certain scenarios, such as complex business logic, long-lived codebases, shared code ownership, and API development. Once you get proficient with unit tests, you can experiment with TDD for yourself.

Advice on TDD

TDD is not mandatory to write unit tests. You can simply go ahead and write your app code first and your unit tests after. However, that approach doesn't ensure that the test itself is running correctly. If your test has a bug and doesn't raise an alarm in case of an invalid function, then the unit test was written for nothing.

TDD offers a three-step approach to ensure that your tests are tested with invalid production code before the production code is written. It makes sense, right?

You may or may not write unit tests for everything—but for the parts that you do, it is advisable to follow the TDD approach.

A.1.3   Unit Testing in Swift

Because Swift is a modern language for today's programmer, it is natural to expect some sort of built-in unit testing mechanism—and that expectation is fulfilled with not one, but two frameworks!

XCTest is the traditional testing framework, integrated into Xcode. It has all the unit testing bells and whistles you could ask for and has been used by programmers for many years. It is widely used for apps, including UI and performance testing.

Swift Testing is a newer, modern framework introduced in Swift 5.9. It offers a lightweight syntax with annotations like @Test and macros like #expect. Swift Testing is mostly designed for pure Swift projects and Swift packages, with a simpler syntax and fewer dependencies.

In this appendix, you will learn about both testing frameworks. Let's follow chronological order and start with XCTest.

A.2   Unit Testing with XCTest

XCTest is Apple's classical testing framework, integrated right into Xcode. It has been the standard unit testing tool of Swift—and its ancestor, Objective-C—for many years. You don't need any dependencies or installations to use XCTest either; everything is included in Xcode. In this section, you will learn how to conduct unit tests using XCTest.

No TDD in Examples

In the upcoming examples, we won't use TDD practices because our purpose is to help you understand the test tools in Xcode. Once you become familiar with the tech and decide to use it in your projects, you can apply TDD.

A.2.1   Preparing a Testable Project

Let's start by creating a new project to house your upcoming unit tests. In Xcode, select **File · New · Project** from the menu, then select **macOS · Framework**, as shown in Figure A.2. Click **Next**.

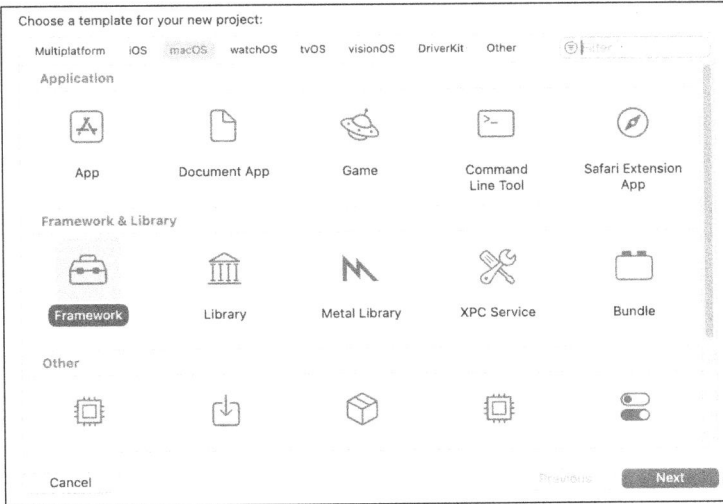

Figure A.2 Selecting Framework Template

In the next window, give a meaningful name to your product, such as **demo1** (see Figure A.3). Select **XCTest** as the **Testing System** (for obvious reasons), then click **Next**.

Figure A.3 Selecting Testing System

Select an appropriate target folder as shown in Figure A.4, then click **Create**, producing the empty project.

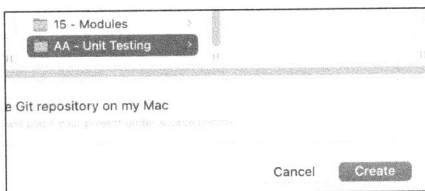

Figure A.4 Selecting Target Folder

Initially, you will have a project structure like that shown in Figure A.5. The **demo1** folder is reserved for your regular Swift code blocks, whereas the **demo1Tests** folder is reserved for your unit tests.

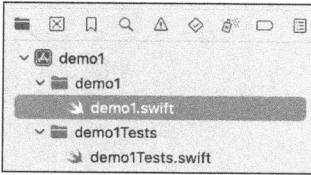

Figure A.5 Folder Structure of Demo Project

Let's put some code into *demo1.swift* now, which will be the subject of your tests. Listing A.1 contains the initial code to apply, but note that it deliberately contains errors. Both `MathHelper.add` and `MathHelper.square` contain incorrect calculations, which you will aim to catch with your unit tests soon.

```
public struct MathHelper {
    public static func add(_ v1: Int, _ v2: Int) -> Int {
        return v1 - v2
    }

    public static func square(_ value: Int) -> Int {
        return value - value
    }
}
```

Listing A.1 Some Incorrect Code in demo1.swift

A.2.2   Coding Unit Tests

Now that you have some testable Swift code at hand, let's write some unit tests! For that purpose, you will edit the *demo1Tests.swift* file. Naturally, you can add further test files to the **demo1Tests** folder, but for this simple case, a single file will suffice.

Swift is kind enough to contain instructions in test files, such as *demo1Tests.swift*, which is shown in Listing A.2. Note that the `demo1` module was imported with the `@testable import` prefix, which obviously indicates that it will be tested here.

```
import XCTest
@testable import demo1

final class demo1Tests: XCTestCase {

    override func setUpWithError() throws {
        // Put setup code here. This method is called before the invocation of
        // each test method in the class.
```

```
    }

    override func tearDownWithError() throws {
        // Put teardown code here. This method is called after the invocation
        // of each test method in the class.
    }

    func testExample() throws {
        // This is an example of a functional test case.
        // Use XCTAssert and related functions to verify your tests produce the
        // correct results.
        // Any test you write for XCTest can be annotated as throws and async.
        // Mark your test throws to produce an unexpected failure when your
        // test encounters an uncaught error.
        // Mark your test async to allow awaiting for asynchronous code to
        // complete. Check the results with assertions afterwards.
    }

    func testPerformanceExample() throws {
        // This is an example of a performance test case.
        self.measure {
            // Put the code you want to measure the time of here.
        }
    }
}
```

Listing A.2 Initial Contents of demo1Tests.swift

The purpose of each function is made clear in the comments if you go through them, but let's address them here nonetheless:

- `setUpWithError` is a function that is automatically called before each test function. If you have initial preparations, such as opening database connections or preparing temporary files, you can place them here.

- `testExample` and `testPerformanceExample` are two sample test functions. In Swift, each test function must begin with the `test` prefix. When you conduct unit tests, only functions starting with `test` are executed. Note that test functions need to have the `throws` suffix because invalid test results will emerge as errors.

- `tearDownWithError` is a function that is automatically called after each test function. If you have concluding operations, such as closing database connections or deleting temporary files, you can place them here.

Targeting the sample `MathHelper` class, Listing A.3 shows the modified version of *demo1-Tests.swift*. Note that we added a test function for each `MathHelper` function.

```
import XCTest
@testable import demo1

final class demo1Tests: XCTestCase {
    func testAdd() throws {
        let sum1 = demo1.MathHelper.add(1, 2)
        XCTAssertEqual(sum1, 3)

        let sum2 = demo1.MathHelper.add(5, 0)
        XCTAssertEqual(sum2, 5)

        let sum3 = demo1.MathHelper.add(100, 100)
        XCTAssertEqual(sum3, 200)
    }

    func testSquare() throws {
        let square1 = demo1.MathHelper.square(4)
        XCTAssertEqual(square1, 16)

        let square2 = demo1.MathHelper.square(0)
        XCTAssertEqual(square2, 0)
    }
}
```

Listing A.3 demo1Tests.swift Containing Meaningful Tests

In this version of the file, testAdd contains unit tests targeting the demo1.MathHelper.add function. After each add operation, you invoke the XCTAssertEqual function, checking if the result of the add equals the expected value:

- If values are equal, then XCTAssertEqual will end silently, and the unit test will pass successfully.
- If values are not equal, then XCTAssertEqual will generate an error and the unit test will fail, demanding your attention.

Following that approach, assert that $1 + 2 = 3$, $5 + 0 = 5$, and $100 + 100 = 200$. Each of those calculations is a test case. In a real-world app, a comprehensive function may have dozens of test cases. It is important to establish a good set of test cases in which regular scenarios are covered—as well as irregular or unexpected scenarios. The more comprehensive your test cases are, the higher the quality of your unit tests will be.

In the second part of the file, similarly, testSquare contains unit tests targeting the demo1.MathHelper.square function.

A.2.3   Executing Unit Tests

Now, let's execute those unit tests—keeping in mind that MathHelper deliberately contains errors to demonstrate invalid unit test results. The easiest way to execute tests is to click the **Product · Test** menu option. Xcode will prepare and execute the tests, showing you the results in the project browser. As shown in Figure A.6, all the tests failed—just like we wanted for this demonstration!

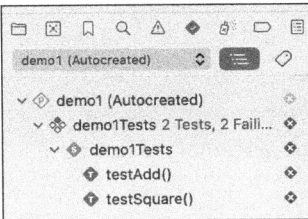

Figure A.6 Failed Unit Test Summary

Failed tests are shown with a red icon. By selecting a failed unit test, you can easily see where and why the test failed, as shown in Figure A.7.

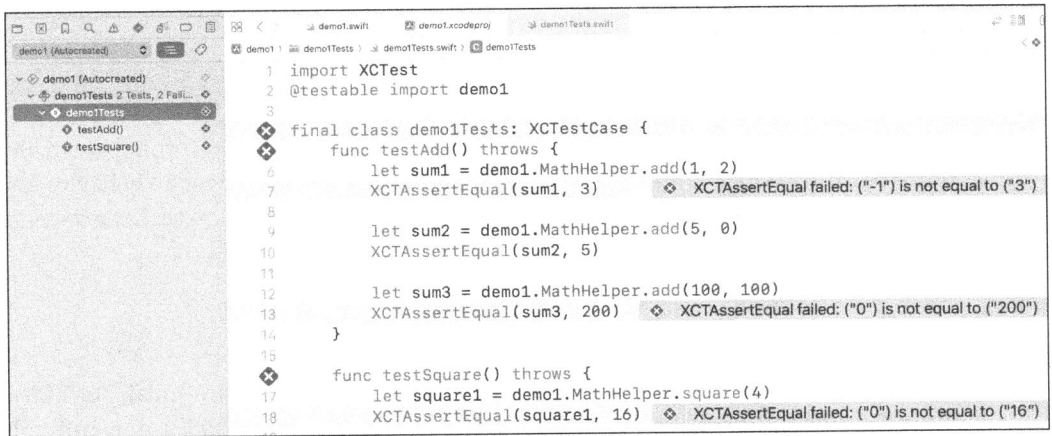

Figure A.7 Details of Unit Test Errors

In this case, Xcode tells us that 1 + 2 was calculated as -1 instead of 3. Likewise, 100 + 100 was calculated as 0 instead of 200, and the square of 4 was calculated as 0 instead of 16.

Now, that's a lot of errors! But it's best to catch them right here, before shipping your code. That is much better than stalling the test team or facing crashes in a published app. Thanks to unit tests, you have the chance to correct the errors right in the source, before things get out of hand!

To that end, Listing A.4 contains the corrected version of *demo1.swift*. This time, add correctly calculates the sum of given values, while square correctly multiplies the given value by itself.

```
public struct MathHelper {
    public static func add(_ v1: Int, _ v2: Int) -> Int {
        return v1 + v2
    }

    public static func square(_ value: Int) -> Int {
        return value * value
    }
}
```

Listing A.4 Corrected Version of demo1.swift

Now, when you reexecute your unit tests, all of them will pass, as indicated in Figure A.8.

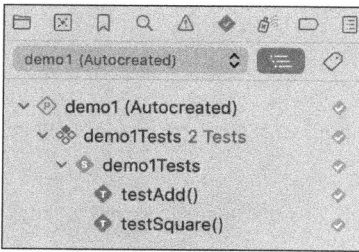

Figure A.8 Passed Unit Test Summary

That's the result you want to see before shipping your code, with everything green. But note that you can trust those green results only if the unit tests cover all behavior and all test scenarios. The "holes" in your tests have a negative correlation with the dependability of the "greenness" of the test results.

A.2.4   Assertions

In our example, we only used XCTAssertEqual to test if two values are equal. But XCtest offers further assertions for your convenience. Table A.1 contains a list of commonly used assertions.

Assertion	Passes the Test...
XCTAssertEqual	...if two values are equal
XCTAssertNotEqual	...if two values differ
XCTAssertTrue	...if a condition is true
XCTAssertFalse	...if a condition is false

Table A.1 Common XCTest Assertions

Assertion	Passes the Test...
XCTAssertNil	...if a value is nil
XCTAssertNotNil	...if a value is not nil
XCTFail	...never; fails the test directly in an unexpected case
XCTAssertThrowsError	...if an expression throws an error
XCTAssertNoThrow	...if an expression doesn't throw an error

Table A.1 Common XCTest Assertions (Cont.)

In unit tests of your projects, you can use any of those assertions as necessary.

A.2.5   Test Coverage

In Swift, as well as any other programming language, test coverage is an important concept in unit testing. It indicates how much of your code is covered by your unit tests. A test coverage of 0% indicates that you don't have any unit tests at all, whereas a test coverage of 100% indicates that you have a unit test for each and every component of your project.

As discussed in the introduction, the ideal coverage target may change from project to project. In a very critical app, where financial transactions are at stake, you may have a coverage target of 100%. In a relaxed app, you may decide that only some critical parts should be covered.

But remember that test coverage doesn't tell you anything about the "quality" of unit tests. You may write a test for each function and reach a coverage rate of 100%, but if your test cases are weak, then the high coverage will be meaningless.

Having that said, Xcode has built-in test coverage measurements; you don't have to do anything extra at all. After running the tests, simply visit the Report Navigator, where test coverage is displayed clear as day (see Figure A.9).

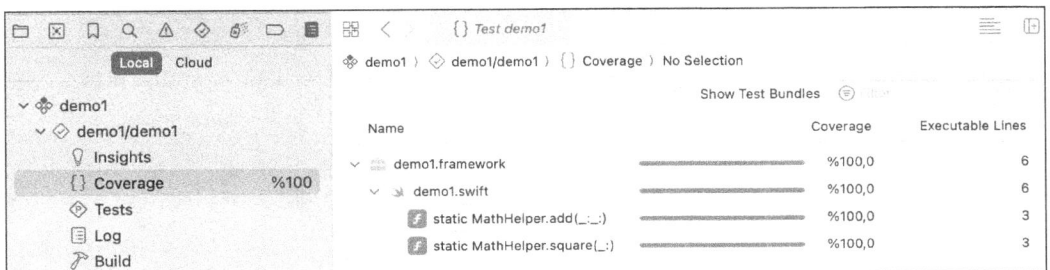

Figure A.9 Test Coverage Rate in Xcode

In this case, we have a coverage rate of 100% because all functions were covered in unit tests. As an exercise, as shown in Listing A.5, let's throw a new function that has no unit tests into *demo1.swift*.

```swift
public struct MathHelper {
    public static func add(_ v1: Int, _ v2: Int) -> Int {
        return v1 + v2
    }

    public static func square(_ value: Int) -> Int {
        return value * value
    }

    public static func doNothing() {}
}
```

Listing A.5 MathHelper with New doNothing Function

If you rerun the tests, the test coverage rate now will reduce, as shown in Figure A.10, just as expected.

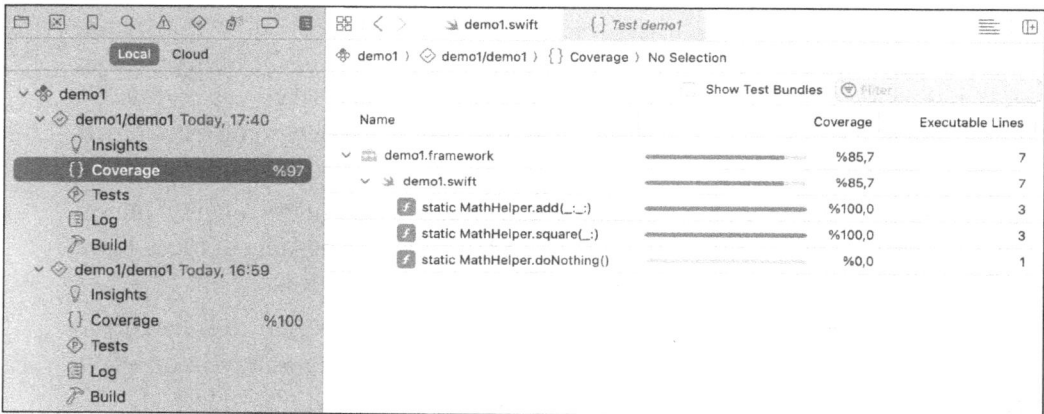

Figure A.10 Reduced Test Coverage Rate

This concludes our look into unit testing with XCTest. Now, let's examine how to conduct unit tests via Swift Testing.

A.3   Unit Testing with Swift Testing

Swift Testing is the new kid on the block. Unlike its ancestor, XCTest, which was originally built for Objective-C and later adopted by Swift, Swift Testing was built from scratch, targeting Swift exclusively. It uses simple annotations and macros, resulting in concise and readable tests.

In this section, you will create a project similar to the one in the former section and see how Swift Testing would be applied.

No TDD in Examples

As in the XCTest section, we won't strictly follow TDD practices for Swift Testing because our purpose is to help you understand the test tools in Xcode. You can apply TDD practices in your projects later on.

A.3.1   Preparing a Testable Project

Let's start by creating a new project to house your upcoming unit tests. In Xcode, click menu option **File · New · Project** and select **macOS · Framework**, as shown in Figure A.11. Click **Next**.

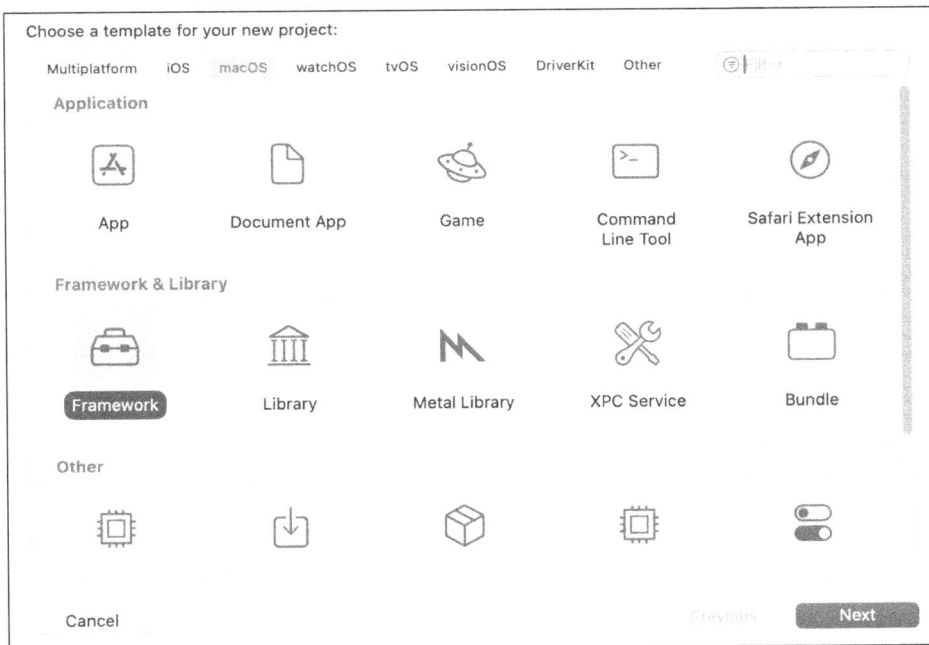

Figure A.11 Selecting Framework Template

In the next window, give a meaningful name to your product, like **demo2**, as shown in Figure A.12. Select **Swift Testing** as the **Testing System** (for obvious reasons), then click **Next**.

Figure A.12 Selecting Test System

Select an appropriate target folder as shown in Figure A.13, then click **Create**, producing the empty project.

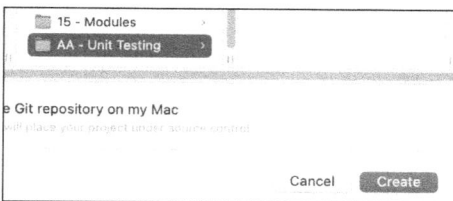

Figure A.13 Selecting Target Folder

Initially, you will end up with a project structure as shown in Figure A.14. The **demo2** folder is reserved for your regular Swift code blocks, whereas the **demo2Tests** folder is reserved for your unit tests.

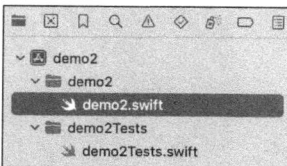

Figure A.14 Folder Structure of Demo Project

As you can see, the structure of the Swift Testing project is the same as the structure of the former XCTest project. Thus, we will proceed likewise—that is, write some testable code and conduct the tests. But this time, the unit tests will be coded using the Swift Testing syntax.

To follow a similar logic, put the code sample from *demo1.swift* in the XCTest project into the *demo2.swift* file of the Swift Testing project, as shown in Listing A.6. Remember that this code snippet deliberately contains errors. Both `MathHelper.add` and `Math-Helper.square` contain incorrect calculations, which you will aim to catch with your unit tests soon.

```
public struct MathHelper {
    public static func add(_ v1: Int, _ v2: Int) -> Int {
        return v1 - v2
    }

    public static func square(_ value: Int) -> Int {
        return value - value
    }
}
```

Listing A.6 Some Incorrect Code in demo2.swift

A.3.2   Coding Unit Tests

Now that we have some testable Swift code at hand, let's write some unit tests! For that purpose, you will edit the *demo2Tests.swift* file. Naturally, you can add further test files to the **demo2Tests** folder, but for this simple case, a single file will suffice.

Swift is kind enough to contain instructions in test files, such as *demo2Tests.swift*, which is shown in Listing A.7. Note that the demo2 module was imported with the `@testable import` prefix, which indicates that it will be tested here.

```
import Testing
@testable import demo2

struct demo2Tests {
    @Test func example() async throws {
        // Write your test here and use APIs like `#expect(...)` to check
        // expected conditions.
    }
}
```

Listing A.7 Initial Contents of demo2Tests.swift

As you can see, the skeleton template of Swift Testing is much simpler than the template of XCTest. The core differences are as follows:

- No subclassing/inheritance is required.
- No setup/teardown functions are present.
- Test functions are marked with the `@Test` annotation instead of by enforcing a naming convention.
- Assertions use macros like #expect and #require.

Now, let's write meaningful tests into *demo2Tests.swift*, as shown in Listing A.8.

```swift
import Testing
@testable import demo2

struct demo2Tests {
    @Test
    func testAdd() {
        #expect(MathHelper.add(1, 2) == 3)
        #expect(MathHelper.add(5, 0) == 5)
        #expect(MathHelper.add(100, 100) == 200)
    }

    @Test
    func testSquare() {
        #expect(MathHelper.square(4) == 16)
        #expect(MathHelper.square(0) == 0)
    }
}
```

Listing A.8 Meaningful Tests in demo2Tests.swift

In the `testAdd` function, we have used the `#expect` macro to validate the results of `MathHelper.add` over various sample values (test cases). Each `#expect` macro checks if the result is as expected:

- If values are equal, then `#expect` will end silently, and the unit test will pass successfully.
- If values are not equal, then `#expect` will generate an error and the unit test will fail, demanding your attention.

Following that approach, we will test *1 + 2 = 3, 5 + 0 = 5*, and *100 + 100 = 200*. In a real-world app, your unit test may (and should) contain as many test scenarios as are needed, ensuring a robust and comprehensive test that includes exceptional and extreme cases.

Similarly, in the second part of the file, `testSquare` contains unit tests targeting the `demo2.MathHelper.square` function.

A.3.3   Executing Unit Tests

Now, let's execute those unit tests—keeping in mind that `MathHelper` deliberately contains errors to demonstrate invalid unit test results. Running tests requires the same action as in XCTest: simply click the **Product • Test** menu option. Xcode will prepare and execute the tests, showing you the results in the project browser. As shown in Figure A.15, all the tests failed—just like we wanted for this demonstration!

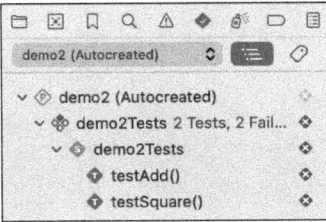

Figure A.15 Failed Unit Test Summary

Failed tests are shown with a red icon. By selecting a failed unit test, you can easily see where and why the test failed, as shown in Figure A.16.

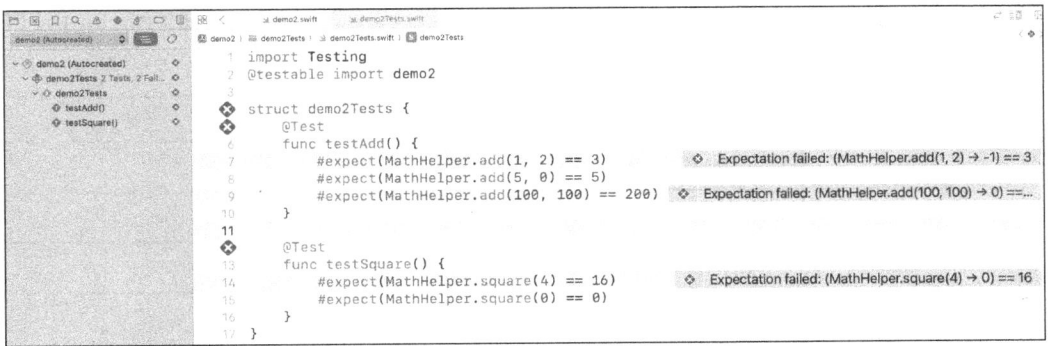

Figure A.16 Details of Unit Test Errors

In this case, Xcode tells you that 1 + 2 was calculated as -1 instead of 3. Likewise, 100 + 100 was calculated as 0 instead of 200, and the square of 4 was calculated as 0 instead of 16.

Just as you did with XCTest, you are going to fix the errors and rerun the tests. Listing A.9 contains the corrected version of *demo2.swift*. This time, add correctly calculates the sum of given values, while square correctly multiplies the given value by itself.

```
public struct MathHelper {
    public static func add(_ v1: Int, _ v2: Int) -> Int {
        return v1 + v2
    }

    public static func square(_ value: Int) -> Int {
        return value * value
    }
}
```

Listing A.9 Correct Version of demo2.swift

Undoubtably, rerunning the unit tests will result in a pleasant list of green results, indicating the success of the tests. Yay!

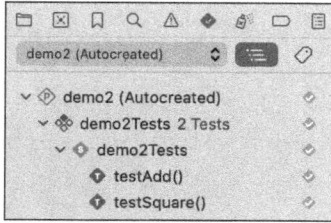

Figure A.17 Passed Unit Test Summary

A.3.4   Test Macros

In XCTest, you had access to various assertions like XCTAssertEqual to validate results in unit tests. Likewise, various macros beyond #expect are available in Swift Testing to help make writing unit tests a little easier. Table A.2 contains a list of commonly used macros.

Macro	Behavior on False
#expect	Fails the test case; proceeds with further test cases
#require	Fails the test case, aborting the test
#assert	Crashes the app

Table A.2 Common Swift Testing Macros

A.3.5   Test Coverage

Built-in test-coverage measurements of Xcode are available for both XCTest and Swift Testing. After running the tests, simply enter the Report Navigator, where test coverage is displayed, as shown in Figure A.18.

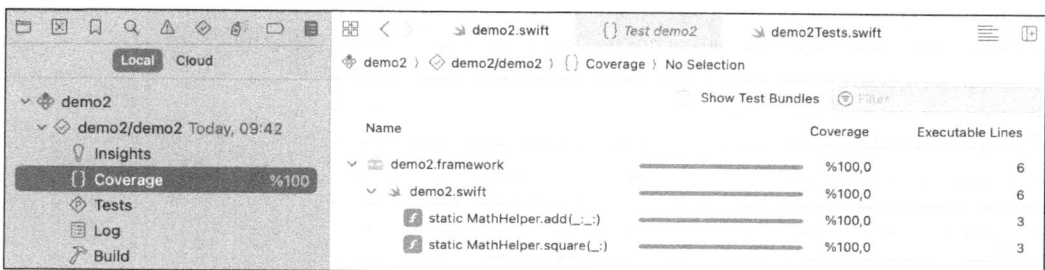

Figure A.18 Test Coverage Rate in Xcode

Further information regarding test coverage, which was explained for XCTest in Section A.3.5, is completely applicable to Swift Testing as well; you can revisit that section if you need a memory refresh.

A.4 Summary

In this appendix, you learned about unit testing and how to write unit tests in Swift.

Unit tests are little tester programs that live alongside productive app code. Every time unit tests are executed, they test your app code through precoded test cases, ensuring that the given inputs produce the expected outputs.

In the programming paradigm called *test-driven development* (TDD), tests and functions are developed simultaneously, ensuring a high test coverage rate and robust tests.

In Swift, there are two options to code unit tests. XCTest is the legacy syntax, offering a traditional approach based on inheritance, naming conventions, and assertions. Swift Testing is the new syntax, offering a modern approach based on annotations and macros.

XCTest is suggested for full app projects in which UI and performance testing is required. Swift Testing is suggested for pure Swift codebases like libraries, where a simpler syntax is desired.

Appendix B
Debugging

Swift naturally provides built-in mechanisms to debug code.
This appendix will explain how to debug your apps.

Debugging is one of the core skills of programming. As you set sail from simple playground examples toward advanced, real-world applications, the importance of this skill increases exponentially.

The term *debugging* originated from an actual insect-related incident. Back in 1947, when computers had vastly different architectures, engineers inspecting a computer error found that the reason for the unexpected malfunction was a moth stuck in a relay. They removed the bug and called the process debugging. Since then, the process of spotting and extracting program errors is commonly called debugging, whether the cause is a real bug or a logical code error.

In today's modern programming environments, debugging basically means running your code in stop-motion, so to speak. Instead of letting the computer process your Swift code with lightning speed, you can make it execute each line sequentially and wait for a green light before the next line. During this stop-motion process, you can comfortably inspect the flow of the code and values of variables, putting your app under the microscope. That helps tremendously when you need to spot bugs or errors in your code or understand the logic of a complex or foreign piece of code.

Most integrated development environments (IDEs) have built-in tools for debugging, and Xcode is no exception. In this appendix, you will learn how to debug your Swift programs in Xcode and pick up some helpful tips and tricks along the way.

B.1 Debugging in Xcode

In this section, you will start your debugging adventures with straightforward basics. You'll learn how to run a step-by-step debug process and how to use breakpoints and inspect variables.

B.1.1 Creating a Sample Project

First, you're going to need a sample project to debug. For this purpose, let's create a simple command line tool containing the code in Listing B.1. This program calculates the area of a circle and a square, then subtracts one from the other to find the difference. That's very straightforward, right?

```swift
func calcSquareArea(side: Double) -> Double {
    let area = side * side
    return area
}

func calcCircleArea(radius: Double) -> Double {
    let area = 3.14 * radius * radius
    return area
}

let area1 = calcSquareArea(side: 10)
let area2 = calcCircleArea(radius: 10)
let areaDiff = area2 - area1
print("Difference: \(areaDiff)")
```

Listing B.1 Sample Swift Code to Debug

When this sample project is executed, it should produce the output shown in Figure B.1.

```
Difference: 214.0
Program ended with exit code: 0
```

Figure B.1 Output of Sample Program

So far, so good! Now you can start debugging that project and see what Xcode has to offer in that regard.

B.1.2   Setting Breakpoints

A *breakpoint* marks a certain line in your code on which you want the program flow to pause. Starting from that point, you can run the program in slow motion, line by line, to observe how it behaves.

To set a breakpoint in Xcode, simply click the row number of the line where you want the breakpoint set. A dark arrow should appear on that line, indicating that the breakpoint is active. Figure B.2 shows an active breakpoint set on line 11.

```
10
11  let area1 = calcSquareArea(side: 10)
12  let area2 = calcCircleArea(radius: 10)
13  let areaDiff = area2 - area1
14  print("Difference: \(areaDiff)")
```

Figure B.2 Active Breakpoint on Line 11

If you reclick the line number, the breakpoint arrow will become a lighter shade, indicating that the breakpoint is temporarily deactivated. Figure B.3 shows such an inactive breakpoint on line 11.

```
10
      let area1 = calcSquareArea(side: 10)
12    let area2 = calcCircleArea(radius: 10)
13    let areaDiff = area2 - area1
14    print("Difference: \(areaDiff)")
```

Figure B.3 Inactive Breakpoint on Line 11

You can toggle breakpoints on and off as you please. Naturally, a Swift code file can contain multiple breakpoints too. Figure B.4 shows a sample with two active breakpoints and one inactive breakpoint.

```
 1    func calcSquareArea(side: Double) -> Double {
 2        let area = side * side
 3        return area
 4    }
 5
 6    func calcCircleArea(radius: Double) -> Double {
 7        let area = 3.14 * radius * radius
          return area
 9    }
10
11    let area1 = calcSquareArea(side: 10)
12    let area2 = calcCircleArea(radius: 10)
13    let areaDiff = area2 - area1
14    print("Difference: \(areaDiff)")
```

Figure B.4 Multiple Breakpoints in Same File

To get rid of a breakpoint completely, you can simply right-click the line number and select **Delete Breakpoint**, as shown in Figure B.5.

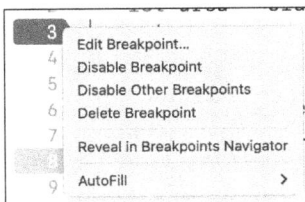

```
 3    Edit Breakpoint...
 4    Disable Breakpoint
 5    Disable Other Breakpoints
 6    Delete Breakpoint
 7
      Reveal in Breakpoints Navigator
 9    AutoFill                      >
```

Figure B.5 Deleting Breakpoint

If you are working through the examples, delete all breakpoints except the initial one at line 11, as shown in Figure B.6. That will be your initial breakpoint for debugging.

```
10
11  let area1 = calcSquareArea(side: 10)
12  let area2 = calcCircleArea(radius: 10)
13  let areaDiff = area2 - area1
14  print("Difference: \(areaDiff)")
```

Figure B.6 Initial Breakpoint to Debug

B.1.3   Starting a Debug Session

To start debugging, all you have to do is to run your app with a breakpoint present in the code. Xcode will start running the app, and as soon as a breakpoint is encountered, execution will be paused, enabling your code and memory inspection. Figure B.7 shows what a debug session looks like.

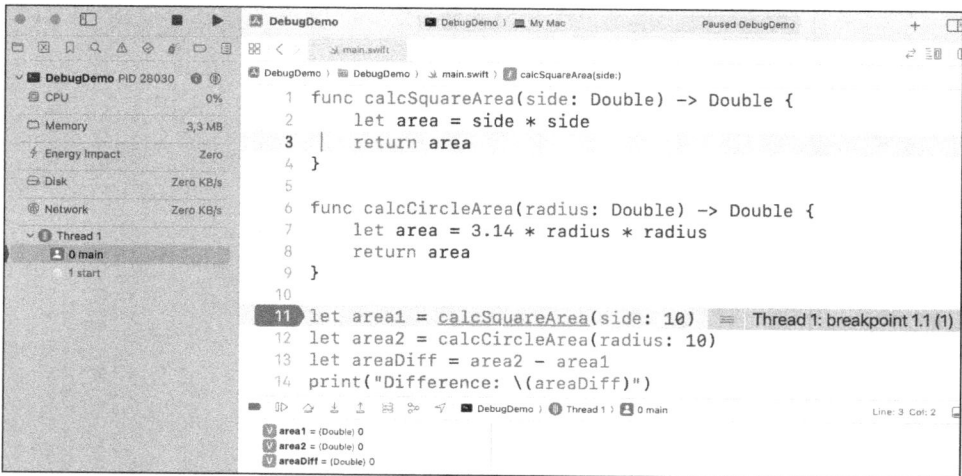

Figure B.7 Initial Debug View of Xcode

In this view, you have multiple areas to monitor the state of your app:

- The left side contains consumption stats, as well as the active threads of the app.
- The right side contains Swift code in execution in the larger area at the top.
- The right side also contains values of variables in scope as well as valuable debug controls in the smaller area at the bottom.

When paused at line 11, the area1, area2, and areaDiff variables are 0 at this time. Because no calculation was conducted yet, they remain at their initial values.

B.1.4   Running Through a Debug

As mentioned before, you can execute the code in stop-motion—that is, one line at a time. To enable this feature, Xcode features debug control buttons, as shown in Figure B.8; these buttons are located right above the variables.

Figure B.8 Control Buttons in Xcode Debugger

Using these buttons, you can control the flow of the program:

- ▯▷ **Continue** will end debugging and keep running the app—until Xcode encounters another breakpoint.

- ⌂ **Step Over** will execute the current line and stop at the next line. If the current line contains a subroutine call, then Xcode will execute the subroutine in the background and move to the next statement.

- ↓ **Step Into** will step into the subroutine called in the current line, enabling a peek into a function.

- ↑ **Step Out** will exit the current subroutine, stepping back into the caller.

In the current state, the cursor is on line 11, as shown in Figure B.7. In this state, click **Step Into** and make Xcode wander into `calcSquareArea`. The debugger will step into that function and wait for your instructions, as shown in Figure B.9.

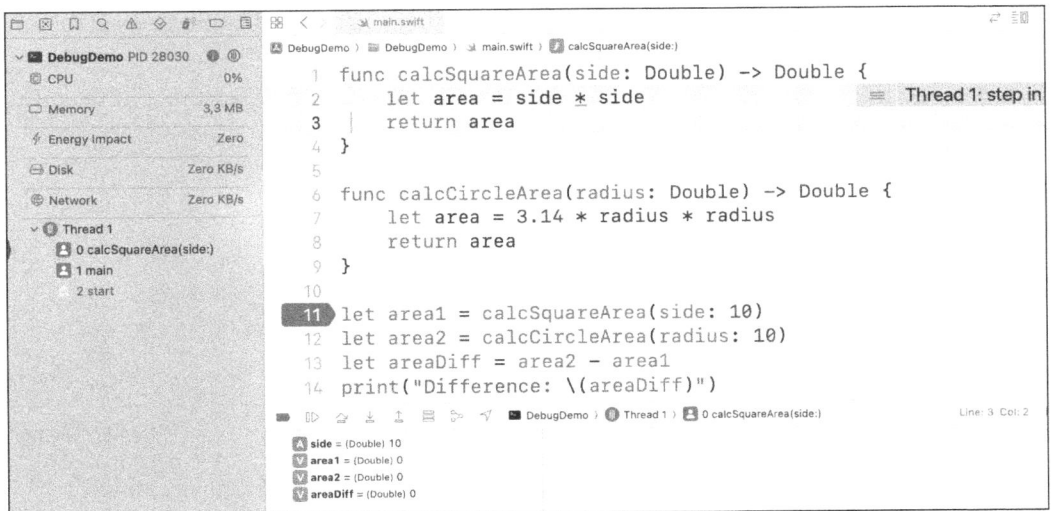

Figure B.9 Stepped Into calcSquareArea

Note that in the thread view on the left, you can see that you have stepped from the main flow into `calcSquareArea`. The deepest stack component is displayed at the top. In the code view, the debugger is pending on line 2, and in the variable view, a new value emerged: **side = 10**. That's because the `side` parameter of the function was sent as 10 by the caller.

Now, click **Step Over**. Xcode will execute the code in line 2 and move to line 3, as shown in Figure B.10.

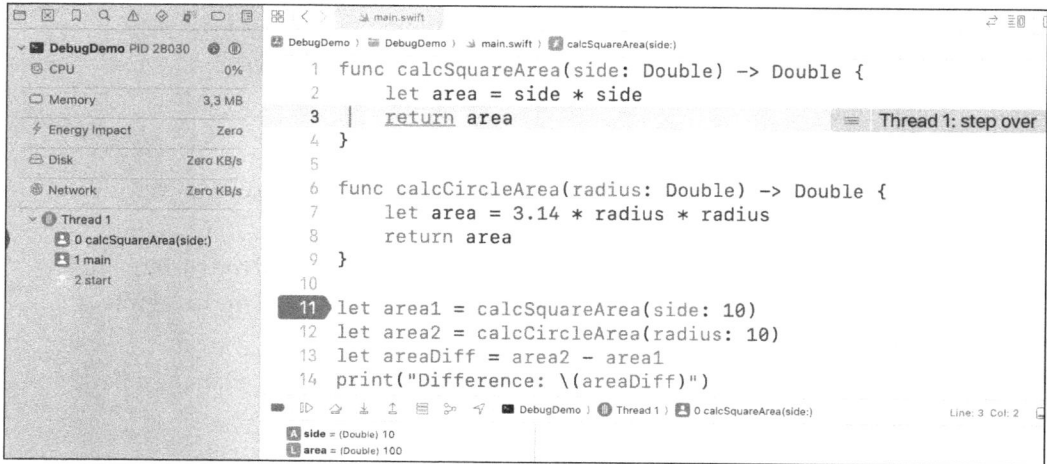

Figure B.10 Stepped Over to Next Line

In the variable view, a new value of **area = 100** has emerged now. That indicates the square's area, which you have calculated.

Clicking **Step Out** will leave this function, bringing you back to the next executable line in the main flow, as shown in Figure B.11.

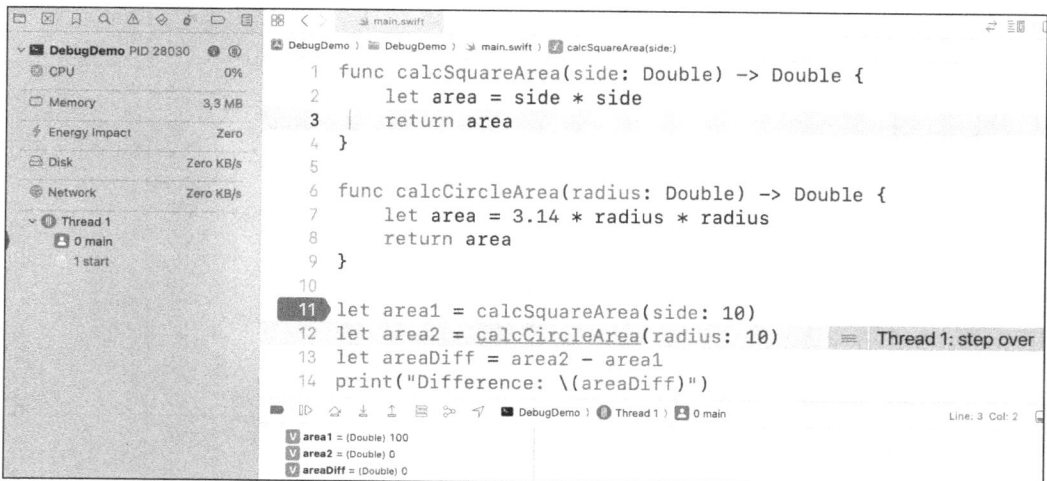

Figure B.11 Stepped Over to Calculation of area2

Assuming that you aren't interested in how the circle's area is calculated, you can click **Step Over** to execute `calcCircleArea` in the background. Xcode will process that function and stop at line 13, as shown in Figure B.12, and will then wait for your instructions.

At this point, both **area1 = 100** and **area2 = 314** are visible in the variable view. The rest of the code seems obvious, so click **Continue** to stop debugging and let Xcode run through the rest of the program. That will produce the expected output of **214**, which was shown in Figure B.1.

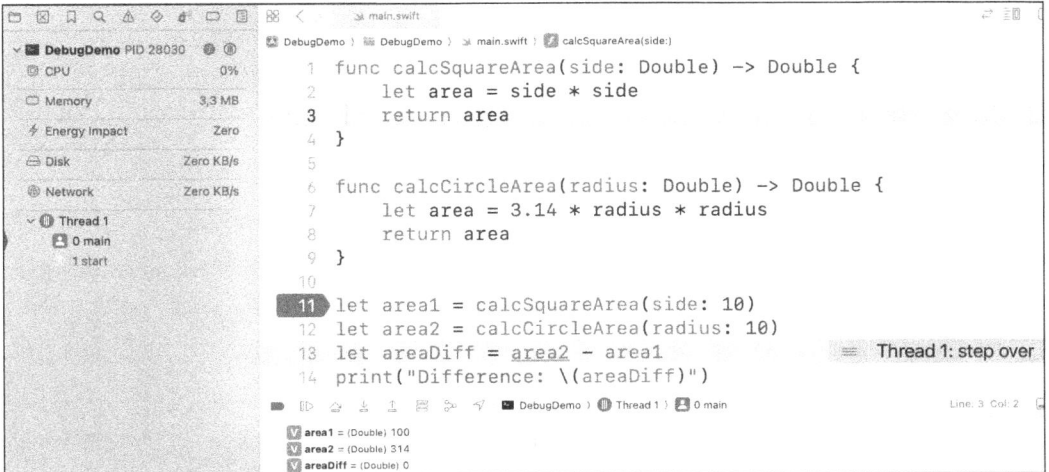

Figure B.12 Stepped Over to Calculation of areaDiff

B.1.5 Changing Breakpoints During Debug

You will be pleased to know that you can set, deactivate, or delete breakpoints during a debug session. As a demonstration, let's restart the app, stopping at the breakpoint on line 11. While Xcode waits, you can click line 8, setting a new breakpoint, as shown in Figure B.13.

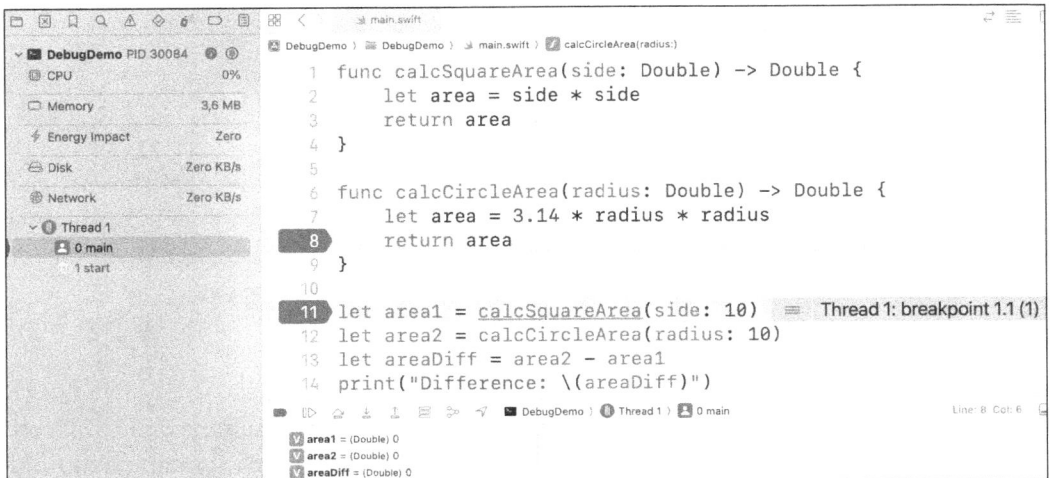

Figure B.13 New Breakpoint During Debug

If you click **Continue** in this state, Xcode will run through the code until it encounters the breakpoint on line 8. When that breakpoint is reached, Xcode will pause once again, as shown in Figure B.14.

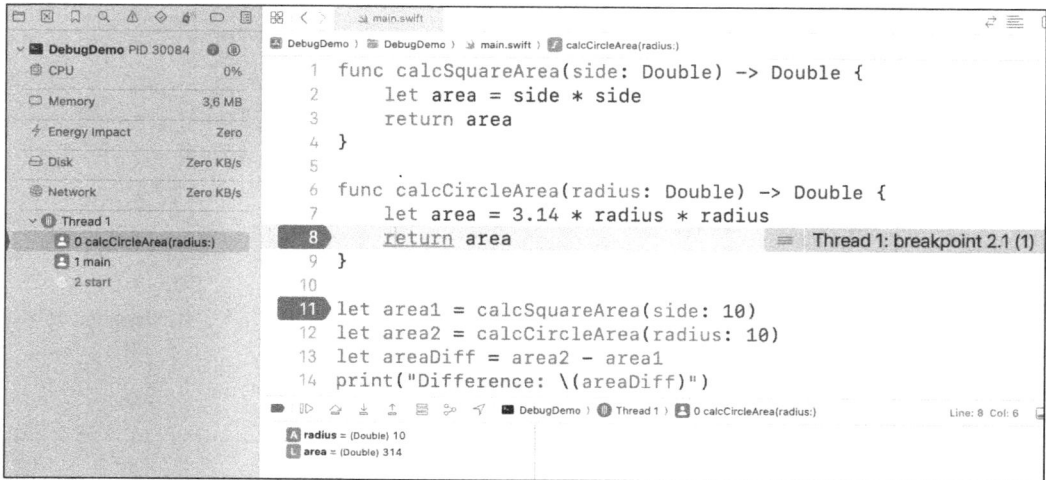

Figure B.14 Xcode Paused at New Breakpoint

That dynamic breakpoint flexibility is invaluable in big code bases in which you might need to chase an insidious bug. Instead of stepping in and out of every single line of code, you can skip seemingly bug-free parts on the go, focusing on suspicious code blocks.

Now that you're familiar with straightforward debugging, let's discuss some advanced features.

B.2   Advanced Debugging Tools

Browsing through the code and stepping in and out of subroutines might be enough to debug simple cases, but sometimes you'll wish you had more options. Xcode offers powerful advanced debugging tools that will help you debug faster and smarter in difficult codebases. In this section, we will introduce some of those tools.

B.2.1   Conditional Breakpoints

A *conditional breakpoint* only stops execution if a specific condition is true. This technique is really useful when you want the app to mostly execute as normal and stop only in an exceptional scenario worth inspecting.

Check the code sample in Listing B.2, in which the app attempts to connect to a resource 300 times.

```
func attemptConnecting(maxTimes: Int) {
    for i in 1...maxTimes {
        print("Attempt \(i)...")
        // Connection attempt code
    }
}
```

```
attemptConnecting(maxTimes: 300)
```

Listing B.2 Connection Attempt Example

Now, what if you want to debug specifically the 100th attempt? Using the traditional method, you can set a breakpoint inside the `for` loop and hit **Continue** repeatedly until i becomes 100.

That doesn't sound very practical, does it? It's a lot of effort and a total waste of time. Besides, if you have to debug the same case again, you must repeat that process.

For such scenarios, in which you want the debugger to stop only in case of a specific condition, Xcode offers conditional breakpoints. After setting the breakpoint on the desired line, you can right-click the breakpoint and select **Edit Breakpoint**, as shown in Listing B.2.

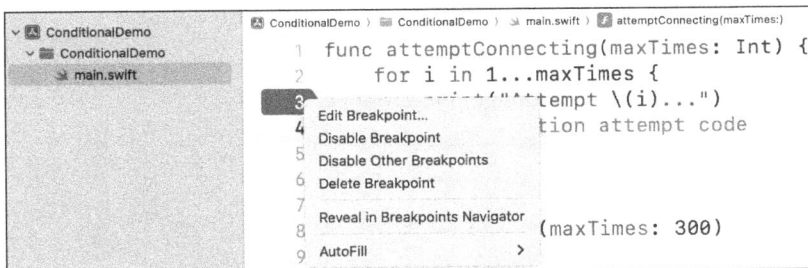

Figure B.15 Editing Breakpoint

In the resulting popup, you can enter any condition for which the breakpoint should function. In our simple example, we will merely enter i == 100 as the condition, as shown in Figure B.16.

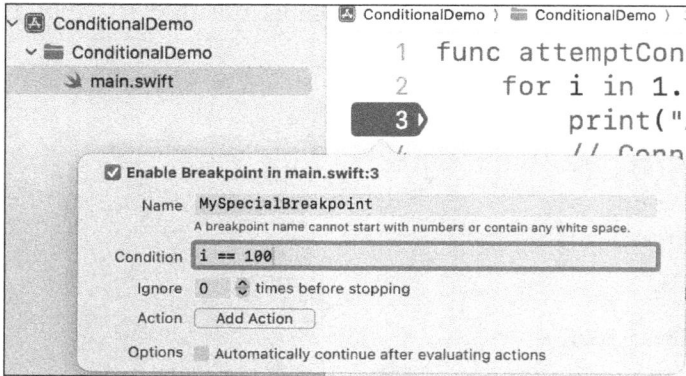

Figure B.16 Entering Breakpoint Condition

You can also give a meaningful name to the breakpoint if you like—for example, **MySpecialBreakpoint**. That habit makes sense especially if you have a lot of breakpoints. You can tell them apart in the breakpoints navigator by their names, as shown in Figure B.17.

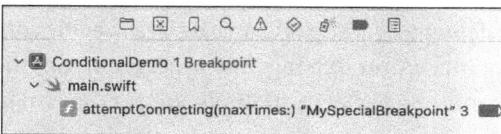

Figure B.17 Breakpoint Names in Project Browser

When you execute the app, it will stop at the third line—but only when i becomes 100. Figure B.18 shows the moment the debugger has stopped.

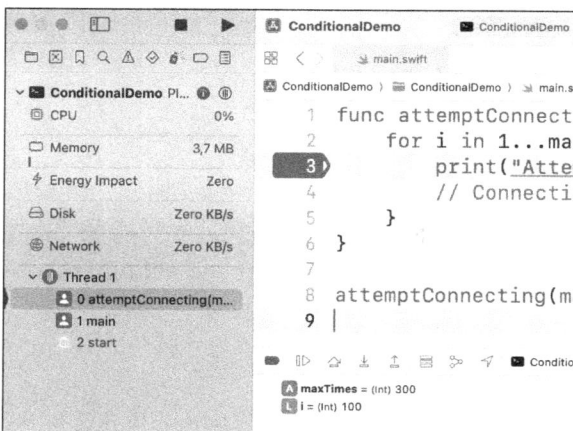

Figure B.18 Breakpoint Stops when i Becomes 100

Next, let's look at symbolic breakpoints.

B.2.2   Symbolic Breakpoints

In Swift debugging, a *symbolic breakpoint* is a powerful tool that allows you to pause the app execution whenever a specific symbol (such as a function or selector) is called. Even if the symbol is not part of your own code, Swift will still stop at it. This approach is particularly useful when you're debugging frameworks, for which you won't necessarily have access to the source code.

Listing B.3 presents a simple app that formats and prints the current date.

```
import Foundation

let formatter = DateFormatter()
formatter.dateStyle = .short
let dateString = formatter.string(from: Date())
print(dateString)
```

Listing B.3 Simple Date Formatter App

Say you want to find out if `DateFormatter.string` is accessing the `[NSDateFormatter stringFromDate:]` Objective-C selector.

If you had that module as an open-source imported package at hand, you could simply set a regular breakpoint at that location and run the program. Sadly, you don't. However, you can resort to a symbolic breakpoint to pause the app on the binary code if that selector is called.

To set the symbolic breakpoint, enter the Breakpoints Navigator, click the **+** button, and select **Symbolic Breakpoint...** from the context menu shown in Figure B.19.

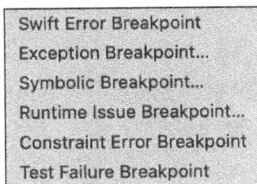

```
Swift Error Breakpoint
Exception Breakpoint...
Symbolic Breakpoint...
Runtime Issue Breakpoint...
Constraint Error Breakpoint
Test Failure Breakpoint
```

Figure B.19 Breakpoints Navigator Context Menu

In the popup window, enter the symbol's name as `[NSDateFormatter stringFromDate:]`. You can also give a meaningful name to that breakpoint, as shown in Figure B.20.

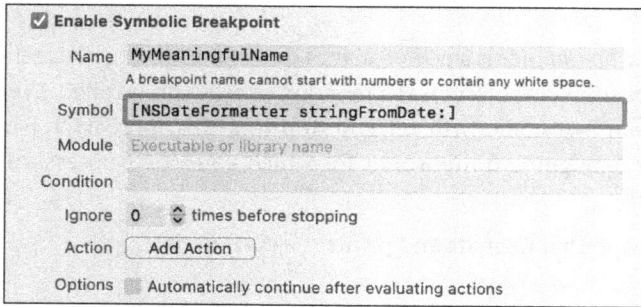

Figure B.20 Entering Symbol's Name

With that breakpoint in place, the execution of the app should stop at the exact point where [NSDateFormatter stringFromDate:] was accessed, as shown in Figure B.21.

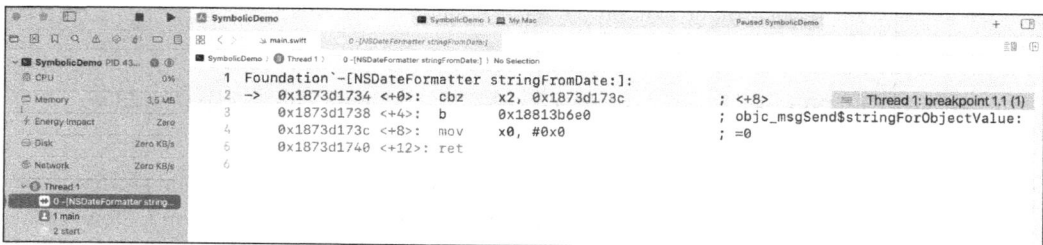

Figure B.21 Debugger Stopped at Symbolic Breakpoint

Because the breakpoint was set on a compiled binary framework, what you see here is not raw Swift code. Instead, you're looking at assembly code that contains machine-level instructions. To be more specific, these are ARM assembly instructions for Apple Silicon chips.

But no matter: The purpose was not reading the Swift code anyway. You understood that [NSDateFormatter stringFromDate:] was accessed and got your answer.

B.2.3   Performance Debugging

From a technical point of view, an app's consistency and reliability are very important features. From the user's point of view, performance is equally important. No one likes an unreliable app, it's true—but no one likes an unresponsive, turtle-speed app either.

To help spot performance bottlenecks and monitor the app's overall footprint, Xcode offers various tools. To discover those tools, use the code sample in Listing B.4. Here, a small loop and big loop are called sequentially.

```
func smallLoop() {
    for i in 0..<100000 {
        let x = i
```

```
    }
}

func bigLoop() {
    for i in 0..<1000000 {
        let x = i
    }
}

smallLoop()
bigLoop()
```

Listing B.4 Small and Big Loops

Obviously, `bigLoop` should run slower than `smallLoop` because it has more iterations within the `for` loop. With such code at hand, let's introduce two significant measurement tools in Xcode: Debug Navigator and Profiling Instruments.

Debug Navigator

While the debug is active, core app metrics can be accessed under the Debug Navigator. As shown in Figure B.22, you can see valuable metrics here such as CPU, memory, and the energy consumption of your app.

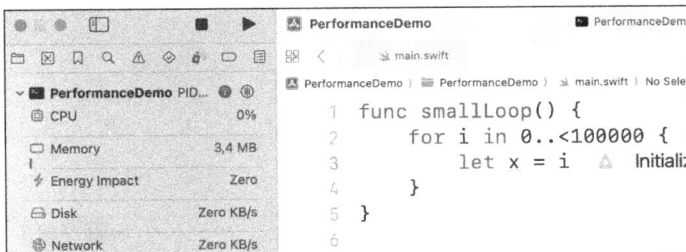

Figure B.22 Debug Navigator Metrics

Such metrics are useful for various use cases. For example, if you develop a mobile app that consumes the battery of an iPhone within an hour, users certainly won't feel motivated to download and use your app. Examining the energy impact could help you to prevent such a problem preemptively. Or in a video game, continuous heavy CPU usage might heat up the computer in an undesirable way—and that's where CPU metrics would help. Different metrics target different scenarios: They might be trivial for small personal projects but become crucial in commercial apps.

Profiling Instruments

For deeper analysis, Xcode offers Profiling Instruments, which are professional-level performance-measurement tools. To access those tools, simply click the **Product • Profile** menu option. A popup like that shown in Figure B.23 will open, presenting many instruments that you can use.

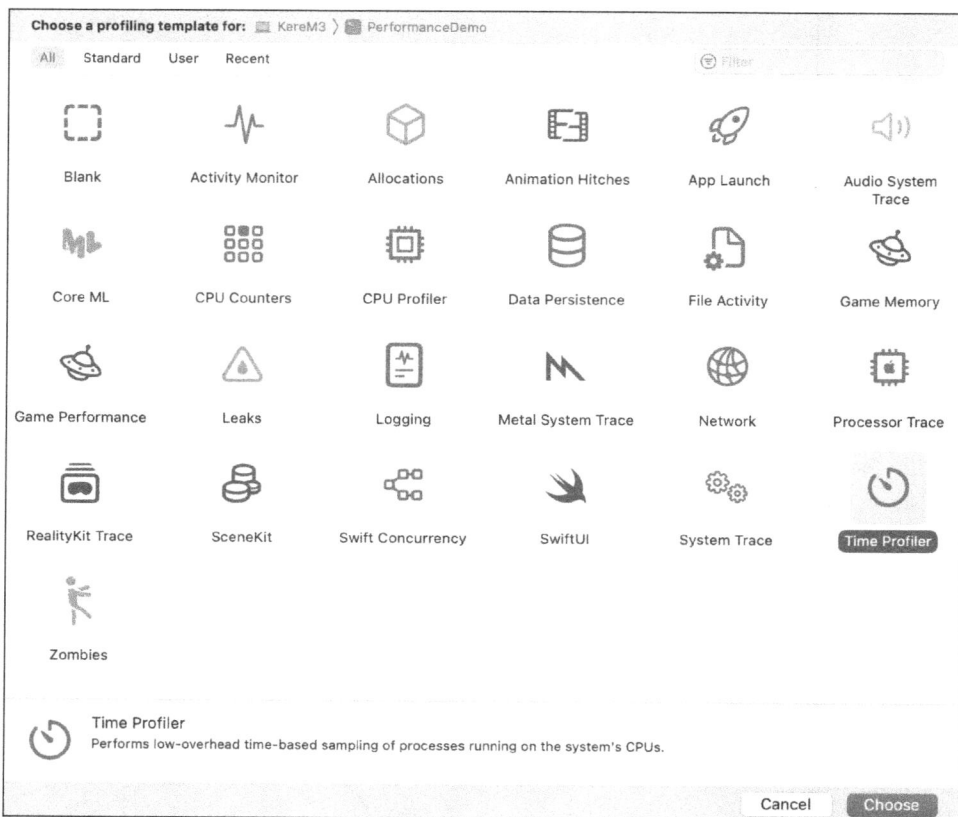

Figure B.23 List of Profiling Instruments

As a demonstration, we'll focus on the **Time Profiler** instrument, which records what your app is doing on the CPU over time. Select **Time Profiler** and click **Choose**. This will open the time profiler window shown in Figure B.24.

Leave all the options in their default states, and click the red **Record** button to let the profiler work. When it's done, the tool will present measurement results, as shown in Figure B.25, where the CPU usage peaks are shown in a timeline of the app's flow. Here, you can navigate through performance peaks, the involved threads, and even the CPU core the app was running on.

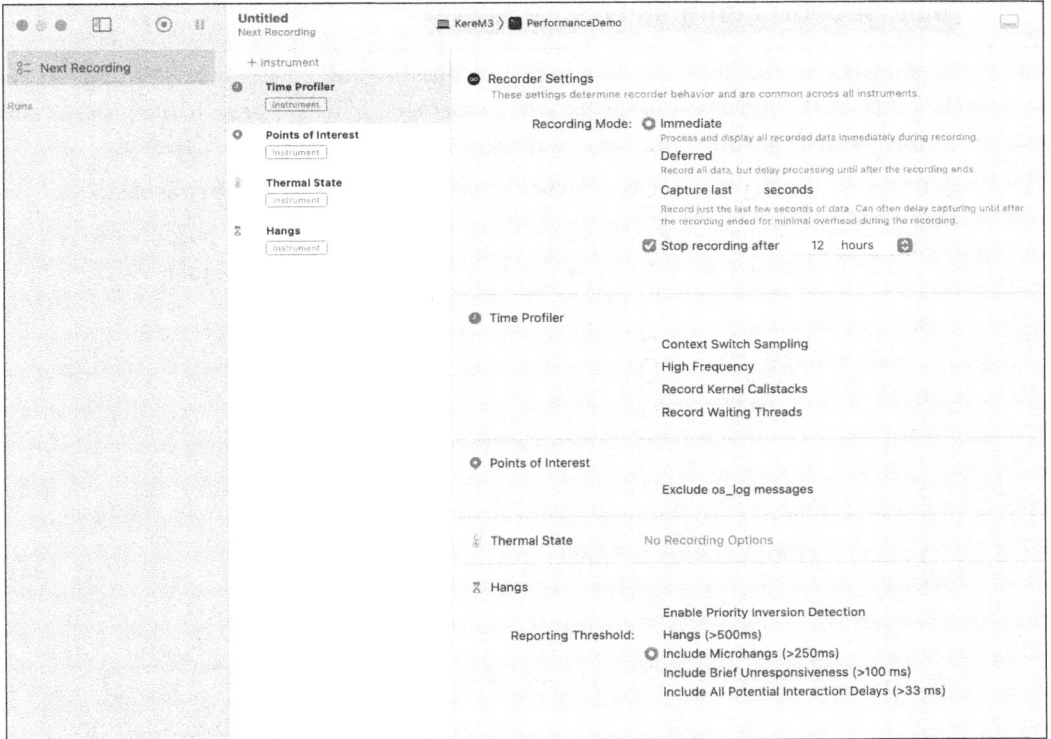

Figure B.24 Time Profiler Window

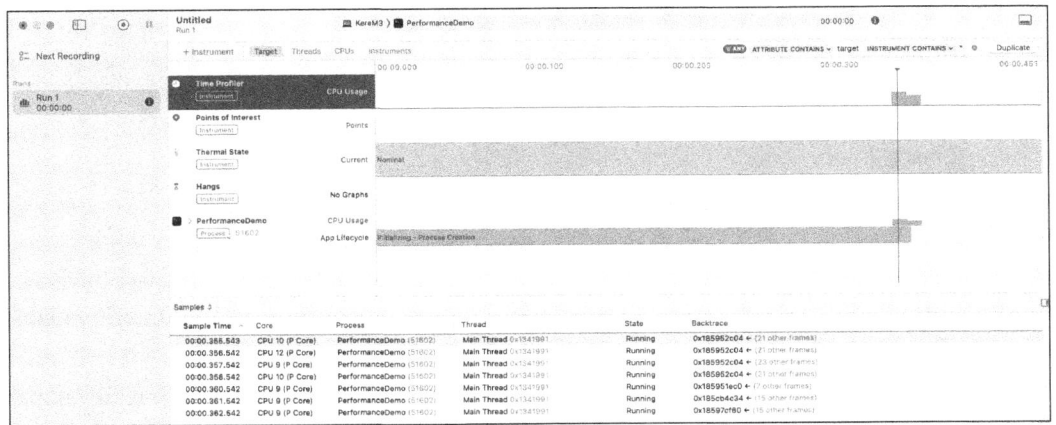

Figure B.25 Time Profiler Results

You are more than welcome to browse the options and explore other instruments, taking notes for your future app projects.

B.3 Support Commands

Despite the availability of various debug techniques and Xcode tools, printing out values may be the most practical choice in some cases. In this approach, the programmer resorts to printing out values of variables to understand the state of the app.

For this, Swift offers several commands to send information to the console.

It's possible to demonstrate all those commands at once. For this purpose, we will use the sample code in Listing B.5, in which each command is called sequentially.

```swift
struct Person {
    let name: String
    let age: Int
}

let person = Person(name: "Kerem", age: 30)

print(person, terminator: "\n\n")
debugPrint(person, terminator: "\n\n")
dump(person)
```

Listing B.5 Demonstration of Debug Support Commands

The output of this demo is shown in Listing B.5. As you can see, each command produces a different output.

```
Person(name: "Kerem", age: 30)

CommandDemo.Person(name: "Kerem", age: 30)

▿ CommandDemo.Person
  - name: "Kerem"
  - age: 30
```

Figure B.26 Output of Support Commands

In the first line, you can see the output of the simple `print` function, which we have used throughout the book. This is the suitable approach for producing a human-readable, simple description of a value.

In the second line, you can see the output of `debugPrint`, which gives a more developer-friendly description, considering the variable's type and scope.

In the final line, `dump` printed out the full contents and structure of the value. That's especially useful for complex objects with many properties.

B.4   Summary

In this appendix, you learned how to debug using Xcode, which arguably has all the bells and whistles one would expect from a modern IDE.

On the basic level, Xcode supports common breakpoint-oriented debugging, in which you can stop and watch the app's flow in stop-motion, as well as see the state of the variables.

On the more advanced level, you have the option to set conditional and symbolic breakpoints, as well as using Profiling Instruments to monitor the app's behavior with a high level of granularity.

Naturally, you can always go back to basics and simply print out values to track the app's behavior. For that purpose, you have multiple commands—like `print`, `debugPrint`, and `dump`—in your toolbox.

The Author

Dr. Kerem Koseoglu is a seasoned software engineer, author, and educator with extensive experience in global projects and commercial applications.

With over twenty years of experience in ABAP, he is also proficient in database-driven development using Python and Swift. As well as the book on Swift in your hands, Dr. Koseoglu is also the author of *Design Patterns in ABAP Objects* (SAP PRESS), *SQL: The Practical Guide* (Rheinwerk Computing), and several best-selling technical books in Türkiye.

Dr. Koseoglu holds a PhD in organizational behavior and is renowned for his work in time management. He conducts a variety of technical and soft-skill trainings. He speaks Turkish, English, and German fluently.

When he's not coding or teaching, you are likely to find Dr. Koseoglu on the stage, playing his bass guitar.

You can visit Dr. Koseoglu's website at *www.keremkoseoglu.com* or contact him via email at *kerem@keremkoseoglu.com*.

Index